SAP S/4HANA aATP

Advanced Available-to-Promise
Optimized Inventory & Fulfilment in Real-Time:
Practical Guides for the Digital Supply Chain

By
Sunil Patil

Introduction: Bridging the SAP S/4HANA aATP Knowledge Gap

Many companies move to SAP S/4HANA to improve their supply chain, but one key question still causes daily stress: "Can we really deliver what we just confirmed to the customer?" Advanced Available-to-Promise (aATP) is SAP S/4HANA's answer to this question, yet in many projects it is not well understood or fully used. This book aims to close that gap and show how aATP can become a simple, practical tool to support business decisions.

In the past, classic ATP in SAP ERP could answer basic availability questions, usually for a single plant and with limited options for allocations and backorders. Today, supply chains are global, inventories are tight, and customers are more demanding, so this old logic often isn't enough. aATP was designed for this new reality, with faster checks, smarter allocation of limited stock, and better ways to reorder priorities when demand changes.

There's still a big knowledge gap around aATP. Many SD, MM, and PP consultants know classic ATP well but aren't sure how aATP works, what to switch on, or how settings impact the full order-to-delivery process. Business users often hear that "aATP is powerful" but don't see clear, simple examples of what it does in their daily work. Project leads and architects may struggle to connect S/4HANA strategy with concrete aATP design and rollout steps. This book brings these different views together and offers a common, easy-to-understand language for everyone working with order promising.

The focus of this book is on real-life situations, not just theory. We'll look at cases such as a critical order when stock is low, priorities between different sales channels, new product launches with limited supply, and multi-plant deliveries. Each situation introduces relevant aATP functions like availability checks, product allocation, backorder processing, supply protection, and alternative-based confirmation and shows how to set them up, run them, and avoid common mistakes.

This book is written for SAP functional consultants, solution architects, business process owners, and technical consultants. Consultants will get a complete view of aATP from basic concepts to advanced options. Architects will learn how to position aATP within the S/4HANA landscape and align it with business goals. Business users in order management, supply chain, and customer service will see what's possible and how to ask for the right solutions. Technical experts will gain the process understanding they need to build stable enhancements and integrations.

The chapters follow a typical S/4HANA journey. We start with the move from classic ATP to aATP, explaining key terms and system behavior in simple language. Then we cover the main building blocks - aATP checks, product allocations, backorder processing, and confirmation strategies - before exploring advanced topics like supply protection, substitution, supply-based confirmation, and business process scheduling. Configuration always pairs with business examples, so you see exactly how settings change confirmation results. Later chapters tackle project essentials: migration from classic ATP, rollout strategies, testing, performance, and real-world best practices.

Author's Motivation-

The reason for this book is simple: In many S/4HANA projects, aATP is seen as important but postponed, configured with minimal thought, or treated as a "black box." Systems go live, but order promising still depends on spreadsheets and manual changes. By explaining aATP in a clear, scenario-driven way, this book aims to change that, so your organization can use aATP with confidence, improve customer commitments, protect scarce supply, and get real value from SAP S/4HANA.

Quick Reference Value-

This book is my promise to help the next generation of consultants and practitioners with clear, future-ready knowledge. It serves as a quick reference guide when you need to review or improve specific aATP areas—even experts forget details when away from processes for a while. Beginners will start their SAP aATP journey with skill and confidence. Use it to refresh knowledge, check your approach, and handle availability check tasks efficiently.

Why This Book is Essential for You-

1. Overcoming the "Knowledge Fade" Challenge
   Stepping away from aATP processes can cause temporary knowledge gaps. This book provides quick concept reviews, checklists, and standard transaction codes to get you back up to speed fast and confidently.

2. aATP Process Clarity in S/4HANA
   S/4HANA transforms availability checking. This book explains the modern aATP system simply, with detailed guidance on key functions and step-by-step configuration.

3. Built for Immediate Use
   No searching required—clear transaction mappings, program references, and master data dependencies make it your instant operational reference.

This isn't a beginner's guide. It's the experienced aATP professional's operational manual—precise, efficient, keeping your skills sharp and projects on track. Especially useful for SD/MM/PP consultants already familiar with availability checks.

Special "Important Notes" Section-

In this book, you'll find a special section titled "Important Notes" - curated Q&A rarely documented elsewhere, plus concise notes for each topic. Drawn from my professional experience and compiled from multiple sources, these provide unique insights with clear explanations.

Screenshots are limited to keep explanations crystal clear and focused.

Key Problem-Solving Coverage-

Valuable notes tackle complex scenarios like:

- Multi-plant availability checks across sales organizations
- Product allocation conflicts during peak demand
- Backorder processing priority rules
- Supply protection for strategic customers

Acknowledgments-

My sincere thanks to colleagues and mentors who guided my SAP journey. Large-scale S/4HANA projects across SD, MM, and beyond gave me the hands-on experience to create this practical aATP guide.

Special thanks to Gary Cole, Paula Politis for their unwavering support and trust in allowing me to work on large-scale, high-impact SAP implementation projects. These opportunities helped me develop strong techno-functional skills and gain deep knowledge across multiple SAP modules, forming a solid foundation for my professional growth. I am truly grateful for your guidance, encouragement, and continued support.

Gary Cole is a hands-on SAP implementation professional with over 25 years of experience delivering global implementations across SAP R/3, SAP ECC, and SAP S/4HANA. His expertise, guidance, and commitment have been invaluable to my professional development.

Finally, to my family and loved ones, and to one special person who means more than words can say, 👑SSP👑 (*dur bhi ho, magar dil ke paas ho*...) - thank you for your endless love, support, trust, and understanding. Your encouragement has been my greatest source of strength throughout this journey.

Table of Contents

Note: The content of this book is based on the SAP AATP version, S/4HANA 2023.

List of S/4 HANA AATP versions-

SAP S/4HANA Advanced Available-to-Promise (aATP) versions have evolved to enhance order fulfillment capabilities with increasingly advanced features. Here is a list of notable aATP versions/releases typically aligned with major S/4HANA releases and their key enhancements:

- S/4HANA 1610 (2016): Initial introduction of aATP with basic advanced ATP functionalities like product allocation, backorder processing, and substitution.
- S/4HANA 1709 (2017): Enhanced integration with supply chain execution and improvements in backorder processing and product allocation.
- S/4HANA 1809 (2018): Introduction of new Fiori apps for aATP, improved user experience, and expanded features in alternative-based confirmation and supply protection.
- S/4HANA 1909 (2019): Functional extensions in supply creation-based confirmation and enhanced scope of checks management.
- S/4HANA 2020: Introduction of new features for packaging size consideration in ATP checks, performance optimization, and further integration with SAP Integrated Business Planning (IBP).
- S/4HANA 2022: Added length-based availability checks for industry-specific requirements and enhanced real-time analytics.
- **S/4HANA 2023: More industry-specific ATP checks, improved supply protection, expanded manual confirmation adjustment capabilities, and better consolidation of shipping locations.**
- S/4HANA 2025 FPS0 (Latest as of mid-2025): Consolidation of shipping locations during alternative-based confirmation, characteristics evaluation in supply creation, enhanced supply protection to prevent overprotection, new rating attributes, and expanded business add-ins for custom modifications.

Each version built upon prior capabilities to provide more comprehensive and flexible order promising functions, tailored to increasingly complex and dynamic supply chains.

This version of history helps enterprises understand the progression of SAP aATP capabilities and plan upgrades or new implementations based on the latest and relevant features available.

SAP S/4HANA aATP 2023 – Book Summary

This book provides a comprehensive deep dive into Advanced Available-to-Promise (aATP) in SAP S/4HANA 2023, showing how order promising evolves from a simple stock check into an intelligent, real-time, constraint-aware fulfilment engine. It explains how aATP replaces the limitations of classic ATP by synchronizing sales, supply, and execution decisions across complex supply chains.

aATP enables real-time, reliable confirmations by considering inventory, planned and confirmed supply, allocations, priorities, and substitutions - all processed in-memory for immediate results.

Core Functionalities Covered-

Feature	Business Value
PAC	PAC performs **real-time availability checks** by simultaneously evaluating on-hand stock, inbound supply, outbound demand, and allocations. Fiori-based configuration allows businesses to define scopes of check across plants, storage locations, and supply categories. **Business value:** Accurate confirmations at order entry with sub-second response times.
PAL	PAL manages **quota-based supply protection** across customers, regions, channels, or products. Allocation sequences and time elements ensure scarce supply is distributed strategically. **Business value:** Prevents over-confirmation and protects strategic customers and markets.
BOP	BOP dynamically reprioritizes demand using **Win, Gain, Redistribute, Fill, and Lose** strategies. Simulation and execution are handled through the **BOP Fiori cockpit**, allowing planners to reallocate supply as conditions change. **Business value:** Maximizes revenue and service levels when supply is constrained.
ABC	ABC supports **product and plant substitutions** using ranking strategies and control profiles. If the requested product or location is unavailable, aATP automatically proposes viable alternatives. **Business value:** Higher order fulfilment rates and reduced lost sales.
Supply Protection	Priority stock reservations protect supply for **VIP customers, key channels, or strategic demand** before general order consumption. **Business value:** Margin protection and improved customer satisfaction for high-value segments.
BPS	BPS enables **supply creation (SBC)** and integrates with **Transportation Management (TM)** to provide realistic, execution-based delivery dates. **Business value:** Commit dates aligned with actual logistics and production capabilities.

Business Impact

aATP transforms order promising by delivering:

- Up to **80% faster order confirmations**
- End-to-end supply chain visibility
- Reliable, execution-ready delivery dates
- Intelligent prioritization that protects margins and key customers

1. The Rising Demand for SAP S/4 HANA aATP

The market demand for Advanced Available-to-Promise (aATP) is rapidly increasing. With the introduction of powerful new capabilities, aATP enables businesses to significantly improve productivity, optimize supply chain decisions, and scale operations more effectively.
Demand for SAP S/4HANA aATP is rising because more companies are moving to S/4HANA, supply chains are under pressure, and standard ATP is no longer enough to promise orders reliably in complex, global environments.

- S/4HANA migration deadlines (ECC mainstream maintenance ending around 2027) push companies onto platforms where aATP is the strategic order-promising engine instead of classic ATP.
- Modern supply chains require real-time, intelligent order confirmation that can deal with shortages, multiple plants, and complex constraints, which aATP is designed to handle.

Historically, several core capabilities now offered in aATP - such as Backorder Processing (BOP), Alternative Base Confirmation (ABC), and Product Allocation (PAL) - were delivered through Global ATP (GATP) in the SAP APO system. APO functioned as a standalone application, interfacing with SAP ECC via the Core Interface (CIF). Because ECC and APO were separate systems, they required continuous data synchronization through queues, batch jobs, and numerous integration programs. This created a highly complex and operationally intensive landscape for both IT teams and business users. As a result of this complexity and the ongoing maintenance effort, many organizations were reluctant to adopt APO-based GATP functionality.

SAP has developed Advanced Available-to-Promise (aATP) as part of S/4HANA, bringing together the key functionalities that were previously handled separately in SAP ECC and SAP APO.

Additionally, SAP has announced that APO will reach its end of life by 2030, further accelerating the need for a modern replacement.
aATP addresses key business problems by helping organizations optimize processes and improve customer satisfaction.

- **Frequent shortages and volatile demand**: aATP helps determine which customers, channels, or regions should receive limited stock, improving service levels for critical segments.
- **Fragmented or manual allocation processes**: Many companies still rely on spreadsheets and custom logic to allocate stock. aATP replaces these with standard, configurable capabilities such as product allocation, supply protection, backorder processing, and alternative-based confirmation.

These are the key capabilities that make aATP attractive for organizations:

- **Real-time, intelligent ATP checks**: aATP considers inventory, planned receipts, and allocation or priority rules instead of relying on a simple "first come, first served" approach.
- **Advanced allocation and fulfillment functions**: Capabilities such as product allocation, supply protection, backorder processing, and alternative-based confirmation enable organizations to prioritize stock, protect key customers, and automatically propose alternative plants or products.

Companies' Interest in aATP is Accelerating (2025–2026) - Interest in advanced Available-to-Promise (aATP) solutions is accelerating in 2025–2026 as organizations shift their focus from basic system migrations to optimization initiatives that deliver measurable ROI. Benefits include improved fill rates, reduced penalties, and higher on-time delivery performance.
At the same time, ongoing supply chain disruptions, geopolitical uncertainty, and Industry 4.0 initiatives are increasing the demand for a centralized, real-time order promising engine that is closely integrated with production planning and logistics.

Industry 4.0 refers to the **fourth industrial revolution**, which focuses on the digital transformation of manufacturing and supply chains. It integrates advanced technologies to create smart, connected, and automated operations.

Adoption approach for aATP-

These are typical adoption patterns for aATP implementations. The exact approach depends on the scale and scope of the business, but in most cases, the following phased strategy is recommended.

aATP is typically adopted in phases during projects. Organizations first go live with basic ATP or a limited aATP scope to minimize risk during S/4HANA migration. In later phases, they expand to advanced aATP capabilities such as product allocation, supply protection, backorder processing, and alternative-based confirmation to drive value realization and optimize service for key products, customers, and multi-plant networks.

Phase 1	Go live with Basic ATP or a minimal aATP scope to reduce risk during the S/4HANA migration. This phase focuses on ensuring system stability and continuity of core order fulfillment processes.
Phase 2	Roll out advanced aATP capabilities as part of value realization or optimization waves. This may include: • Product allocation and supply protection for key product and channel combinations • Backorder processing for strategic customers • Alternative-based confirmation across multi-plant networks

2. Why aATP Is the Future

aATP (Advanced Available-to-Promise) is considered the future of order promising in SAP. Unlike classic ATP, which performs simple stock checks and is tied to ECC-era processes, aATP transforms order fulfillment into an intelligent, rule-based, real-time engine. It is designed to handle volatile demand, constrained supply, and complex networks, supporting priority-based confirmations and being AI-ready for modern S/4HANA supply chains.

Strategic role in S/4HANA-

- S/4HANA is now the strategic ERP, with ECC mainstream maintenance ending around 2027, so new investments in order promising and fulfillment innovation are concentrated in aATP rather than classic ATP.
- aATP is positioned by SAP as the standard for "next-generation order promising," tightly embedded in S/4HANA's in-memory platform, analytics, and Fiori UX, which means future enhancements and roadmap focus will land there first.

Advanced capabilities vs classic ATP-

- aATP offers capabilities like product allocation, supply protection, backorder processing, and alternative-based confirmation that allow businesses to allocate stock by priority, guard key customers, and propose best alternatives automatically; these go far beyond the simple first-come-first-served logic of classic ATP.
- The engine works with real-time inventory and order context, improving feasible confirmations, protecting service levels, and reducing manual allocation and firefighting in shortage situations.

Fit with 2026+ trends-

- Current SAP trends emphasize clean-core S/4HANA, cloud adoption, AI, and integrated analytics, and aATP aligns with this by being configuration-driven, analytics-enabled, and ready to consume AI-based insights (for example, in forecasting or risk scoring) as these capabilities mature.
- As companies modernize order-to-cash, the focus shifts from basic transactional order entry to intelligent, automated fulfillment, and aATP is specifically called out as the mechanism

to make more precise, data-driven fulfillment decisions in real time.

Business value going forward-

- By allocating "the right stock to the right demand" and protecting key customers and products, aATP helps improve delivery performance and avoid penalties, which directly supports revenue and margin in increasingly constrained supply chains.
- Because almost all new S/4HANA transformations now include a roadmap to unlock advanced capabilities after go-live, aATP becomes a central lever in post-migration value realization, rather than a niche add-on.

Overview of why aATP is the next-generation order promising solution.

Here are a few suggestions for Transforming Order Promising with aATP in S/4HANA.

1. From basic checks to intelligent promises

Classic ATP mainly answers, "Is there enough stock on this date?" using relatively simple, often batch-based checks against current inventory and planned receipts. aATP instead performs real-time, in-memory checks in SAP S/4HANA, evaluating inventory, production, and transportation constraints together to give more reliable and realistic confirmations.

2. Core capabilities that make aATP future-ready

The aATP functionalities in S/4HANA 2023 provide a complete, real-time order promising solution, covering allocation, substitutions, backorders, delivery prioritization, and scheduling effectively consolidating ECC ATP and APO capabilities.

aATP bundles several advanced functions that directly address today's fulfillment challenges. The most important are:

- Product Allocation (PAL)
- Backorder Processing (BOP) - with multiple strategies (Win, Gain, etc…)
- Alternative-Based Confirmation (ABC)
- Supply Protection
- Release for Delivery

- Supply-Based Confirmation (SBC)
- Business Process Scheduling (BPS)

3. Designed for volatile, constrained supply chains

Global supply chains now face frequent disruptions, regulatory pressures, and fast-changing customer expectations, which make static, one-dimensional ATP logic insufficient. aATP supports constraint-based confirmations (including capacity and transport limits) and supply-protection strategies, so key products, customers, and markets retain access to scarce inventory even in shortages.

4. Tight integration with planning and S/4HANA

Because aATP is embedded in SAP S/4HANA and tightly integrated with SAP Integrated Business Planning (IBP), it links long-term planning with real-time execution. Planning systems define allocation strategies and simulate scenarios, while aATP enforces those strategies at order entry, creating a closed loop between plan and actual fulfillment.

5. Strategic business impact

By moving to aATP, companies typically improve service levels, reduce stockouts for strategic customers, and avoid costly delivery failures and penalties. The result is a more resilient, responsive, and customer-centric order-to-cash process, which is why many SAP customers and SAP itself position aATP as the next-generation standard for order promising.

With SAP S/4HANA, aATP is now embedded directly within the same system. This brings several significant advantages:

- Unified Architecture: Master data and transactional data reside in one system, eliminating the need for CIF integration.
- Lower Complexity: No queues, no cross-system synchronization, and no additional technical overhead.
- Cost-Effective: Businesses only incur a minimal additional license cost to activate aATP.
- High Performance: Built on the HANA in-memory platform, aATP delivers real-time insights and rapid processing.

3. Opportunities for Consultants

For consultants with backgrounds in SD, MM, PP, or APO, aATP represents a "low-hanging fruit," as many of its concepts build upon existing ATP and gATP foundations. This makes the learning curve manageable and allows consultants to quickly improve their skills and become proficient in the new aATP functionalities.

While this perspective is primarily relevant for SAP consultants and implementation teams, experienced SCM end-users can also benefit by understanding how aATP enhances order fulfillment processes. For end-users, the focus shifts from learning system configuration to leveraging practical benefits, process improvements, and new functionalities in day-to-day operations.

Given the ongoing migration of companies from ECC/APO to S/4HANA, the adoption rate of aATP is growing rapidly. As a result, there is a strong and rising demand in the market for consultants with solid aATP skills. Consultants with SAP S/4HANA aATP expertise are in strong demand today due to widespread migrations from classic ATP and the growing need for resilient supply chain solutions amid global disruptions. Opportunities span implementations, optimizations, and integrations across multiple industries, including Manufacturing, Automotive, High-Tech & Electronics, Pharmaceuticals & Life Sciences, Consumer Products & Retail, and Chemicals & Process Industries.

In short, SAP aATP is applicable wherever accurate order promising, stock allocation, and real-time supply chain visibility are essential. It is particularly valuable in industries with complex production, high-value products, regulated supply, or volatile demand.

Active roles include aATP Leads designing Backorder Processing (BOP) and Product Allocations (PAL), often requiring 4-5 years of hands-on S/4HANA experience plus SD module knowledge. Key tasks involve conducting workshops, debugging Fiori apps, and managing integrations with SD, MM, PP, and EWM.

Companies seek consultants with proven, hands-on aATP experience from end-to-end S/4HANA implementations, particularly those involving migrations from classic ATP, APO/gATP, or ECC systems. Nice-to-have skills include variant configuration, Agile methodologies, and industry domain knowledge in environments facing supply volatility.

Demand is expected to surge with the 2025 S/4HANA FPS0 releases, which enhance aATP-IBP integration and support cloud adoption, alongside ongoing ECC-to-S/4 conversions. SAP's push for best-practice optimizations in volatile supply chains positions aATP specialists for leadership in end-to-end projects, feature ramp-ups, and advisory roles. In the long term, growth is anticipated in entitlements management and AI-driven promising, sustaining high employability.

4. Supply Chain Network and the Role of aATP

To understand where Advanced Available-to-Promise (aATP) fits within an organization, it is essential to first understand the broader supply chain network and how demand flows across it.

What Is a Supply Chain Network-

A supply chain encompasses the full set of activities required to deliver a product to a customer from raw material sourcing all the way to final delivery. Although it can be represented with a simple linear flow, real-world supply chains are complex, involving multiple partners, facilities, and processes.

A basic supply chain flow typically involves:

1. Customers
2. Retailers or Sales Channels
3. Distribution Centers (DCs)
4. Manufacturing Units (Plants)
5. Suppliers (Raw materials or finished goods)

*In any supply chain, **requirements begin with the customer and are fulfilled by suppliers**.*

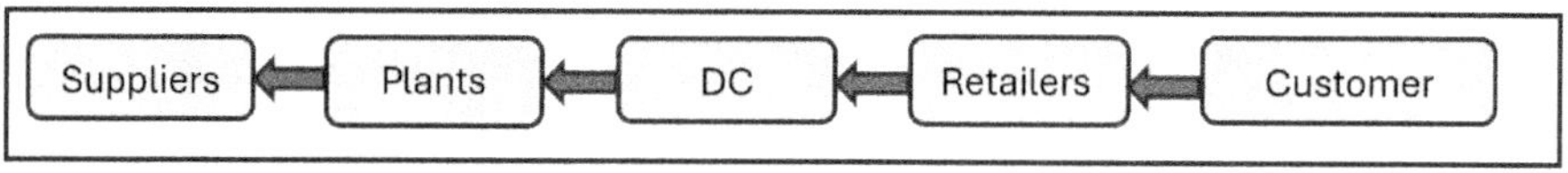

The ultimate objective of any supply chain is straightforward: *Deliver the right product to the customer, on time, and in full quantity.* However, to achieve this consistently - especially in a volatile market - organizations must operate an optimized and agile supply chain supported by modern planning and execution tools.

Example: How Demand Flows Through the Network-

Imagine a customer who needs Product FG10101, Quantity 100 units:

1	**Customer → Retailer** The customer places the request at the nearest retailer. The retailer checks its available stock to see if the full quantity can be supplied.
2	**Retailer → Distribution Center (DC)** If stock is insufficient, the retailer passes the demand to the DC. The DC checks its inventory availability.
3	**DC → Manufacturing Unit (Plant)** If the DC cannot fulfill the demand, the request moves to the manufacturing unit. The plant checks: ○ If the finished product is available, or ○ Whether production is required.
4	**Manufacturing Unit → Supplier** If raw materials are not available for production, the plant sends demand to its suppliers.
5	**Supplier → MU → DC → Retailer → Customer** Suppliers deliver raw materials to the plant. The plant produces the product and sends it to the DC. The DC supplies the retailer, and the retailer finally fulfills the customer order.

This flow demonstrates how demand travels backward through the supply chain, while goods move forward from suppliers to customers.

Where Does aATP Fit in the Supply Chain Process-

aATP sits at the point where customer demand enters the supply chain. It acts as the intelligent gatekeeper that decides:

- *Can we commit this demand?*
- *When can we deliver?*
- *From which location should inventory be sourced?*
- *Do we need to reschedule or re-prioritize existing orders?*

aATP plays a significant role on three critical points:

1	**Order Processing**	When sales orders or customer requests are received.
2	**Inventory Allocation**	Ensuring stock is prioritized for the right customers or channels.
3	**Supply Alignment**	Connecting demand with the downstream supply chain (DCs, plants, suppliers).

By providing accurate, real-time commitments, aATP ensures the entire supply chain functions smoothly, reducing shortages, improving customer service, and aligning operational planning with business priorities.

Planning Tools for the End-to-End Supply Chain-

To manage a complete supply chain, companies rely on powerful planning tools. Popular solutions include:

- Anaplan
- Blue Yonder (JDA)
- Kinaxis RapidResponse (KX)
- SAP APO (on-premises)
- SAP IBP (cloud-based)
- SAP PP/DS (Embedded in S/4 HANA)
- SAP DDMRP
- MRP Live

In the ERP world, SAP covers nearly 70% of the global supply chain planning market, making SAP solutions the dominant choice for enterprises.

In SAP S/4HANA 2023, aATP acts as the real-time order-promising engine, while IBP, PP/DS, MRP Live, and DDMRP provide planning inputs across different horizons. Together, they replace classic ECC and APO planning landscapes with embedded, resilient, and real-time planning architecture.

SAP Supply Chain Planning Landscape-

SAP divides end-to-end supply chain planning into five major components:

1. Demand Planning (DP)
2. Supply Network Planning (SNP)
3. PP/DS – Production Planning & Detailed Scheduling
4. TP/VS – Transportation Planning & Vehicle Scheduling
5. gATP – Global Available-to-Promise

1.**Demand Planning (DP)** - is a fundamental forecasting tool that analyzes historical data to predict future sales. It helps estimate demand for the upcoming months—typically over a 6-month to 1-year horizon. By knowing the expected demand in advance, organizations can better prepare for production, procurement, and delivery. Demand Planning is a long-term planning process that uses past trends and patterns to generate reliable sales forecasts.

2. Supply Network Planning (SNP) – Once demand is forecasted, proper supply planning within the supply chain network is required, which is handled through Supply Network Planning. SNP considers both the predicted demand and ongoing sales to create a rough supply plan. It determines where to produce, how much to produce, and how to distribute products within the supply chain network.
For example, if an organization has 2–3 manufacturing facilities, SNP will suggest how much to produce at each facility and which distribution centers (DCs) to supply. All recommendations are based on the forecasted demand.
Supply Network Planning is considered a mid-term planning process, bridging the gap between long-term demand planning and short-term execution.

3. Production Planning and Detailed Scheduling (PP/DS) – Once the supply plan is generated through SNP, it needs to be converted into an actual production plan, which is handled by PP/DS. The plan received from SNP is executed here and transformed into detailed production schedules.
PP/DS determines the sequence of production, resources to be used, and timing of operations to ensure efficient execution. It operates at a detailed level to manage short-term production planning and scheduling.
PP/DS is considered short-term planning, bridging the supply plan with actual shop floor execution.
In SAP S/4HANA, PP/DS (Production Planning and Detailed Scheduling) is referred to as embedded PP/DS (ePP/DS) because:

- "Embedded" = Fully integrated directly into the S/4HANA core (no separate APO system needed)
- "ePP/DS" = Explicitly calls out the embedded version to distinguish from classic APO PP/DS

4. Transportation Planning and Vehicle Scheduling (TP/VS) – TP/VS manages the movement of products within your supply chain network, optimizing the transportation of goods from one node to another. It ensures that deliveries are planned efficiently to reduce transportation costs and improve resource utilization.

For example, if you need to deliver 1,000 units to a customer and you produce 500 units today and another 500 units tomorrow, TP/VS can help schedule a single delivery of 1,000 units instead of two separate shipments. This optimizes transportation, reduces costs, and improves overall logistics efficiency.

TP/VS is part of short- to mid-term supply chain planning, focusing on effective transportation and delivery scheduling.

5. Global Available-to-Promise (gATP) – gATP is a front-end tool that interacts with customers to capture their requirements and based on that information, proposes delivery dates. It checks product availability across the supply chain, considers ongoing and planned orders, and provides realistic delivery commitments to customers.

gATP is used for short-term operational planning and ensures that promised delivery dates are feasible based on actual stock and supply conditions.

APO and IBP:

SAP APO	SAP APO is the older, on-premises solution for supply chain planning, which is nearing end-of-life (sunset) around 2027–2030. SAP has stopped active development of APO.
SAP IBP	The modern, cloud-based replacement for APO, offering advanced planning capabilities across demand, supply, inventory, and sales & operations. IBP provides real-time analytics, scenario simulation, and better integration with SAP S/4HANA.

MRP and Demand Planning-

Material Requirements Planning (MRP) is an execution-level planning tool within the ERP system.

- MRP generates receipts (production proposals, purchase requisitions) to cover requirements (sales orders, planned independent requirements).
- Planned Independent Requirements (PIRs) are typically created by Demand Planning.
- Before MRP is executed, these PIRs represent forecasted demand.

Here is a clean, professional, and comparison table showing how key supply chain planning components map across S/4HANA, SAP APO, and SAP IBP.

Comparison of Supply Chain Capabilities across S/4HANA, APO, and IBP-

Process / Function	SAP S/4HANA	SAP APO (On-Premises)	SAP IBP (Cloud)
Demand Planning	*Managed via embedded functionalities or via IBP integration*	**Demand Planning (DP)**	**IBP for Demand**
Supply Network Planning	*Not available within S/4; managed by IBP*	**SNP – Supply Network Planning**	**IBP Supply / IBP S&OP**
Detailed Production Planning	**ePPDS (embedded PP/DS)**	**PP/DS**	*Not part of IBP; executed in S/4*
Transportation Planning	**SAP TM (Transportation Management)**	**TP/VS – Transportation Planning & Vehicle Scheduling**	*Managed at strategic level via IBP S&OP; execution in SAP TM*
Available-to-Promise	**aATP – Advanced ATP**	**gATP – Global ATP**	*ATP logic not in IBP; planning supports ATP in S/4*

This illustration shows which components were moved from SAP APO to SAP S/4HANA and SAP IBP.

S/4 Hana	SAP APO (on Premise)	SAP IBP (Cloud)
	Demand Planning →	IBP Demand
	SNP →	IBP Supply/IBP S&OP
ePPDS	← PP/DS	
TM	← TP/VS	
aATP	← gATP	

SAP APO components have been migrated (redistributed) between SAP S/4 HANA and SAP IBP as follows:

1	**Demand Planning (DP)**	Migrated/transitioned to SAP IBP for Demand.
2	**Supply Network Planning (SNP)**	Migrated/transitioned to SAP IBP for Supply and SAP IBP for S&OP.
3	**Production Planning and Detailed Scheduling (PP/DS)**	Moved into SAP S/4HANA as embedded PP/DS (ePPDS).
4	**Transportation Planning/Vehicle Scheduling (TP/VS)**	Moved into SAP S/4HANA Transportation Management (TM) module.
5	**Global ATP (gATP)**	Replaced by Advanced ATP (aATP) in SAP S/4 HANA.

5. Overview of SAP aATP - (Advanced Available-to-Promise)

aATP is SAP's advanced availability check functionality that helps businesses make realistic delivery promises to customers. **Advanced Available-to-Promise (aATP)** is SAP S/4HANA's modern, intelligent availability-checking engine. It enables businesses to make *accurate and reliable delivery commitments* by evaluating product availability in real time. aATP goes far beyond the classical ATP check by incorporating dynamic, rule-based logic and advanced supply chain insights.

When there is significant demand vs. supply imbalances, aATP functionalities are used to manage and prioritize orders effectively. Advanced ATP (aATP) is not mandatory in SAP S/4HANA implementations; classic ATP remains the default availability check, while aATP can be activated optionally via business function (or the corresponding IMG switch) when clients face demand-supply imbalances requiring advanced features such as PAL, BOP, Supply Protection, or Supply-Based Prioritization (SBP).

aATP is not mandatory for all SAP S/4HANA clients and should be used only when there is demand–supply imbalances. Its functionalities (such as PAL, BOP, ABC, etc.) are not activated by default; they are enabled selectively based on business needs via the relevant business function.

Key points-

Default:	• *Classic ATP handles standard availability checks.*
Selective activation:	• *Use aATP for constrained scenarios (product launches, shortages).*
Business-driven:	• *Configure only relevant features to avoid unnecessary complexity.*

*Here's a **clear explanation** of when clients use **classic ATP** (default) versus **aATP** in SAP S/4HANA:*
Simple businesses can continue using classic ATP for standard availability checks. Complex businesses with constrained supply should implement aATP to leverage advanced features such as PAL, BOP, ABC, Supply Protection, and Supply-Based Confirmation. Both approaches effectively manage supply and demand, but aATP provides greater flexibility, fairness, and prioritization when inventory is limited.

- ***Simple businesses:*** *Can continue using classic ATP.*
- ***Complex businesses with constrained supply:*** *Should implement aATP for advanced features.*
- *Both systems allow companies to* ***manage supply and demand****, but aATP adds* ***flexibility, fairness, and prioritization*** *when stock is limited.*

1. Classic ATP (default)-

Who can use it:	• Clients with simple supply-demand scenarios, where stock availability is usually sufficient and there is no significant demand-supply imbalance.
Features:	• Basic availability check (stock vs. order demand) • No advanced allocation rules • No supply protection or prioritization
Use case example:	• A company with stable inventory levels and predictable orders, e.g., a small-to-medium distribution business.

2. Advanced ATP (aATP)-

Who should use it:	• Clients with complex supply-demand imbalances, high-volume orders, or limited stock for strategic allocation.
Features:	• Product Allocation (PAL) – control distribution to customers, regions, or channels. • Backorder Processing (BOP) – prioritize orders when stock is insufficient. • Supply Protection / Fair Share – protect stock for key customers or markets. • Check against transportation capacity, production, and planned receipts.
Use case example:	• A company with limited stock, high demand, multiple priority customers, or global distribution where fair allocation is critical.

Key Point for Implementing aATP Functionalities:

aATP is essential only when client demand significantly exceeds supply that is, when demand is higher than available stock (e.g., product launches, shortages, or capacity constraints). It ensures fair allocation using PAL/BOP instead of classic ATP's first-come-first-served approach. Some trigger scenarios as below.

- New high-demand products where stock must be allocated based on customer priority.
- Supply chain disruptions requiring order prioritization and segmentation.
- Stable supply chains: aATP can be skipped to keep processes simple.

With Advanced Available-to-Promise (aATP), the system evaluates several key factors to determine whether a customer order can be fulfilled on time:

Current Stock Levels:	• Available quantities at the requested plant or warehouse.
Inbound Supply:	• Planned receipts from purchase orders, production orders, stock transfers.
Existing Commitments:	• Other open sales orders or reservations that might consume available stock.
Product Allocation:	• Allocated quantities reserved for specific customers or sales channels to manage scarce resources.
Alternative Sources:	• Availability from alternate plants or substitute materials when the primary source is insufficient.
Backorder Processing:	• Prioritization and rescheduling of existing orders when supply or demand changes.
Transportation and Production Constraints:	• Consideration of lead times and scheduling relevant delivery dates.
Customer and Business Rules:	• Customer priority, segmentation, and special agreements influencing confirmation.

Based on these evaluations, aATP provides realistic confirmation proposals indicating confirmed quantities and delivery dates, potentially splitting deliveries if full fulfillment is not possible on the requested date. This comprehensive evaluation ensures precise, reliable order promising aligned with actual supply chain capabilities in SAP S/4HANA.

- **Current stock levels**
- **Planned production** (e.g., planned orders, production proposals)
- **Inbound supply** such as Purchase Order, Production Orders, Stock Transfer Orders
- **Product allocations** to ensure fair and strategic distribution of limited inventory.
- **Substitution options**, including product, plant, and location substitutions.

aATP is a commitment to the customer to deliver the product on time.

Example-
Suppose we receive an order from our customer for Material FG10101 with a quantity of 5,000 units and a requested delivery date of November 20th. When the customer sends us their Purchase Order (PO), it is converted into a Sales Order in the system with the product, quantity, and requested delivery date.

During order confirmation, the system performs the availability checks to determine whether the requested delivery date can be met. In addition, the system also checks warehouse capacity.

Customer requests for 5,000 units of product FG10101 for delivery on November 20th - aATP will check real-time availability and supply elements.

- aATP may confirm all 5,000 units for November 20th, or
- It may split the confirmation, for example:
    - 4,000 units on November 20th
    - 1,000 units on December 5th

This confirmation depends on stock, planned production, inbound supply, allocations, and other aATP calculations.

In short, aATP calculation in SAP S/4HANA works as follows:

1	*Collect Demand:*	The system gathers all relevant requirements such as Sales Orders, Stock Transport Orders (STOs), and other customer or internal demands.
2	*Check Supply (inbound and outbound):*	It evaluates the current stock situation, including available warehouse inventory, open production orders, purchase orders, planned receipts, and in-transit stock—to determine what supply can cover the demand.
3	*Apply Business Rules & Priorities:*	aATP considers factors like product allocations, backorder processing rules, customer importance or profitability, and strategic priorities that influence how limited stock should be distributed.
4	*Confirm Realistic Quantities & Dates:*	Using the combined view of demand, supply, and business priorities, the system provides reliable confirmations for quantities and delivery dates that can realistically be met.

AATP Calculation Flow Diagram-

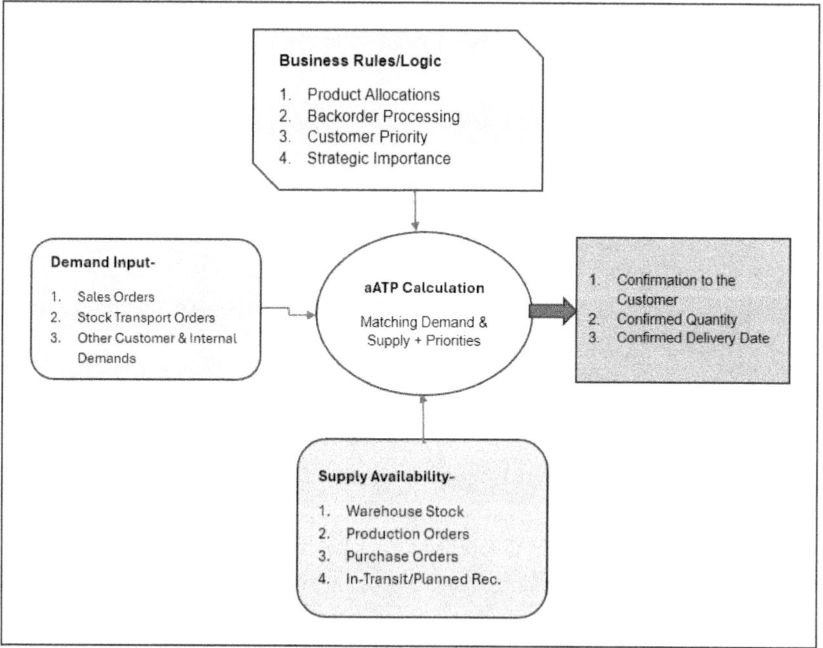

SAP S/4HANA aATP follows the strategy of being **Simple, Fast, Flexible, Intelligent, and Integrated**.

aATP is called an Intelligent Order Promising solution because it uses real-time data, advanced algorithms, and business rules to deliver the most accurate and optimized confirmations for order quantities and delivery dates.

It is considered *intelligent* because exceptions are handled automatically through rules-based logic, back-order processing, and priority-driven decision making.

aATP is also fully integrated into the end-to-end SAP S/4HANA landscape, connecting with all key modules such as Sales and Distribution, Manufacturing, Logistics Execution and Supply Chain Planning, ensuring seamless and consistent availability checks across the entire supply chain. aATP also integrates with multiple planning systems - both SAP (like IBP and S/4 PP/DS) and non-SAP (such as Oracle, Kinaxis, or custom APS via OData/REST APIs) - enabling synchronized allocation strategies and real-time checks across diverse landscapes.

6. aATP Key Concepts

aATP transforms SAP from a stock-focused promise engine to a smart, supply-aware, and capacity-aware order fulfillment system, giving realistic delivery dates, prioritization, and improved customer satisfaction. aATP ensures that sales orders are confirmed realistically, balancing customer demand with supply availability and production capacity. It closes the gap between sales promises and what the supply chain can actually deliver, which is essential for reliable order fulfillment.

1. Before committing a delivery date to the customer, the system checks all the aATP parameters mentioned above, such as supply availability, demand inputs, and applicable business rules and logic, to ensure an accurate and reliable delivery commitment.

2. Schedule Line Confirmations in aATP-
The Sales Order confirmation is determined at the Schedule Line level. The schedule line displays both:

- The requested delivery date (as given by the customer), and
- The confirmed delivery date (calculated by aATP)

After the aATP logic is executed, the confirmed date reflects real-time stock availability, supply elements, production plans, allocations, and warehouse capacities.
The Schedule Line is one of the most important components of a Sales Order because it clearly shows the customer's request and the system's realistic confirmation based on availability.

3. Core aATP Components and Key Functions-

a. **Product Availability Check (PAC)**: Verifies whether the requested quantity can be delivered on the requested date. Determines what quantity of requested material can be confirmed by considering current stock, receipts in transit, planned production, and concurrent demands.

b. **Product Allocations (PAL):** Controls how much product each customer/region can receive during shortages. Controls how limited stock is allocated to various customers or channels, supporting fair and strategic allocation.

c. **Backorder Processing (BOP):** Automatically reallocates limited stock based on priority rules (e.g., VIP customers first). Prioritizes and reschedules existing sales orders based on business rules and changes in stock or supply situation.

d. **Alternative-Based Confirmations (ABC):** Suggests automatic substitutions like:

- A different storage location
- A different plant
- A different product
- A different delivery date

When requested stock is not available in the chosen plant, aATP can confirm delivery from alternate plants or sources.

ABC checks hierarchy:

1. Storage locations (same plant first)
2. Plants (alternate plants)
3. Products (substitutes)
4. Dates (earlier/later delivery)

ABC includes "different delivery date" because it adjusts dates when stock isn't available on request date. Key strategies: FULL_CONFIRMATION (exact date), MAX_ON_TIME (max qty on request date), MAX_EARLIER (earliest possible date). Without date flexibility, many real-world substitutions fail.

e. **Supply Protection:** Reserves future inventory for strategic customers or sales channels. SuP ensures that important customers always receive their promised stock, even when overall inventory is limited.

f. **Release for Delivery**

aATP can run a delivery release check to ensure that all constraints (stock, allocation, protection, capacity) are fulfilled before releasing an order for delivery. It ensures that only valid, confirmed orders move to the shipping stage.

g. **Supply Based Confirmation**

Used when supply does **not exist today** but will be created in the future. The system considers:

- Planned production
- Purchase orders
- Planned orders
- Inbound deliveries

h. **Business Process Scheduling**

Business Process Scheduling (BPS) is a flexible, configurable framework in SAP S/4HANA for scheduling logistical activities across business documents like sales orders, deliveries, and stock transports.

aATP can confirm an order **based on future supply creation**, ensuring realistic promised dates.

4. aATP vs. Classic ATP-

- aATP provides more sophisticated allocation, simulation, and fulfillment capabilities, including customer prioritization and multi-level supply chain checks.
- In S/4HANA, classic ATP (ECC logic) and aATP can coexist, assigned at the material level, but only one can be active for a given material in an order.
- Classic ATP is suitable for straightforward order processing, while aATP is designed for complex fulfillment, strategic customer management, and advanced supply chain needs in modern enterprises.

Here is a comparison table highlighting the key differences between Classic ATP (Available-to-Promise) and Advanced ATP (aATP) in SAP S/4HANA, focusing on features, capabilities, and business benefits:

Feature/Aspect	Classic ATP (Basic ATP)	Advanced ATP (AATP)
Availability Check	Real-time stock and receipt check	Multiple supply chain constraints included
Confirmation Level	Single-level, basic confirmation	Priority/customer-based and multi-level
Backorder Processing	Limited (mass change)	Advanced BOP: Prioritization, rescheduling
Product Allocation	Not available	Full-feature product allocation
Alternative-Based Confirmation	Not available	Supports alternative plant/source logic
Segmentation/Customer Priority	Not supported	Fully configurable prioritization

Simulation & Analytics	Minimal	Scenario simulation and analysis tools
User Experience (UI/Fiori)	Classic GUI only	Modern Fiori apps and user interfaces
Integration with Other Apps	Standard SD/MM	Deep integration with sales, EWM, PP, TM
Extensibility & Innovations	Not evolving further	Continually enhanced with new features
Use Case	Simple fulfillment, standard processes	Complex, multi-channel/order scenarios

5. Document Types-

In SAP S/4HANA, aATP (Advanced Available-to-Promise) can be applied to several types of documents that involve customer demand and delivery commitments. These documents are where the system checks real-time availability and confirms quantities and delivery dates.

aATP checks primarily occur in the following processes:

- Sales orders
- Deliveries
- Stock transport orders
- Production orders
- Backorder processing (BOP)
- Product allocation (PAL)
- Alternative confirmations (such as location, product, or mode substitution) purpose

These multi-level aATP checks ensure realistic, accurate, and flexible availability commitments across sales, procurement, and production processes. They help organizations manage supply constraints effectively and provide reliable confirmations to customers.

6. Consultant's Role in aATP Setup for Business-

A consultant's role in SAP S/4HANA aATP (Advanced Available-to-Promise) implementation is both strategic and hands-on, guiding clients through design, configuration, and optimization to fit specific business requirements.

A consultant is responsible for designing and configuring how aATP checks and confirmations operate. This includes customizing rules for customer prioritization, product allocation, scheduling, and exception handling to align with the client's specific business processes and service-level expectations.

7. Benefits of aATP-

aATP provides advanced, intelligent availability checking that is essential for modern supply chains. It helps businesses meet complex customer demands, improve fulfillment accuracy, and optimize delivery commitments within SAP S/4HANA.

Key Points-

1. aATP **replaces classic ATP** for the assigned materials.
2. SAP S/4HANA Advanced ATP (aATP) is offered in both on-premise and cloud deployments, with identical core concepts but some advanced features delivered first or exclusively in the cloud.
3. The availability check is assigned at the material level (in the MRP3 view, field "Availability Check"), so one material may use classic ATP while another uses AATP.
4. The same material can have different availability checks for different plants because, in SAP, the availability check is maintained at the **plant level** for a material. This means:
    - Material FG10101 in Plant 1000 can use classic ATP.
    - The same Material FG10101 in Plant 2000 can be used as aATP.

   The setting in the MRP3 view allows plant-specific configuration, so the availability check method can differ across plants for the same material.
5. During Sales Order creation, aATP is triggered based on the Checking Group assigned in the Material Master, and the Checking Rule assigned to that group determines the detailed logic and extensions (BOP, PAL, ABC, SBC, BPS) for the availability check.
6. aATP works with scheduled lines in sales orders or stock transport orders to determine confirmed quantity and delivery dates.

7. **Checking rule determined in the sales order (aATP check)-** The determination of the checking rule (such as checking rule "A") during sales order creation in SAP S/4 HANA is based on a combination of system-internal logic and master data settings:

In SAP ECC & S/4 HANA aATP, the checking rule is *NOT taken from master data.*
Instead, it is determined by the application (transaction) you are executing. So, the source of checking-rule determination is the *transaction type / application area.*

Debug Confirmation: In /h debug mode during VA01 -> ATP check -> checking rule "A" appears automatically (not from material master).
Sales Order Analysis:
VA03 → Item → Environment → Availability
Log shows: "Checking rule A / Checking group 02 used"
Hard Proof: Checking rule never comes from material master—always transaction-driven:

- Sales Order = A
- Delivery = B
- Invoice = No ATP

SAP determines checking rule internally by hard-coded transaction logic, not configuration.

Examples of standard SAP checking rules:

Business Process / Transaction	**Checking Rule**	**Source of Rule**
Sales Order (VA01 / VA02)	**A**	Hard coded in SD application
Delivery (VL01N / VL02N)	**B**	Hard coded in LE application
Production Order (CO01 / CO02)	**PP**	Hard-coded in PP application
Stock Transfer Order	**Varies**	Determined by MRP / STO logic
Backorder Processing (BOP)	**aATP Rules**	Determined by aATP framework

- The checking rule is primarily predefined and hardcoded in SAP for different transaction types. For example, checking rule "A" is the standard rule for normal sales orders, while other rules like "AE" or "AW" apply to special cases like make-to-order or consignment orders.
- The system derives the checking rule based on the transaction context—mostly the sales order type and sales document category.
- The checking group (also called availability check group) comes from the material master, typically from the MRP 3 or Sales views. This group controls which set of availability check rules apply to the material.
- The combination of checking rule and checking group together defines the Scope of Check configured in transaction OVZ9. This scope determines which stock categories, receipts, and requirements are considered during the availability check.
- During sales order creation, the system automatically assigns the checking rule according to the sales document type and material characteristics, combines it with the checking group from the material, and uses this combination to perform the aATP check.
- For example, for a standard sales order of a make-to-stock material, the system uses checking rule "A" + checking group from the material, triggering the appropriate AATP availability logic.

This means the checking rule is not explicitly a user input but is determined by SAP based on sales order type and transaction logic, in combination with material master settings, to ensure the correct advanced availability check settings apply seamlessly during order processing.

Here is the explanation of common SAP checking rules – A, AE, B, BE, and BO – and their purposes in availability checks:

Checking Rule	Purpose / Usage
A	Standard checking rule for make-to-stock (MTS) sales orders. It considers stock, receipts, and issues to confirm material availability. Used in typical sales order processing.
AE	Used for make-to-order (MTO) or customer-specific production scenarios. It incorporates special logic reflecting ordered production and associated supply planning.
B	Applied in delivery processes, such as outbound deliveries and stock transport orders. It usually has stricter criteria since delivery confirmation needs firmer inventory availability.
BE	Used in special stock scenarios, often for rush orders, consignment, or returns processes. Contains customized availability check logic suitable to business needs.
BO	Associated with backorder processing, specifically for re-allocation and reprioritization of open orders when stock shortages occur. Helps in managing priority and fairness in confirmations.

8. Scope of Check in ATP v/s, aATP

The scope of check in SAP S/4HANA is a critical configuration that determines exactly which stock types, receipts, and requirements the system considers during an availability check (ATP or aATP). It works as a set of rules that define how the system evaluates available quantities to confirm customer orders, deliveries, production, or transfers.

There are effectively two levels for defining “scope of check”: the classic SD/PP scope of check and the aATP scope of check, and they sit in different IMG nodes and are maintained with different transactions. (OVZ9 and OVZ9A).

The key difference between ATP and aATP scope of check is that AATP is configured based on **categories**, allowing more flexible and advanced availability checks.

The scope of check is defined using a combination of the Checking Group (from material master) and Checking Rule (based on transaction type, e.g., sales order or delivery). This combination determines which elements (such as unrestricted stock, purchase orders, reservations, planned orders, safety stock, etc.) are taken into account during the availability check.

a. Classic ATP scope of check

In classic ATP (the "old" SD/PP ATP), the scope of check is defined in the availability check control and tied to the checking group and checking rule. Technically this is where you decide which stock types, receipts, and requirements are considered (e.g. safety stock, stock in transfer, purchase orders, planned orders, sales orders, deliveries, reservations, etc.).

Key Points-

- Maintained with the classic availability-check transactions OVZ9 (for example via the scope-of-check transaction that is linked to checking rule and checking group).
- IMG path is under Cross Application Components → Advanced Available-to- Promise (aATP) → Product Availability Check (PAC) → Configure Scope of Availability Check
- Result is a scope that controls classic ATP quantity calculation (plant / storage-location level, ATP categories, etc.).

b. aATP scope of check

Advanced ATP (aATP) has its own scope definition that works on ATP categories and supply/demand segments and is configured in the aATP-specific IMG area. Here you define which ATP categories (stocks, receipts, and requirements) are relevant for the aATP run for a given scenario or document type, independent of the old OVZ9-style scope.

Key Points:

- Accessed via the aATP customizing transaction OVZ9A – "Configure Scope of Advanced Availability Check".
- IMG path is under Cross Application Components → Advanced Available-to- Promise (aATP) → Product Availability Check (PAC) → Configure Scope of Advanced Availability Check
- Used by aATP checks (PAC, product allocation, supply protection, etc.), not by classic SD ATP.

9. How Classic ATP and Advanced ATP Coexist

Classic ATP and aATP can coexist in the same S/4HANA system, but they do not run on the same material/plant at the same time; you decide per material-plant where classic ATP is used and where aATP takes over. This allows a gradual transition: stable or simple scenarios can stay on classic ATP, while complex, high-priority or constrained products move to aATP.

Both run in parallel- controlled by material master activation (OVZ2 -> Advanced ATP flag).

- aATP scope of check is evaluated when an availability check is routed through the aATP engine; in those cases, the aATP configuration (ATP categories and segments) is leading, while the classic scope mainly needs to stay consistent so that classic checks in the same process chain do not contradict aATP.
- Activation is done at material/plant level (for example via configuration like OVZ2 – Availability Check 02), where a flag determines whether the material uses classic ATP or aATP.
- Because the engines work differently (temporary quantity assignments, product allocation logic, etc.), SAP requires a clear decision per material/plant rather than mixing both checks for the same combination.
- Many companies keep standard or low-complexity materials on classic ATP and move critical, allocated, or highly constrained materials to aATP to use PAL, BOP, and ABC.
- Over time, as experience and license coverage grow, more materials are migrated from classic ATP to aATP until classic ATP is used only for residual or legacy cases.
- From a user standpoint, both are triggered from similar SD processes (sales order, delivery, etc.), but the underlying check logic differs depending on whether the material/plant is configured for classic ATP or aATP.
- This setup lets businesses adopt aATP incrementally without disrupting all existing ATP processes at once, while still leveraging S/4HANA's advanced capabilities where they add the most value.

10. Scope of Check in SAP – How It Works

The scope of check defines which stock, receipt, and requirement elements the SAP system considers when performing an availability check (ATP).

- **Configuration:** - In SAP S/4HANA, the scope of check configuration differs for Classic ATP and Advanced ATP. Classic ATP scope of check is maintained in transaction code OVZ9, whereas Advanced ATP scope of check is configured in OVZ9A using ATP categories.

During configuration, you define what should be included or excluded in the availability check, including:

- Available stock
- Planned receipts (purchase orders, production orders)
- Sales and delivery requirements
- Safety stock
- Reservations and dependent requirements
- Requirement types and check rules

Execution: - When an availability check is performed (for example, during sales order creation, delivery, or production order processing), the system uses the scope of check that is mapped to the checking group and checking rule. Based on this configuration, the system dynamically evaluates all relevant stock, as well as future supply and demand elements defined in the scope, to determine whether availability exists, the confirmed quantity, and the confirmation date.

Example: - For a standard sales order, the scope of check may include unrestricted-use stock, inbound purchase orders, and scheduled production, while excluding quality inspection stock and planned orders. For delivery processing, the scope of check is typically more restrictive, considering only stock that is physically available.

The objective is to align system confirmations with real-world supply chain execution and business priorities, ensuring that commitments made to customers are realistic and achievable.
As a result, the scope of check is a central component of S/4 HANA's advanced order fulfillment logic, enabling availability checks that are precise, flexible, and driven by business requirements.

11. Transactions OVZ9/OVZ9A flags map to specific ATP tables-
In Classic ATP, OVZ9 flags control which MRP elements are read from tables like MARD, EKET, AFPO, and VBEP. In Advanced ATP, OVZ9A maps to ATP categories that drive HANA-optimized reads from MATDOC and aATP-specific tables, enabling advanced fulfillment logic. The flags and indicators you see in the Scope of Check configuration in transaction **OVZ9/OVZ9A** in SAP S/4HANA map to specific internal tables that control which stocks, receipts, and requirements are considered during an availability check (ATP or aATP).

In SAP ATP (Classic or Advanced), flags in the scope of check (T441V for Classic ATP, aATP categories for Advanced ATP) control what stock, receipts, and requirements are considered during the availability check. They act as on/off switches that dynamically determine which elements are included or excluded in the calculation.

a. Key OVZ9 flags (classic ATP)-
OVZ9 as "classic ATP scope of check control" tied to table T441V and used to decide which stock, receipts, and issues are included.
These are defined per Checking Group / Checking Rule combination and saved in scope-of-check table entries (T441V). Exact technical field names differ by element, but conceptually they map to indicator fields in T441V and related ATP macros.

ATP macros in aATP define the calculation logic used during availability checks, controlling how supply and demand are evaluated and how confirmations are determined.
Note: These macros are internal ATP evaluation routines, triggered via aATP macros and ATP categories. SAP aATP macros are embedded within the aATP framework as non-configurable evaluation routines and cannot be modified directly.

b. Key OVZ9A flags (aATP scope of check)-
OVZ9A as "aATP scope of check" that reuses the same idea (scope of check per checking group/rule) but for aATP-active groups, with ATP category usage and aATP-specific flags maintained only when Advanced ATP is active in OVZ2.

OVZ9A is the S/4HANA transaction used for aATP scope-of-check customization. It is linked to checking groups that are marked as 'Adv.ATP active' in OVZ2 (field 'Adv.ATP'). The technical data is stored within the scope-of-check framework and builds on the same category/indicator concept as T441V but extended to support aATP categories and advanced logic.

- ATP category usage: Controls which ATP categories (stock/receipt/issue types) are relevant for aATP for a given checking group/checking rule; configuration is only possible if the checking group has Adv.ATP = A (active).

- Element inclusion flags: For each category (for example stock, purchase orders, production orders, sales orders, deliveries), on/off indicators determine whether that category flows into the aATP calculation.
- aATP-specific extensions: New categories and flags relevant for aATP scenarios (such as special supply elements) are controlled here, ensuring that aATP reads the correct set of elements distinct from classic ATP.

12. Different scope of check for SD/MM/PP-

In SAP S/4HANA, there are different scopes of check for Sales and Distribution (SD), Materials Management (MM), and Production Planning (PP). Although the fundamental concept of scope of check—to define which stocks, receipts, and requirements are considered during availability checks—is consistent, the specific configuration and elements considered vary depending on the module and business process.

a. Sales and Distribution (SD): The scope of check is configured to include stock types and requirements relevant to fulfilling customer orders. This typically considers unrestricted stock, stock in transit, purchase orders, production orders, sales orders, deliveries, and backorder processing with focus on customer order fulfillment.

b. Materials Management (MM): The scope of check in MM often focuses on procurement and inventory management aspects. It considers stock, purchase requisitions, purchase orders, and planned receipts for managing material availability from a procurement perspective.

c. Production Planning (PP): In PP, the scope of check is tailored for production related availability, such as components for production orders, planned orders, safety stock, and material reservations necessary for production scheduling and execution.

How It Works-

- The scope of check is defined in OVZ9 using combinations of Checking Group (from material master) and Checking Rule (transaction type).
- Different checking rules are assigned or designed for SD, MM, and PP to reflect these process-specific needs.

- For example, standard checking rules like A and AE are often used in SD, while PP can use specific rules like PP for production order checks.
- The flags set within scope of check determine which types of stocks, receipts, and requirements are included or excluded, thus tailoring availability checks according to module business logic.

These are **checking rules encountered in SD, MM, and PP**:

Module	Checking Rule	Typical Use / Description
SD (Sales & Distribution)	*A*	Sales Orders – standard ATP check for order confirmation
	AE	Sales Orders for make-to-order stock
	AW	Consignment sales orders (in some setups).
	B	Deliveries – ATP check at delivery stage
	BE	Deliveries for make-to-order stock
	BO	Backorder processing scenario in SD
	BW	Delivery for consignment or special scenarios
	BV	Delivery for returnable packaging (custom).
MM (Materials Management / STO)	*RP*	Stock Transport Orders – ATP check for inter-plant transfers
PP (Production Planning / Manufacturing)	*PP*	Production / planned orders – component and finished goods availability

The basic aATP configuration is controlled through the node 'Advanced Available-to-Promise (aATP)' under 'Cross-Application Components,' including the activation of the relevant functionalities.

- ***Classic ATP Configuration Node- Transaction Code OVZ9***

SPRO→ IMG→ Cross Application Components → Advanced Available-to-Promise (aATP) → Product Availability Check (PAC) → Configure Scope of Availability Check

- ***AATP Configuration Node- Transaction Code OVZ9A***

SPRO→ IMG→ Cross Application Components → Advanced Available-to-Promise (aATP) → Product Availability Check (PAC) → Configure Scope of Advanced Availability Check

13. Material Availability Date - MAD Calculation-

At the time of sales order creation, when ATP is calculated, the system first determines the Material Availability Date (MAD) and then adds the relevant additional days (such as picking, packing, loading, and transportation) to arrive at the final confirmed delivery date.

MAD (Material Availability Date) in SAP S/4HANA sales orders is the calculated date when sufficient material stock becomes available for picking and packing to meet the requested delivery date (RDD).

a. MAD (Material Availability Date) is the date on which sufficient material is available in the plant/storage location so that picking and packing can start. Essentially, it tells SAP: *"When can we actually start fulfilling this order?"*

At the time sales order creation, SAP:

- Determines MAD using ATP (stock + receipts – issues)
- Adds additional lead times to determine the Confirmed Delivery Date (CDD)

Confirmed Delivery Date = MAD + Picking + Packing + Loading + Transportation

Note- In SAP, Customer RDD is the requested delivery date specified by the customer - the target date for goods arrival at their location - It drives ATP scheduling and planning but may differ from the actual receipt date confirmed after shipment.

b. Scheduling Logic Overview- SAP uses two types of scheduling in sequence:

During the ATP check at sales order creation, SAP calculates the Material Availability Date (MAD) by first performing backward scheduling. If backward scheduling fails, the system then performs forward scheduling to determine the earliest possible MAD and confirmed delivery date.

Step	Scheduling Type	Purpose
1	Backward Scheduling	Try to meet the **Requested Delivery Date (RDD)**
2	Forward Scheduling	Used if backward scheduling fails

c. Backward Scheduling (Primary Logic)

During MAD calculation, backward scheduling is always attempted first to try to confirm the customer's requested delivery date (RDD). Backward scheduling subtracts lead times (transportation, loading, picking/packing, material staging) from RDD to derive MAD, the date when materials must be available. ATP then checks cumulative quantities (stocks + inbound receipts like purchase orders) from MAD backward; if sufficient, it confirms on MAD, reserving inventory on a first-come, first-served basis.

Example - Backward Scheduling at Sales Order Data

- Requested Delivery Date (RDD): **15 Oct**
- Transportation time: **2 days**
- Loading time: **1 day**
- Picking + Packing time: **2 days**

Step-by-Step Calculation

Step	Calculation	Date
RDD	Customer Request	15 Oct
Transportation	15 - 2	13 Oct
Loading	13 -1	12 Oct
Picking/Packing	12 - 2	10 Oct (**MAD**)

So, in this case, during the ATP check, if stock is available on 10 Oct, the results will be:

• MAD = **10 Oct**

• Confirmed Delivery Date = **15 Oct**

d. Forward Scheduling (Fallback Logic)

During MAD calculation, backward scheduling is always attempted first to try to confirm the customer's requested delivery date (RDD). If backward scheduling fails because sufficient stock is not available on the calculated MAD, the system then performs forward scheduling. In forward scheduling, the system:

- Finds the earliest date on which the material becomes available.
- Adds the relevant logistics lead times (picking, packing, loading, transportation) forward to calculate a new confirmed delivery date.

Example - Backward Scheduling Fails → Forward Scheduling at Sales Order Data

- Requested Delivery Date (RDD): **15 Oct**
- Material available only on: **14 Oct**
- Picking + Packing: **2 days**
- Loading: **1 day**
- Transportation: **2 days**

Backward Scheduling Attempt

- Calculated MAD = **10 Oct**
- Stock available on **10 Oct** = No

Forward Scheduling

Step	Calculation	Date
Material Availability	Earliest Stock	14 Oct
Picking/Packing	14 + 2	16 Oct
Loading	16 + 1	17 Oct
Transportation	17 + 2	19 Oct

So, in this case, during the ATP check, if stock is NOT available on 10 Oct, the results will be:

- MAD = **14 Oct**
- Confirmed Delivery Date = **19 Oct**

e. MAD Calculation Example: Backward & Forward Scheduling

Customer orders on Jan 1, 2025, with Requested Delivery Date (RDD) = Jan 10. Lead times from master data: Transit=2 days, Loading=1 day, Pick/Pack=2 days, Staging=1 day.

Backward Scheduling (Stock Available)

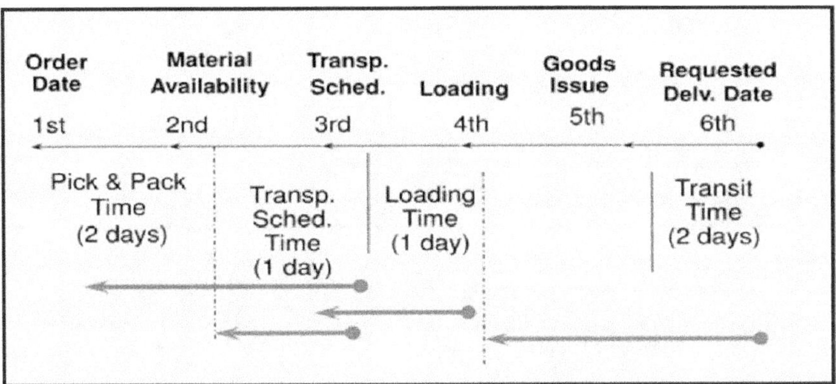

Step	Activity	Duration	Date Calculation	Result Date
1	RDD	-	Jan 10	Jan 10
2	- Transit Time	-2 days	Jan 10 - 2	Jan 8 (Goods Issue)
3	- Loading Time	-1 day	Jan 8 - 1	Jan 7 (Loading Date)
4	- Transp. Sched. Time	-1 days	Jan 7 - 1	Jan 6 (Transp. Sched Time)
5	- Pick/Pack Time	-2 days	Jan 6 - 2	Jan 4 (Transp. Planning)
6	= MAD	-1 day	Jan 4 - 1	Jan 3 ← ATP checks stock here

* Result: 100 units available on Jan 3 → Confirm delivery Jan 10.

Forward Scheduling (No Stock on Jan 3 MAD)

Forward scheduling is used if backward scheduling fails, e.g., stock is not available on Jan 3. Start from earliest feasible date (today or stock availability)

- Order creation date = Jan 1
- Suppose stock is available only on Jan 5

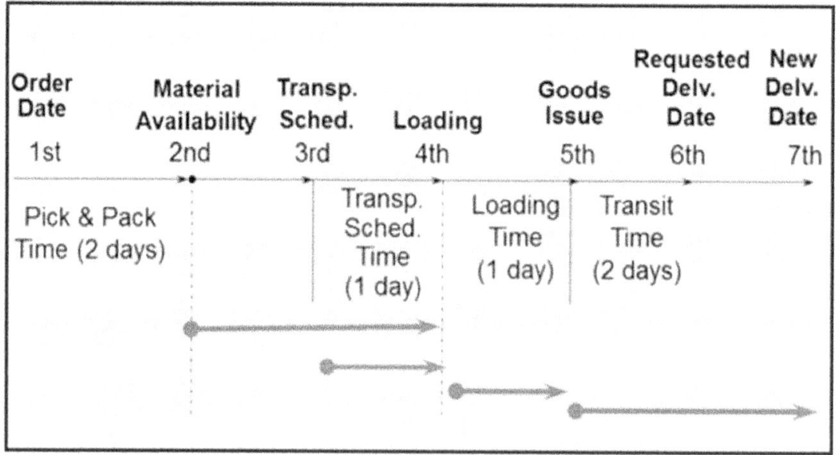

Step	Activity	Duration	Date Calculation	Result Date
1	Today	-	Jan 1	Jan 1
2	+ Staging	+1 day	Jan 5 + 1	Jan 6 (New MAD)
3	+ Pick/Pack Time	+2 days	Jan 6 + 2	Jan 8
4	+ Transp. Sched. Time	+1 day	Jan 8 + 1	Jan 9
5	+ Loading Time	+1 day	Jan 9 + 1	Jan 10
6	+ Transit Time	+2 days	Jan 10 + 2	Jan 12 (Confirmed Delivery)

* Logic: Backward fails (past/no stock) → Forward from current date proposes earliest feasible dates, optimizing pull to minimize gaps.

f. Key Dates in Sales Order (Schedule Line)

Key dates in sales order (schedule line) are MAD, GI Date and Delivery Date

Date Field	Meaning
MAD	The date when sufficient stock is available for picking and packing.
GI Date	Goods issue date (after picking/packing/loading). the date when goods are issued from the warehouse.
Delivery Date	Customer confirmed delivery date. the confirmed date on which the customer is expected to receive the goods.

Key Terms-

1. MAD (Material Availability Date)

- The date when enough material is available in the plant or storage location to start picking and packing.
- MAD Calculated by ATP (Available-to-Promise) in SAP:
  ATP = STOCK + RECEIPTS - ISSUES
- Essentially, it tells SAP: *"When can we actually start fulfilling this order?"*

2. CDD (Confirmed Delivery Date)

- The date SAP confirms for delivery, considering all additional lead times after material is available.
- Formula:
  CDD = MAD + Picking + Packing + Loading + Transportation
- This is what SAP communicates as the expected delivery date to the customer.

3. Customer RDD (Requested Delivery Date)

- This is the date the customer wants the goods to arrive.
- SAP uses it as a target in ATP checks, but the actual confirmed date (CDD) may differ if stock or resources are insufficient.

Flow in Sales Order (Step-by-Step):

1	*Sales Order Creation*	• Customer RDD is entered. • SAP checks ATP to see when material is available (MAD).
2	*Lead Time Addition*	• SAP adds internal processing times: picking, packing, loading, transportation -> gives CDD.
3	*Confirmation*	• SAP confirms the delivery date (CDD) to the customer. • The customer may get a date earlier, later, or on their requested date depending on material availability and lead times.

Example:

- Customer requested **RDD = 8-Jan**
- SAP confirms **CDD = 10-Jan** because MAD + processing + transport = 10-Jan

Parameter	Days
MAD	Jan 05
Picking	1
Packing	1
Loading	1
Transportation	2
CDD	Jan 10

In SAP Sales Orders, the Customer Requested Delivery Date (RDD) refers to the date the customer wants the goods to arrive at their location, not the date to leave your warehouse. Here's the distinction:

Term	What it Represents
RDD (Customer Requested Delivery Date)	Target date for the goods to **arrive at the customer's site**. Drives ATP scheduling in SAP.
Planned Goods Issue / Shipping Date	The date the goods need to **leave your warehouse** so they can reach the customer by the RDD.
CDD (Confirmed Delivery Date)	The actual **confirmed arrival date** at the customer site after considering material availability and lead times.

Example:

- Customer RDD = 20-Jan
- Transportation time = 3 days
- Picking, packing, and loading = 2 days

Step 1: Determine Goods Issue / Shipping Date

- Goods need to reach the customer by 20-Jan, and transportation takes 3 days. So, we subtract transportation time:
  Shipping Date = RDD – Transportation (20-Jan – 3 days = 17-Jan)

Step 2: Check if warehouse processing fits

Picking + Packing + Loading = 2 days

- That means the warehouse needs to start processing 2 days before the shipping date: Start Picking/Packing = 17-Jan – 2 days = 15-Jan

Step 3: Confirm the flow

Date	Activity
15-Jan	Material available (MAD), start picking/packing/loading
17-Jan	Goods leave warehouse (shipping date / goods issue)
20-Jan	Goods arrive at customer (RDD)

The goods should leave the warehouse on 17-Jan to reach the customer on 20-Jan. So, in short: **RDD is arrival at customer**, and SAP calculates backward to figure out when the goods must leave your plant.

14. Lead Times - The lead times used in MAD and confirmed delivery date calculation come from the following SAP objects.

Lead Times in SAP for MAD and CDD Calculations-

Lead Time	Correct SAP Source / Object	Notes
Picking / Packing	Shipping Point	Usually linked to the shipping point and internal warehouse processes. Picking and packing times can also be in shipping conditions or route determinations.
Loading	Shipping Point	Loading time is generally maintained at the shipping point level.
Transportation	Route	Transportation or travel time comes from the route assigned to the delivery, often including planned transit time.
Planned Delivery Time	Material Master – MRP2	PDT is purely internal lead time, stored in MRP2, and it represents the time from order creation until goods are ready for delivery.
Transit Time	Route	Time for goods to physically move from plant/warehouse to customer location; derived from the route (can also include shipment scheduling).

Planned Delivery Time (PDT) during Sales Order ATP check:
Planned Delivery Time represents the time required to make the material ready for shipment after a sales order is received. It is used in the calculation of the Confirmed Delivery Date (CDD) and for MAD alignment. For externally procured materials or services, it indicates the number of calendar days needed to obtain them.

15. Plant Determination in SAP Sales Orders (Line-Item Level)

When a sales order line item is created, SAP determines the plant from which the material will be supplied. The sequence depends on order data, material master, customer, and configuration.

Sequence / Logic for Plant Determination-
During ATP calculation, SAP first needs a delivering plant. The plant is determined automatically (e.g. CMIR-> customer master -> material master). Only once a plant exists does the system perform the ATP check using parameters such as planned delivery time and other MRP/availability data. If no plant is determined automatically, the user must enter a plant manually without a plant, the system cannot calculate ATP.

Sequence	Determination Method	Source / Notes
1	*Customer-Material Info Record*	• If a delivery plant is maintained for that customer–material combination, the system takes that plant with highest priority. • Field: Customer-Material Info Record → Delivery Plant.
2	*Customer-specific sales area rule*	• In the standard S/4HANA SD plant determination, the second step is the ship-to customer master's Delivering Plant field
3	*Material Master*	• If there is no plant in CMIR and customer master, the plant from material master (Sales: Sales Org. 1 → Delivering Plant) is used.
4	*Manual entry*	• Entered by user at sales order line item. Manual entry in the sales order item (overrides any of the above)

16. Important Notes in aATP-

1. MAD (Material Availability Date) is ATP-relevant and aATP checks. Partial confirmations can result in multiple MADs-
   - Stock
   - Planned receipts (POs, production orders)
   - Allocations & BOP (if active)
2. SAP always tries backward scheduling first; forward scheduling is used only if stock is not available.
3. In SAP, the Material Availability Date (MAD) is the key ATP date. The final confirmed delivery date is calculated as MAD plus logistics lead times
4. ATP Integration - ATP then checks cumulative quantities (stock + receipts) from MAD backward; if available, it confirms on MAD before adding post-MAD lead times for the final delivery

proposal. If MAD falls in the past or stock lacks, forward scheduling shifts from current date or earliest receipt.

In SAP sales orders, the Material Availability Date (MAD) calculation relies on key input fields that drive backward scheduling from the Requested Delivery Date (RDD).

5. Primary Sales Order Fields-
    - Requested Delivery Date (RDD): Entered at header or item level (VBKD-EDATU); serves as the starting point for subtracting lead times to derive MAD.
    - Shipping Point: Determines calendars and loading times (VBEP-VSTEL); pulled from item or route.
    - Route: Specifies transportation lead time (VBEP-ROUTE); influences transit duration subtraction.
6. Master Data Dependencies - Material Master provides picking/packing/handling time (MARC-ANZSN, MARC-LADEZ), while shipping point and plant contribute staging durations via configuration. Customer master and sales area data indirectly affect via route determination.

17. Fixed Date & Qty

The ‘Fixed Date & Quantity’ checkbox in SAP sales order schedule lines (VBEP-FIXMG / FIXDAT) locks the confirmed quantities and dates after the ATP check, making them non-reschedulable.
ATP confirmations are flexible by default; Fixed Date & Quantity is required to explicitly lock customer-accepted commitments and prevent rescheduling or reallocation.

Business Interpretation

- **System ATP/aATP** → Planning promise (dynamic, flexible)
- **Fixed Date & Quantity** → Contractual promise (locked, firm)

Even with the advanced capabilities of aATP, the system **requires explicit confirmation from the business** to convert a planning promise into a contractual promise.
The “Fixed Date & Quantity” indicator is stored in the schedule line table VBEP, not in VBAP.

- VBAP → Sales order header/item data (overall item details)

- VBEP → Sales order schedule line data (dates, quantities, delivery schedule, and indicators like FIXMG/FIXDAT)

FIXMG/FIXDAT belongs to VBEP, because it applies at the schedule line level, controlling rescheduling for each delivery date and quantity.

- FIXMG (Fixed Quantity) - Locks the confirmed quantity for a schedule line.
- FIXDAT (Fixed Date) - Locks the confirmed delivery date for a schedule line.
- Together, the flag ensures that once ATP confirms a quantity/date, the system cannot reschedule it, even if stock or demand changes later.

The Fixed Date & Quantity flag locks confirmed schedule line quantities and dates, preventing ATP or aATP from rescheduling them, thus ensuring committed deliveries are maintained.

Behavior in Classic ATP-

- During ATP check, the system calculates available quantity and confirms the schedule line.
- If FIXMG/FIXDAT is set:
    - Confirmed quantity/date cannot be reduced or shifted automatically during subsequent rescheduling.
    - The system ignores stock shortages or other orders that would normally adjust this line.
- Useful for critical customer orders that must not be delayed or reduced.

Behavior in Advanced ATP (aATP)-

- aATP extends the concept with categories, rules, and macros, but the principle is similar:
    - Once a schedule line is confirmed and FIXMG/FIXDAT is set, the line is protected from rescheduling or reallocation.
    - Even if aATP performs product allocation, backorder processing, or fair-share distribution, these lines are excluded from changes.
- Key difference: aATP uses categories and rules to control flexibility for other lines, but fixed lines remain strictly non-adjustable.

Fixed Date & Quantity (FIXMG / FIXDAT) – Manual vs Automatic
The Fixed Date & Quantity flag can be set manually in sales order schedule lines or automatically via configuration, copy control, or enhancements like BAdIs and user exits, ensuring confirmed quantities and dates remain protected.

The Fixed Date & Quantity (FIXMG / FIXDAT) flag in SAP can be managed both manually and automatically, depending on business requirements and system configuration. Here's the detailed explanation:

- Manual marking: Most common for ad-hoc critical orders.
- Automatic marking: Achieved via configuration, copy control, or enhancement (BAdI/user exit).
- Ensure consistency and automation in protecting confirmed quantities/dates across orders.

Manual Management - In the sales order schedule line screen (VA02/VA01), users can manually select the checkbox for:

  - FIXMG (Fixed Quantity)
  - FIXDAT (Fixed Date)

This is common when: Critical orders must not be rescheduled and Customer-specific commitments are made.

Automatic Management - The flag can also be set automatically using SAP configuration or enhancements:
a. Through Copy Control or Item Category

- In some sales order item categories, system logic can default FIXMG/FIXDAT based on rules (e.g., standard items, rush orders).

b. Through User Exits / BAdIs

- Using BAdI BADI_SD_SCHEDULE or user exits in sales order processing, you can automatically set FIXMG/FIXDAT based on:
  - Customer group
  - Material type
  - Priority of order
  - Special conditions in the order

c. Integration with ATP/aATP

- When ATP or aATP confirms a schedule line, system can automatically set FIXMG/FIXDAT if configured to:
    - Treat certain confirmed lines as non-reschedulable
    - Ensure protection during backorder processing or fair-share allocation

Comparison table for Fixed Date & Quantity (FIXMG / FIXDAT) – Manual vs. Automatic.

Aspect	Manual	Automatic
How it's set	User selects the checkbox in the sales order schedule line (VA01/VA02).	System sets the flag via configuration, copy control, or enhancements (BAdIs / user exits).
Typical Use Case	Critical or high-priority orders where confirmed quantity/date must not change.	Standardized rules for certain materials, customers, or order types; ensures protection during ATP/aATP or backorder processing.
Flexibility	Fully controlled by user; can override defaults.	Consistent and automated; reduces manual errors.
Integration with ATP/aATP	Protects confirmed schedule line from rescheduling once set.	Automatically marks lines as non-reschedulable based on ATP/aATP confirmation or business rules.

Check 'Fixed Date & Quantity' in SAP sales orders when the customer explicitly accepts the ATP-confirmed quantities and dates after the availability check.

- When checked after the customer accepts ATP-proposed dates (MAD-derived), it transfers fixed requirements to MRP/MD04, preventing Backorder Processing (BOP) or rescheduling from altering them.
- When unchecked, dates remain flexible—the system reschedules to the requested delivery date (RDD) if better supply becomes available, showing demand in MD04 without a fixed commitment."

Controlled by standard configuration based on sales area level-
The behavior is controlled through standard SAP configuration at the sales area level using Transaction Code OVZJ. This configuration standardizes system behavior and overrides manual user decisions, ensuring consistent handling across all sales orders created within the same sales area.

In this case, the "Fixed Date and Quantity" option has been applied for the sales area 1710 / 10 / 00. As a result, during the ATP (Available-to-Promise) check, once the system confirms the quantity, it marks the corresponding sales order schedule line as fixed.

This ensures stable and predictable confirmations for sales orders in the defined sales area.

Configuration Path for OVZJ (Availability Control)-
SPRO → SAP Reference IMG → Sales and Distribution → Basic Functions → Availability Check and Transfer of Requirements → Availability Check with ATP Logic or Against Planning → Carry Out Control for Availability Check (Transaction Code: OVZJ)

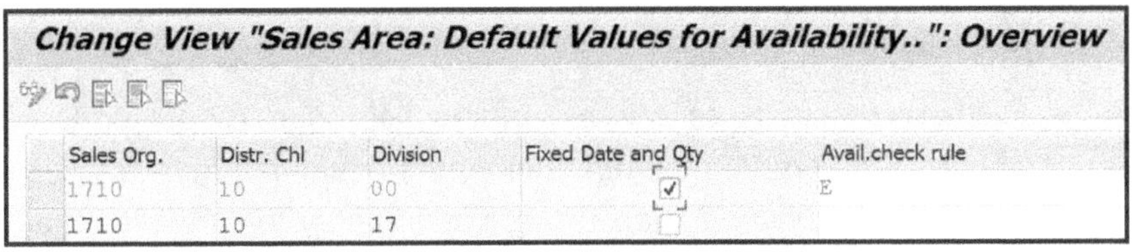

Change View "Sales Area: Default Values for Availability..": Overview

Sales Org.	Distr. Chl	Division	Fixed Date and Qty	Avail.check rule
1710	10	00	☑	E
1710	10	17	☐	

Scenario: "Fixed Date and Quantity" (Sales Area 1710/10/00)

1. **Full stock available** → ATP confirms full quantity:
   - Schedule line is marked fixed
   - Quantity is reserved
   - No rescheduling happens in future planning runs
2. **Partial stock available** → ATP confirms only part of the requested quantity:
   - The confirmed quantity is fixed in the schedule line
   - The unconfirmed (back ordered) quantity remains open
   - System may create a delivery proposal or back-order for the remaining quantity depending on configuration (e.g., rules for partial confirmations)
   - The fixed confirmed quantity won't be rescheduled, but the unconfirmed quantity can still be rescheduled

3. **No stock available** → ATP cannot confirm any quantity:
    - No schedule line is created for the fixed quantity because nothing can be confirmed
    - The sales order remains in back order or unconfirmed status
    - When stock becomes available, ATP will attempt confirmation according to the availability check rules

Notes-

- The "Fixed Date and Quantity" only applies to the quantity that ATP confirms at the time of order creation.
- It does not force the system to confirm quantity that isn't available.
- So, if stock is insufficient, the system still follows standard ATP logic for the unconfirmed part, but whatever does get confirmed is protected from rescheduling.
- OVZJ = fixed behavior at sales area level only - by default, OVZJ settings work at the sales area level (Sales Organization + Distribution Channel + Division). If you need different behavior based on other combinations, you have a few options in SAP, but it depends on what "other combination" means.
- Other combinations can be achieved through BAdIs or User-Exits. For very flexible requirements, you can implement a BAdI in the ATP check, which allows you to apply logic based on any combination, such as customer, material, plant, sales organization, etc. Within the BAdI, you can force the schedule line to be fixed or adjust the confirmed quantity.
- Even with fixed settings, SAP supports manual rescheduling rules. However, to automate behavior based on other combinations, enhancements using BAdIs are usually required.

Key Points-

1. The flag is set at the schedule line level (VBEP) in the sales order document.
2. Sales Order Schedule Line (VBEP) - This is the main place where FIXMG/FIXDAT is marked. It locks the confirmed quantity and delivery date for that particular schedule line.

3. Sales Order Item (VBAP) - The flag is not stored here, because VBAP is at the item level, and the fix applies per schedule line, not per item.

4. Effect on Other Documents -

- Delivery Document (LIKP/LIPS):
    - When a delivery is created from the sales order, the confirmed quantities/dates from the schedule line are copied to the delivery line.
    - FIXMG/FIXDAT does not exist as a field in the delivery tables, but the delivery is indirectly affected because the system cannot reduce or delay quantities/dates beyond the fixed value during delivery creation.

- Subsequent Documents:
    - Purchase requisitions or production orders created via ATP/aATP respect the fixed schedule line because the ATP check considers FIXMG/FIXDAT when confirming availability.
    - The flag itself is not marked in those documents, only the confirmed quantities/dates are protected.

5. FIXMG/FIXDAT is only marked in the sales order schedule line (VBEP). It ensures committed quantities and dates remain non-reschedulable throughout the order fulfillment process.

6. *The Fixed Date & Quantity flag is marked in the sales order schedule line (VBEP) and locks confirmed quantities/dates; while it is not stored in downstream documents, its effect is carried over to deliveries and ATP/aATP checks.*

7. Configuration Impact: Set during sales order entry post-ATP (manually or automatically via user exits); affects requirement transfer but does not impact the initial MAD calculation.

8. Protects confirmed quantity/date from ATP/aATP or backorder rescheduling.

9. Useful for high-priority, time-sensitive, or partially available orders.

Scenarios –

Here are some common business scenarios where Fixed Date & Quantity (FIXMG / FIXDAT) is used in SAP sales orders:

Fixed Quantity & Date is used to lock customer-accepted ATP confirmations, preventing rescheduling or reallocation and ensuring stable execution and planning.

1. Critical Customer Orders

- Scenario: A key customer requires strict delivery of high-value or time-sensitive goods (e.g., electronics launch, perishable items).
- Use of FIXMG/FIXDAT: Lock confirmed quantity and delivery date to prevent rescheduling even if stock situations change.

2. Promotional or Campaign Orders

- Scenario: Orders tied to marketing campaigns or seasonal promotions (e.g., Black Friday, Diwali sale).
- Use: Ensure delivery aligns with the campaign date; ATP/aATP cannot move dates forward or backward.

3. Partial Stock Availability

- Scenario: Part of the order is available now, part arrives later via procurement or production.
- Use:
    - Fix the schedule lines for available quantity to guarantee delivery.
    - Protect the later schedule lines from being shifted during Backorder Processing (BOP).

4. Pre-confirmed Customer Commitments

- Scenario: Customer agrees to ATP-proposed MAD-based delivery after confirmation call or email.
- Use: Check FIXMG/FIXDAT to transfer confirmed requirements to MRP/MD04, preventing automatic rescheduling or allocation changes.

5. Supply Chain Planning Stability

- Scenario: Company wants to avoid frequent changes in production or procurement plans.
- Use: Fix schedule lines for key orders to reduce fluctuations in MRP, BOP, or aATP allocations.

Example: Using Fixed Date & Quantity

- Customer: ABC Corp
- Material: ABC1001
- Ordered Quantity: 100 units
- Requested Delivery Date (RDD): Jan 15, 2025 (Today is Jan 10)
- Stock Availability: 50 units in stock today (Jan 10), 50 units arriving via PO on Jan 14
- ATP Check Result (MAD): Jan 14

Example: Using Fixed Date & Quantity

Schedule Line	Confirmed Qty	Confirmed Date	FIXMG/ FIXDAT	Remarks
1	50	Jan 10	Checked	Stock already available; fixed to prevent rescheduling
2	50	Jan 14	Checked	PO arrival; fixed to prevent BOP or rescheduling

Effect of FIXMG / FIXDAT-

- Once FIXMG / FIXDAT is checked, the confirmed quantities and dates are locked. They cannot be rescheduled, even if stock becomes available earlier later.
- If the flag is not checked, SAP may reschedule the second 50 units to an earlier date if stock arrives sooner, or during Backorder Processing (BOP).
- This ensures that customer commitments are honored, maintaining consistent order fulfillment during ATP / aATP checks and BOP runs.

Why Fixed Quantity & Date Is Required – ***Business Reasons***

The Fixed Quantity & Fixed Date (FIXMG / FIXDAT) checkbox is required to protect confirmed customer commitments and to stabilize planning and execution. Here’s the business reasoning clearly explained as below.

Business Reasons	Explanation
To Honor Customer Commitments	• Once a customer explicitly accepts ATP-confirmed quantities and dates, the business must ensure they are not changed later. • FIXMG/FIXDAT guarantees that confirmed delivery promises are honored.

To Prevent Backorder Processing (BOP) Changes	• During ATP/aATP backorder processing, SAP may:
	  ○ Reallocate stock.
	  ○ Shift dates.
	  ○ Reduce quantities.
	• Fixed lines are excluded from BOP, preventing critical orders from being deprioritized.

To Stabilize MRP & Supply Planning	• Without FIXMG/FIXDAT, sales demand remains flexible and may be rescheduled.
	• When this checkbox is selected:
	  ○ Requirements are transferred to MRP (MD04) as firm demand.
	  ○ Production and procurement plans remain stable and predictable.

To Avoid Continuous Rescheduling	• If better supply becomes available later, SAP may:
	  ○ Pull-in dates
	  ○ Split or merge schedule lines
	• FIXMG/FIXDAT stops automatic rescheduling, reducing operational noise.

To Protect High-Priority or Time-Sensitive Orders	• Used for:
	  ○ Strategic customers
	  ○ Promotional or launch dates
	  ○ Contractual SLAs
	• Ensures priority orders are not impacted by other demand.

No. Without the Fixed Date & Quantity flag, the system does *not* fully protect the confirmed dates and quantities. Why the system alone is not enough (without FIXMG / FIXDAT)

1. ATP confirmation is not a firm commitment by default - Standard ATP/aATP confirmations are dynamic and SAP assumes dates and quantities can change if supply or priorities change. So even though ATP confirmed something earlier, SAP may recalculate and change it later.

2. System optimizes globally, not per customer commitment - SAP's goal is to optimize overall supply fulfillment, not to protect individual promises. During BOP or rescheduling Higher-priority orders may take stock and Earlier RDDs may be favored. Without FIXMG/FIXDAT, your order is negotiable.

3. FIXMG / FIXDAT is the explicit "business lock" - This flag tells SAP that this confirmation is final and contractually accepted and without this flag, SAP assumes flexibility.

Comparison: With vs Without the Flag-

Business interpretation- In SAP, ATP represents a planning promise, while Fixed Date & Quantity represents a contractual promise. SAP deliberately separates these two concepts to allow supply-chain optimization unless the business explicitly chooses to lock the commitment.

- System ATP -> Planning promise
- Fixed Date & Quantity -> Contractual promise

Aspect	Without FIXMG/FIXDAT	With FIXMG/FIXDAT
ATP confirmation	Temporary / flexible	Firm / locked
Backorder Processing	Can change	Excluded
Rescheduling	Allowed	Not allowed
MRP requirement	Changeable	Firm
Customer commitment	Implicit	Explicit

No, there is no other standard ATP-safe way to fully lock both the date and quantity without using the Fixed Date & Quantity (FIXMG / FIXDAT) flag.

There are limited alternative ways to protect or "reserve" date and quantity in SAP, but none provide the same clean, ATP-native locking behavior as Fixed Date & Quantity (FIXMG / FIXDAT). SAP deliberately maintains FIXMG/FIXDAT as the primary and explicit locking mechanism.
Below is a clear and honest explanation of what other options exist, why they are different, and why FIXMG/FIXDAT continues to be required.

There is no standard ATP mechanism that fully locks both date and quantity without FIXMG / FIXDAT.
Other methods may influence or indirectly protect confirmations, but they do not guarantee immutability across ATP/aATP, Backorder Processing (BOP), and rescheduling.

Alternatives & Why They Are Not Equivalent

1. Delivery Creation (VL01N)

How it Works	• Creating an outbound delivery reserves stock physically.
Why it's Not Equivalent	• Only works when stock already exists. • Does not protect future receipts. • Too late for ATP commitments.
Use Case	• Execution-level protection, not planning-level locking.

2. Manual Reservations (MB21 / MRP Reservations)

How it Works	• Creates inventory reservations.
Why it's Not Equivalent.	• Not ATP-relevant. • ATP ignores MB21 reservations. • Breaks SD–MM integration logic.
Use Case	• Internal consumption, not customer commitments.

3. Product Allocation / aATP Allocation

How it Works	• Limits available quantity per customer, region, or period.
Why it's Not Equivalent.	• Controls *distribution*, not *confirmation locking*. • BOP can still reshuffle inside allocation limits.
Use Case	• Fair-share, not fixed promises.

4. High-Priority Requirements / Requirement Classes

How it Works	• Influences BOP prioritization.
Why it's Not Equivalent	• Still reschedulable. • Priority ≠ immutability.
Use Case	• Preference, not protection.

5. BOP Exclusion Rules

How it Works	• Excludes certain orders from backorder processing.
Why it's Not Equivalent	• Only applies during BOP runs. • ATP rescheduling may still change confirmations.
Use Case	• Operational protection, not full ATP locking.

6. MRP Firming (Planned Orders / Production Orders)

How it Works	• Firms supply elements.
Why it's Not Equivalent	• Firms *supply*, not *sales demand.* • ATP can still reassign supply.
Use Case	• Supply stability, not demand locking.

In SAP S/4HANA aATP 2023, there is no alternative standard mechanism that fully replaces the Fixed Date & Quantity flag for locking ATP-confirmed date and quantity. Other features like product allocation, BOP rules, and custom logic can stabilize commitments but do not guarantee immutability across ATP, BOP, and rescheduling the way FIXMG/FIXDAT does.

However, there *are* complementary approaches that can help protect or stabilize commitments around dates and quantities, depending on business requirements. These approaches do not provide a 100% equivalent ATP-native lock like FIXMG/FIXDAT - but they *help achieve similar business outcomes* in certain scenarios.

FIXMG/FIXDAT remains the only standard way in SAP to explicitly freeze both quantity and date at the ATP confirmation level.
Legal & Commercial Accuracy: Only the business—not the system—can decide when a promise becomes legally and commercially binding."

Alternatives in SAP S/4HANA aATP 2023 (Not a Full Replacement)
While none *fully replaces* FIXMG/FIXDAT, the following tools help stabilize commitments or control fulfillment behavior:

Alternatives	Limitation
Product Allocation • Controls how available supply is distributed across customers, regions, or channels. • Help ensure that specific customers or groups get a guaranteed share of available quantity.	• Allocation protects quantity distribution but does not lock specific dates.
Backorder Processing (BOP) Rules • This allows you to define priorities and sorting rules for reshuffling orders. • Can be used to optimize allocations toward preferred orders.	• BOP works *within available supply* and still can reschedule dates and quantities unless FIXMG/FIXDAT is set.
Alternative-Based Confirmation (ABC) • Let SAP propose multiple fulfilment options with different dates or sources. • Improve customer experience and decision-making.	• Does not *lock* a specific date/quantity — it just *presents options.*
Custom Enhancements or Rules (BAdIs / User Exits) • You can automate behavior around FIXMG/FIXDAT (e.g., automatically set it for certain items). • Or implement logic to *block changes in confirmation* via custom checks.	• These are *workarounds or governance enforcers*, not true ATP-level locks.
Order Change Prevention at Delivery/Shipment Level Examples: • Locking delivery dates • Restricting changes to confirmed quantities	• Not ATP-native; acts after ATP has already confirmed.

Why There Is No Alternative ATP Lock-

SAP's ATP/aATP model is designed for supply-chain optimization, not automatic contractual locking:

- ATP is intended to be *flexible, dynamic, and optimized* across demand and supply.
- FIXMG/FIXDAT is the *only explicit exception* where the business says:

"Do not change this confirmation."
Without that explicit instruction, SAP assumes:

- "Optimized planning > rigid commitments"
- So, the system reschedules to make planning efficient.

This design is intentional and global across SAP's supply-chain logic.

Feature	Fixed Date & Qty	Other Mechanisms
Locks quantity	Yes	(no)
Locks date	Yes	(no)
Prevents ATP/aATP rescheduling	Yes	(partial or no)
Works across BOP	Yes	(no)
SAP standard without enhancement	Yes	(partial)

This is a very common question, and the distinction is important to understand. Even with SAP S/4HANA aATP, the Fixed Date & Quantity (FIXMG/FIXDAT) flag still plays a critical role. Here's the detailed explanation:

Even with aATP, Fixed Date & Quantity is required to lock ATP confirmations because aATP optimizes supply dynamically, whereas FIXMG/FIXDAT converts a planning promise into a contractual, unchangeable commitment.

Note-
BOP strategies like WIN control allocation priority but cannot replace Fixed Date & Quantity, which is the only SAP mechanism to lock confirmed ATP/aATP dates and quantities as a firm, contractual commitment.

BOP strategies like WIN cannot replace Fixed Date & Quantity (FIXMG/FIXDAT). Here's why, explained clearly:

<table>
<tr><td>What aATP Does</td><td><ul><li>It provides real-time availability check across multiple sources (plant stock, production, inbound deliveries, stock in transit, alternative locations, etc.) & it optimizes order fulfillment dynamically.</li><li>aATP can propose dates and quantities based on available supply, rules, and priorities.</li><li>aATP is designed to optimize planning and supply, not to make a contractual commitment.</li></ul></td></tr>
<tr><td>What BOP WIN Does</td><td><ul><li>WIN stands for “When-Needed / Importance / Net Availability” strategy in Backorder Processing (BOP).</li><li>It prioritizes orders during stock shortages:<ul><li>Determines which orders get fulfilled first</li><li>Sorts orders based on rules (priority, requested delivery date, order type, etc.)</li></ul></li><li>Helps optimize allocation of limited stock.</li><li>WIN only controls prioritization, not locking confirmed dates or quantities.</li></ul></td></tr>
<tr><td>What FIXMG/FIXDAT Does</td><td><ul><li>Locks schedule line quantity and delivery date after ATP/aATP confirmation.</li><li>Prevents:<ul><li>Rescheduling by ATP/aATP</li><li>Reallocation during BOP</li><li>Changes from supply adjustments or reprioritization</li></ul></li><li>FIXMG/FIXDAT guarantees a contractual commitment, independent of stock availability or BOP logic.</li><li>FIXMG/FIXDAT is the explicit business instruction: “This delivery promise is final and cannot change.”</li></ul></td></tr>
<tr><td>Business Interpretation</td><td><ul><li>BOP WIN = optimization mechanism (who gets stock first)</li><li>FIXMG/FIXDAT = commitment mechanism (locked, unchangeable promise)</li></ul>You can use WIN in combination with FIXMG/FIXDAT to optimize fulfillment of other flexible orders without impacting the fixed orders.</td></tr>
<tr><td>Why WIN Cannot Replace FIXMG/FIXDAT</td><td><ul><li>WIN influences how stock is allocated among competing orders.</li><li>FIXMG/FIXDAT ensures an order’s confirmed quantity/date cannot change, even during BOP.</li><li>Without FIXMG/FIXDAT, WIN may still reschedule or reduce the order if stock is reallocated.</li></ul></td></tr>
</table>

18. Safety Stock-

Safety Stock is the minimum level of inventory you want to keep as a buffer to protect against stock-outs. It is not meant to be used in normal order fulfillment but acts as a "reserve" to prevent stock-outs in situations such as the following:

- Unexpected customer demand spikes
- Supply delays or late deliveries
- Production uncertainties

This describes the use of safety stock in Classic ATP and Advanced ATP (aATP).

Classic ATP (Basic ATP)	Advanced ATP (aATP)
• Safety stock is simple and static. • Purpose: Prevents the system from confirming stock below the minimum level. **Behavior:** 1. If safety stock is active in the ATP scope of check, only stock above the safety stock is confirmed. 2. Safety stock cannot be used automatically; it can only be manually overridden in exceptional cases. • Limitation: Cannot do complex rules like allocating stock to multiple customers or product lines.	• Safety stock is part of more flexible, rule-based availability checking. • Purpose: Protects inventory like classic ATP, but allows more sophisticated control. **Behavior:** 1. Safety stock is considered when confirming sales orders or delivery proposals. 2. Can define product allocations, product priorities, and rules for releasing safety stock. 3. Supports backorder processing, partial confirmations, and future stock reservations. • Advantage: You can protect stock but still allow planned exceptions automatically, based on rules.

Stocks go below threshold only via **manual or exceptional consumption**. Replenishment restores the buffer; normal orders cannot consume it when ATP is active.

Safety stock in the ATP scope of check can be either **active** (system reserves it, reducing confirmable stock) or **inactive** (system ignores it, allowing full stock to be promised).

ATP Scope of Check-

Active	• ATP considers safety stock means only stock above safety stock is confirmed.
	  ○ ATP will only promise quantities above the safety stock.
	  ○ Helps prevent over-committing stock that should be reserved for unexpected demand.

Inactive	• ATP ignores safety stock means all stock can be promised.
	  ○ ATP ignores safety stock in the calculation.
	  ○ System can commit all available stock, even if that would bring inventory below safety stock.
	  ○ Essentially, ATP might overpromise, since it doesn't "protect" the safety stock.

When stock goes below the safety stock level, MRP is triggered to fulfill and restore the safety stock.

Process:
Stock dropping below safety stock → MRP RUN → Planned Order → Goods Receipt → Safety Stock restored.
Planned order or purchase requisition generation depends on procurement type:

- **In-house production (manufactured material):** MRP creates a **Planned Order** → Production Order → Goods Receipt.
- **External procurement (purchased material):** MRP usually creates a **Purchase Requisition (PR)** → Purchase Order (PO) → Goods Receipt.

When Safety Stock Comes into Picture.

Sales Order Creation (ATP Check)	• Safety stock is considered if ATP check includes it. • Normal orders: Only stock above safety stock is confirmed. • Emergency orders: Safety stock can be manually released if needed.	Example: • Stock = 100, • Safety Stock = 50 • Order Qty = 60 ATP confirms only 50 unless safety stock is manually released.

Purchase Requisition (PR)-	• PR is generated by MRP to replenish stock.
	• Safety stock comes into the calculation indirectly:
	  o MRP calculates net requirements = demand – available stock + safety stock
	  o Safety stock ensures MRP plans to maintain buffer after meeting demand.
		Example:
		Safety Stock = 50, Forecasted demand = 80, Stock = 40
		Net requirement = 80 – 40 + 50 = 90 → PR quantity 90

Purchase Order (PO)-	• Safety stock is not directly used in PO, but it is the reason PR or planned order was created. • When creating PO from PR, the planned quantity already considers safety stock.	

MRP Run	• Safety stock is actively used in MRP calculations:
	  o Net requirements = Demand – Stock + Safety Stock
	  o Ensures that after fulfilling demand, safety stock level is maintained
	• Helps avoid shortages and plan timely procurement or production.
		Example:
		Stock – 40
		Safety Stock – 50
		Forecast Demand – 80
		MRP Requirement – 90

Summary Table-

Process	How Safety Stock is Used
Sales Order / ATP	Reserved buffer: only stock above safety stock is promised
MRP	Net requirement calculation includes safety stock to maintain buffer
PR	Generated quantity considers safety stock indirectly via MRP
PO	Reflects quantity planned to maintain stock above safety stock
Emergency Orders	Safety stock can be manually released if needed

This is a recommendation / best practice for using safety stock functionality in Classic – ATP and Advanced (aATP).

1. In **Classic ATP**, activate the safety stock check in the material master (MRP3 view → *Check Safety Stock*) and ensure the scope of check supports it.

2. For more advanced and flexible control, consider **Advanced ATP (aATP)**, which supports rule-based protection of safety stock, product allocations, and backorder processing.

Example-

1	*Setup Assumptions*	• Material: MAT100 • Current stock: 100 units • Safety stock: 50 units • No planned receipts • Basic ATP check includes safety stock

2	*Scenario: Sales Order Request Customer requests 60 units.*	**ATP Calculation-**
		• Available stock = 100
		• Safety stock = 50
		• ATP quantity = Stock – Safety Stock = 100 – 50 = 50 units available to promise
		Result-
		• Only **50 units** can be confirmed in the sales order.
		• **10 units cannot be confirmed** because confirming them would violate safety stock.
		• System may either:
		  o Confirm partially (50 units), or
		  o Suggest a later date if replenishment is expected (depending on configuration).

3	*Scenario: Sales Order Request Customer Requests 40 Units*	• Requested quantity = 40 • ATP quantity = 100 – 50 = 50 • Result: Full quantity 40 units confirmed, stock after order = 60 (still above safety stock).

Key Points-

1. If you want full order confirmation regardless of safety stock, you would deactivate safety stock in ATP (as in your current setup).
2. Safety stock is the minimum inventory level you want to keep reserved to prevent stock-outs.
3. ATP with safety stock active means: the system will not promise quantities that would bring stock below safety stock.
4. Safety stock is used only when confirming a sales order or delivery. It is not consumed until the order is confirmed.
5. ATP only allows confirmation above safety stock. Safety stock itself is never reduced by ATP confirmation — it's the protected buffer.

Safety stock is considered at the point of confirming availability for a sales order:

- During **sales order creation (VA01/VA02)** → ATP check calculates confirmable quantity
- During **delivery creation (VL01N/VL02N)** if ATP re-check is active
- Safety stock is **subtracted from available stock** before confirming the order
- Safety stock is **never directly consumed by sales orders**.
- It's a **buffer to prevent over-committing** and protect against stock-outs.

7. gATP Overview-

gATP (Global Available-to-Promise) was a component of SAP APO (Advanced Planning and Optimization) designed to provide complex, global availability checks for order promising across distributed supply chains. It enabled enterprises to confirm delivery quantities and dates by evaluating multiple locations, sources, and constraints, beyond the basic ATP found in SAP ECC.

SAP Advanced Planning and Optimization (APO) was a separate SCM server for advanced supply chain planning, replacing basic ECC MRP. SAP APO launched in 2002 because ECC MRP was too basic for complex global supply chains.

APO gATP was the predecessor to today's S/4HANA aATP. Same concepts, embedded power—no separate APO learning required.

Global ATP (gATP) was the APO-based, advanced ATP engine that SAP provided on the separate SAP SCM/APO system to overcome the limits of classic ECC ATP, especially for multi-plant, rule-based and allocation-driven confirmations. In S/4HANA, gATP is not continued; its role is being replaced by Advanced ATP (aATP) embedded in S/4HANA together with SAP Integrated Business Planning (IBP) for planning functions that used to sit in APO.

Key Concepts of gATP

Global Scope:	Checks availability not only at a single location but across multiple plants, distribution centers, and even alternative sources of supply, offering a comprehensive supply chain view.
Rule-Based Checks:	gATP can use sophisticated rules to propose alternative plants, substitute products, and consider various fulfillment strategies when the preferred source or product isn't available.
Advanced Allocation:	Incorporates product allocation mechanisms to manage scarce stock, ensuring strategic customers or channels are prioritized based on predefined business rules.
Backorder Processing (BOP):	Allows for advanced, automated re-promise and reallocation of stock when supply situations change, re-prioritizing open orders as needed.
Integration with Planning:	gATP was tightly linked with other APO components, allowing dynamic interaction with forecasting, supply

	network planning, and production scheduling to provide the most accurate promise dates.
Scheduling and Transportation:	Can perform availability checks alongside transport and production scheduling, distinguishing it from less advanced ATP methods.

Typical gATP Process Flow

1. Customer order triggers a global availability check.
2. System analyzes all possible stock, open supply, and production across the network.
3. If insufficient stock exists at the requested site, gATP proposes alternatives (different plants or products).
4. Confirms the optimal delivery proposal based on rules, allocations, and priorities.

Some key details about gATP, including its history, why it is being replaced, and what is replacing gATP now.

How gATP came into the picture	• gATP was delivered as part of SAP APO to offer:
	  • Multi-location ATP (checking several plants/DCs).
	  • Rule-based ATP, product and location substitution, multi-level ATP, and capable-to-promise (CTP).
	  • Sophisticated backorder processing and product allocation tightly integrated with planning.
	• ECC/basic ATP could only check one plant and had simpler logic, so gATP became the go-to solution for complex global supply chains.

Why gATP is being replaced	• SAP's APO stack is being phased out; planning pieces are split into:
	  • IBP for demand, supply and allocation planning.
	  • PP/DS and related modules moved into S/4HANA for detailed scheduling.
	• In S/4HANA, SAP built aATP natively in the S/4 core as the strategic successor to gATP, with:
	  • Simplified, in-memory architecture on HANA and faster response.
	  • Tighter integration with S/4HANA Sales, Manufacturing and Logistics, and a Fiori-based user experience.

What replaces gATP now	• Advanced ATP (aATP) in S/4HANA takes over order-side ATP capabilities:

	• Product Availability Check (PAC) as the core ATP engine. • Advanced Backorder Processing with prioritization "buckets". • Product allocation (PAL in S/4 + allocation planning in IBP). • Supply protection and intelligent plant/product substitution (without static substitution rules like gATP needed). • SAP IBP complements aATP by providing the planning layer that used to live in APO/gATP (demand, supply, allocation planning).
Current Status (2026)	APO is end-of-life and fully replaced by S/4HANA embedded functionality: Maintenance Timeline: • APO 5.1 -> Mainstream ended 2018 • APO -> Extended maintenance → **2027 final end** *****Migrate NOW** to avoid risks/costs***
Where APO Stands Today	• Legacy systems: Still running in ~10% of SAP customers (mostly large enterprises) • Migration deadline: 2027 (hard stop) • No new development: Pure maintenance mode *Embedded successors: All APO power now lives inside S/4HANA core (no separate server)*

Conceptually, gATP introduced the "advanced, global, rules-based" ATP layer on top of classic ATP; in S/4HANA that role is taken over by aATP + IBP, so new implementations are encouraged to use aATP rather than deploying APO gATP.

gATP vs. aATP (in SAP S/4HANA)-

gATP (Global Available-to-Promise) was a core component of SAP APO and provided advanced availability checking, supporting global, decentralized, and complex supply chain processes. It offered features far beyond classic ATP, such as rule-based availability, substitution, allocation, and global order promising.

With SAP S/4HANA, GATP is largely replaced by AATP (Advanced ATP). While both solutions provide advanced availability logic, AATP uses a modern, in-memory architecture built directly into S/4HANA. This delivers:

- Real-time processing
- Simplified configuration
- Better integration with logistics and supply chain modules
- A more streamlined user experience

gATP historically enabled sophisticated, customer-centric, and flexible order fulfillment for global supply chains. AATP continues this vision but with a redesigned, faster, and more flexible framework optimized for today's real-time business requirements.

SAP APO (including gATP) mainstream maintenance ends in 2027, with extended support available until the end of 2030. After that, all gATP functionality will transition to SAP S/4HANA Advanced ATP (aATP), the strategic successor. The gATP sunset timeline is as follows:

- Mainstream Maintenance: Ends Dec 31, 2027 for SAP SCM/APO 7.0.
- Extended Maintenance: Available through Dec 31, 2030 (customer-specific, costly).
- Post-2030: No further support; mandatory migration to S/4HANA aATP + IBP.

Functionality Migration to aATP - All core gATP capabilities are embedded in S/4HANA aATP:

- Product Availability Check (PAC) replaces multi-level ATP.
- Advanced Backorder Processing (BOP) with prioritization replaces gATP BOP.
- Supply Protection handles location/product substitution dynamically.
- Product Allocation (PAL) integrates with IBP for planning-based allocations.
- Capacity Check/CTP via PP/DS embedded in S/4HANA.

Why the Transition-

- Embedded Architecture: aATP runs natively on HANA in S/4 core - no separate APO system needed.
- Fiori UX: Modern apps replace APO transactions (e.g., /SAPAPO/ATPQ for simulation).
- Real-time: In-memory processing eliminates CIF latency between ECC-APO.

Key Point –

While 2030 marks the end of APO/gATP support, the functionality lives on in aATP, which covers 95%+ of gATP scenarios with simplified configuration (no complex rules tables like APO 9V/9E).

APO Rules 9V/9E were complex gATP time-series allocation tables requiring manual characteristic maintenance and CIF transfers.
aATP replaces them with simpler PAL objects/sequences configured directly in S/4HANA - no CIF, no rules tables needed.

CIF (Core Interface) is SAP APO's standard integration layer that transfers master data (materials, BOMs, routings) and transactional data (sales orders, planned orders) between ERP/ECC and APO systems.

8. Activating aATP in SAP S/4HANA

Advanced Available-to-Promise (aATP) enhances order fulfillment by providing real-time, accurate availability checks across sales, procurement, and production processes. Activating aATP in SAP S/4HANA enables these real-time checks for improved order fulfillment. The core steps to activate aATP in SAP S/4HANA are as follows:

To activate Advanced Available-to-Promise (aATP) in SAP S/4HANA, follow these key steps:

1. Activate Business Functions-
Use transaction code "SFW5 or the SAP IMG path to activate the business functions "S4H_AATP" and
optionally SCM_APO_ON_ERP for integration purposes. This enables the underlying AATP engine and related capabilities.

- S4H_AATP (SAP S/4HANA Advanced ATP)
- SCM_APO_ON_ERP (if integration with APO is needed

This step enables the core AATP capabilities in the system.
To enable Advanced Available-to-Promise (aATP) in SAP S/4HANA, the first step is to activate the business function ***S4H_AATP****. This business function unlocks the core aATP capabilities—including Product Availability Check, Backorder Processing (BOP), Alternative-Based Confirmation (ABC), and Supply Protection.*

Activation is performed via the ***SAP S/4HANA Switch Framework (SFW5)****. Once enabled, the system exposes all relevant aATP configuration nodes in the IMG and makes the functionalities available for use across sales, procurement, and production processes.*
Activation in the ***Switch Framework (SFW5)*** *confirms that your system is* ***licensed and enabled for AATP****.*

If anything is missing or not yet active, the ***SAP Basis team*** *will complete the activation through SFW5.*
Because this involves system-level switches and technical setup, the activation must be performed by the Basis team.

- Path: SAP Customizing Implementation Guide > SAP NetWeaver > Business Functions > Activate Business Functions -> Activate S4H_AATP to enable core AATP capabilities.
  Tcode – SFW5

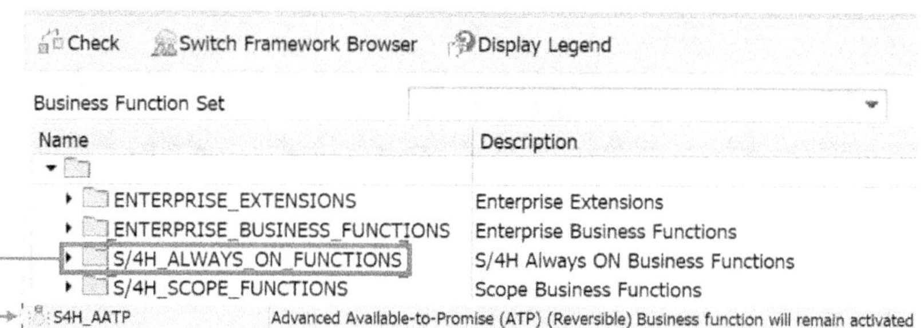

2. Fiori App-

AATP (Advanced Available-to-Promise) in SAP S/4HANA is fully accessible through Fiori apps. In fact, SAP requires using Fiori for AATP because AATP is not supported in SAP GUI (except for a few monitoring transactions).

- You cannot perform full AATP functions (like BOP, ABC, PAC, or supply check) from GUI.
- SAP requires Fiori apps for all new aATP capabilities.

You can open the Fiori Launchpad directly from SAP GUI by using the command - /n/UI2/FLP

aATP relies on several Fiori apps to perform and manage its key functionalities.
Therefore, as a first step, you need to ensure that all relevant aATP Fiori apps are added to the user's Fiori Launchpad Home Page.
This allows users to easily access and operate features such as Product Allocation, Supply Protection, ABC configuration, BOP variants, and monitoring tools.

After adding the required aATP Fiori apps to the Launchpad, the next step is to ensure that the necessary backend services and business roles are active and assigned to the user.

To know exactly which OData services, business catalogs, business groups, and business roles are required for a particular app, you can use the:
"SAP S/4HANA Fiori Apps Library".

The Fiori Apps Library provides complete information for each Fiori app, including:

- Required OData services
- Required UI services
- Business catalogs
- Business roles
- Technical app details
- Implementation prerequisites

How to Check Requirements for Any App

1. Go to the SAP Fiori Apps Library.
2. Search for the app name (e.g., *Monitor Product Availability*, Configure BOP Segment, *Schedule BOP Run*, etc.).
3. Open the app details.
4. In the "Implementation Information" and "Configuration" sections you will find:
   - Required OData services
   - Required Backend services
   - Required Business Catalogs
   - Required Business Roles
5. Your Basis and Security (Authorization) teams will activate the services and assign the required roles to the user.

This can be accessed through ***Google*** *– "sap Fiori apps library".*

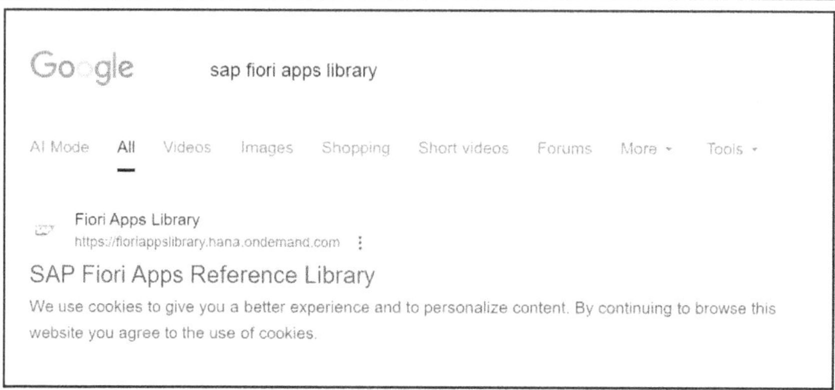

Find the app **"Configure BOP Segment"**; the details are available under the **"Implementation Information"** → Configuration.

Configure BOP Segment-

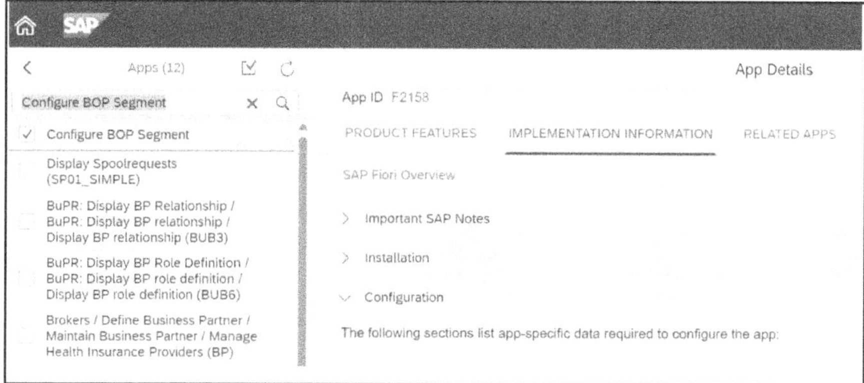

SAPUI5 Application –

SAPUI5 Application

The ICF nodes for the following SAPUI5 application must be activated on the front-end server:

Component	Technical Name	Path to ICF Node	SAP UI5 Component
SAP UI5 Application	ATP_ABOPSEGS1	/sap/bc/ui5_ui5/sap/atp_abopsegs1	sap.i2d.atp.configurebopsegment
	ATP_ABOPVARS1 *	/sap/bc/ui5_ui5/sap/atp_abopvars1	sap.i2d.atp.configurebopvariant
	ATP_MONBOPRUNS1 *	/sap/bc/ui5_ui5/sap/atp_monbopruns1	sap.i2d.atp.monitorboprun

* Added automatically due to dependencies

To activate the service (Technical Name), use transaction code **SICF**.

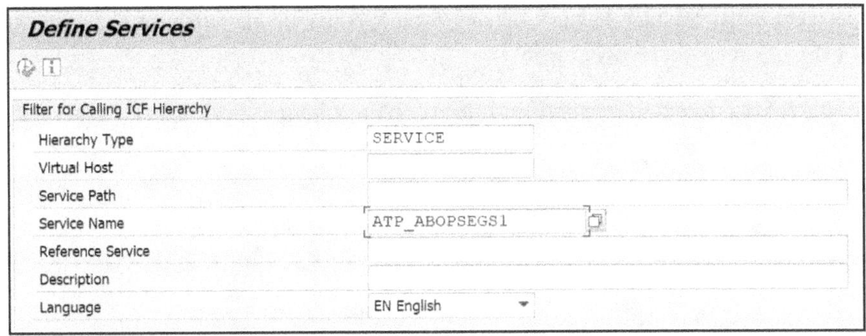

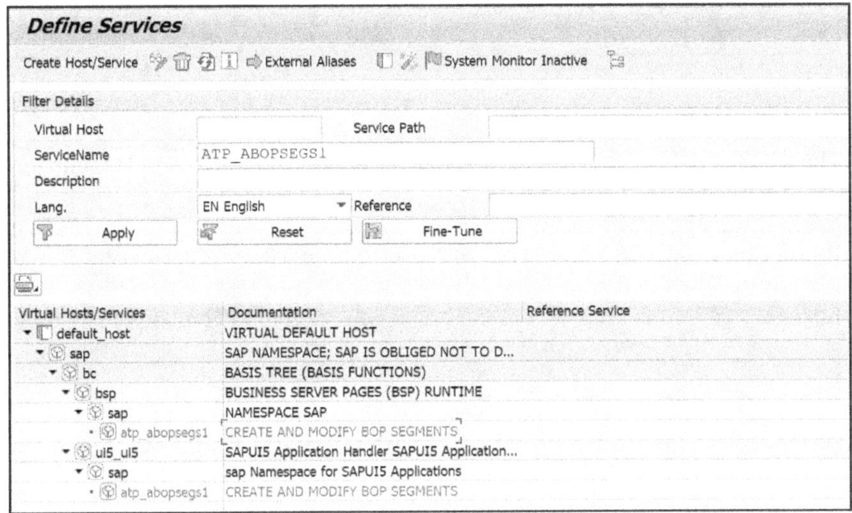

*If the service is not activated (shown as **greyed out**), right-click on it and select **Activate Service**. In this way, you can activate all the required services one by one.*

OData Service-

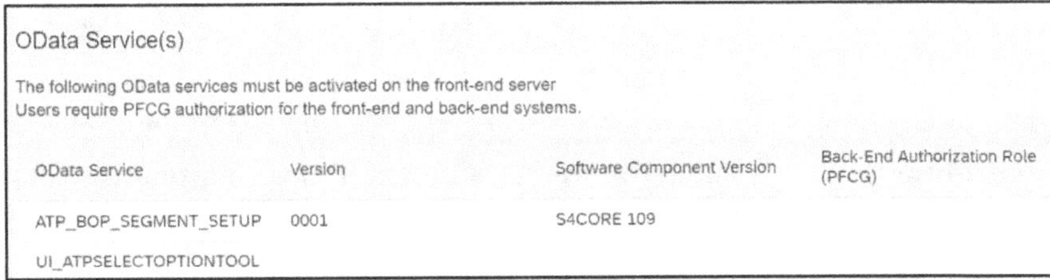

OData Service(s)

The following OData services must be activated on the front-end server
Users require PFCG authorization for the front-end and back-end systems.

OData Service	Version	Software Component Version	Back-End Authorization Role (PFCG)
ATP_BOP_SEGMENT_SETUP	0001	S4CORE 109	
UI_ATPSELECTOPTIONTOOL			

To activate the OData service, use transaction code - **/N/IWFND/MAINT_SERVICE**. This activity is typically handled by the **UI5 team** or the **Basis team**.

Use **"Add Service"** to add and activate the OData service.

Activate and Maintain Services

Filter | Add Service | Delete Service | Service Details | Load Metadata

Refresh Catalog | OAuth | Soft State | Processing Mode | Add to Transport

Service Catalog

Type	Technical Service Name	Vers..	Service Description

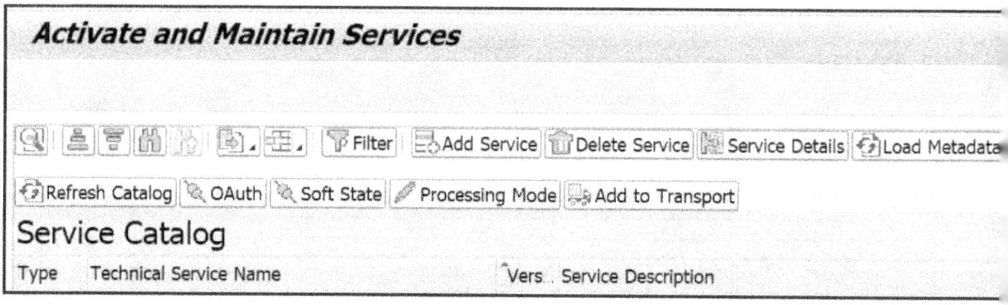

Business Role-

The required business role must be assigned to your user ID through transaction SU01.

Business Role(s)

Business Role	Business Role Description
SAP_BR_ORDER_FULFILLMNT_MNGR	Order Fulfillment Manager

These three activities are mandatory to access and use a Fiori app:

1. Activate SAP UI5 applications – Tcode "SICF".
2. Add and activate the required OData services – Tcode "/N/IWFND/MAINT_SERVICE".
3. Assign the relevant business role to the user ID – Tcode "SU01".

3. Basic Configuration & Master Data Setup in SAP S/4 HANA aATP

To set up Basic Advanced Available-to-Promise (aATP) in SAP, certain configurations and material master data must be maintained in the system. These configurations primarily define the Scope of Check, which determines how SAP evaluates stock, planned receipts, and requirements to confirm product availability for sales orders. Proper master data setup ensures that the system can execute the Scope of Check correctly for aATP processing.

1. ***Availability Check (Checking Group) and Activate AATP-***

   In transaction OVZ2, activate the AATP functionality for the relevant availability check groups. This defines which availability check groups use AATP (e.g., 01, 02). Note that in S/4HANA, daily requirements (01) are typically not active for AATP; focus mainly on group 02 or custom groups.

   01 - Daily requirement Check

   02 - Individual Requirement Check

- In aATP for S/4HANA, we usually use Availability Check 02 (Individual Requirement) and need to activate the 'Advanced ATP' column.
- Defines the rules for how availability is checked for materials during sales order processing.
- It determines whether the system considers stock, planned receipts, and other factors.
- Normally, we don't create a new checking group and use the standard 01 or 02 based on the requirement. This checking group is assigned to the Material Master record in the MRP3 view (field: Availability Check)
- A Checking Group is a code (e.g., 01, 02) assigned to a material, which specifies the Availability Check to be applied.
- Availability Check is the actual logic or procedure used to verify stock, receipts, and planned orders for confirming availability according to the configured scope of check
- SAP provides separate checking groups for classic ATP and Advanced ATP (aATP). Standard checking groups like 01 and 02 are delivered for aATP and are visible only in the aATP-specific customizing. These values do not appear in the classic ATP configuration, as aATP uses its own enhanced availability-check framework.
- *Configuration Path: SPRO → SAP Reference IMG → Cross-Application Components → Advanced Available-to-Promise (aATP) → Define Availability Checking Group → Define Availability Checking Group*

2. ***Assigning Availability Check Group to Materials-***
   In the material master (MM02-MRP3 View), assign the activated availability check group for AATP to the materials in the appropriate plant views. This ensures that the material will trigger AATP during sales order or production processes.

3. ***Checking Rules (OPJL)-***

   Configure checking rules relevant to your business scenario via transaction OPJL.

   Checking rules like "A," "AE," and others control how supply and demand elements are considered per process. This controls how the availability check is performed in combination with availability check groups.
   - *Defines which combinations of product, location, and date are considered during* the availability check.
   - It links the availability check to the specific scenario (e.g., sales orders, stock transport orders).
   - Normally, we use the following standard checking rules based on the application: These are standard, in-built rules provided by SAP and are used according to the specific application scenario. The values of the Checking Rule (e.g., A, B) represent the application for which it is used A for Sales Order, B for Delivery.
     - A – Sales Order
     - B – Delivery
     - BO – Backorder Processing
     - PP – Production Planning (PP) Checking Rule

4. ***Define Scope of Check***

   Using transaction **OVZ9A**, configure which stock types, receipts, and requirements are part of the availability check per checking group and checking rule combination. This tailors ATP logic to business needs.
   The Scope of Check in aATP is determined by two key components: the Checking Group, which comes from the material master, and the Checking Rule (Business Scenario), which is

derived from the application. Together, these two elements define how the availability check is executed

- Purpose of the Scope of Check-The Scope of Check determines which stock, receipt, and requirement elements are considered during the ATP check. In other words, it defines what the system should look at when confirming material availability.
- The Scope of Check is always determined by the combination of:
  Checking Group + Checking Rule.
- In aATP terminology, the Availability Check is represented by the Checking Group, while Checking Rules are referred to as Business Scenarios.
- In the SAP configuration, two separate Scope of Check nodes are provided: one for Classic ATP and another for Advanced ATP (aATP). Each operates independently because aATP uses enhanced, category-based availability-check logic.
- The Scope of Check is structured in two parts: a Header Level, where global availability-check parameters are defined, and a Category Level, where the specific stock, receipt, and requirement categories to be considered are maintained.
- At the Header Level, we can define parameters such as Future Supply, Delayed Supply, Quantity Distribution, Check Horizon, Special Scenarios, and other global controls.
- At the Category Level, we can specify which Stock Categories to include—for example, Unrestricted Stock (S1), Quality Inspection Stock (S2), Blocked Stock (S7), and Safety Stock (SH). The same approach applies to Receipt and Requirement categories, where we can activate or deactivate specific categories based on business needs.
- The main difference between the Classic ATP Scope of Check and the aATP Scope of Check is the use of categories. In aATP, Stock, Receipts, and Requirements are configured using detailed category settings, whereas in Classic ATP the Scope of Check is maintained on a single screen with limited parameter options.

- In addition, custom categories can be created and activated within the Scope of Check, allowing businesses to tailor stock, receipt, and requirement category behavior according to specific requirements.
- During sales order creation or change, the system retrieves the Checking Group from the material master and the Checking Rule from the application. Based on this combination, it determines the appropriate Scope of Check to calculate and confirm the availability date

5. ***Mandatory Fields in Material Master for aATP***

For aATP to work correctly, certain fields in the Material Master must be maintained. The Checking Group is one of the mandatory fields, as it determines which availability check logic will be applied. In addition to the Checking Group, the MRP views and Strategy Group are also mandatory to ensure proper calculation of stock, receipts, and requirements during the aATP check.

Strategy Group: - The Strategy Group in the Material Master determines the **Requirement Class** for the material. At the Requirement Class level, you control whether **ATP is activated** for that type of requirement. Based on this activation, the system decides if the respective Requirement Class is **eligible for ATP checks** during sales order processing.
The Requirement Class controls whether a particular requirement is eligible for ATP checks. Only requirements with ATP activated at the Requirement Class level will be considered during availability checks.

6. ***Configure Schedule Line Categories and Requirement Class***

To ensure that AATP is triggered properly during sales order processing, both the schedule line categories and the requirement class must be configured to support availability checks.

Schedule Line Categories:

For AATP, the availability check must be active at the schedule line level. This setting ensures that the system executes the Advanced ATP logic when processing a sales order. By enabling availability checks and requirements transfer in the schedule line category, AATP can perform real-time confirmation of quantities and dates based on the checking group, checking rule, and advanced capabilities such as Backorder Processing (BOP), Product Allocation (PAL), and Alternative-Based Confirmation (ABC).

Requirement Class:
The requirement class determines whether a sales order item or dependent requirement is considered in planning and ATP. For AATP to function, the requirement class must allow availability checks. If this setting is disabled, AATP will not run—even if the checking group and checking rule are correctly defined. Ensuring the requirement class is ATP-relevant allows the order requirement to participate fully in AATP logic.

In SAP aATP, when determining whether to perform an ATP check, the Schedule Line level takes higher precedence than the Requirement Class.
Here's how it works:

- Requirement Class: Defines whether ATP is generally active for a type of requirement (e.g., sales order, delivery).
- Schedule Line / Item Level: You can override the Requirement Class setting at the schedule line level. If ATP is deactivated at the schedule line, the system will not perform ATP for that requirement, even if the Requirement Class allows it.

The system always checks the most specific level first (schedule line). If no override exists, it falls back to Requirement Class, and finally to the material-level defaults.
Schedule Line settings > Requirement Class settings > Strategy Group / Material settings.

In SD, Requirement Class settings are only effective if they are allowed by the Schedule Line Category.
For example:

- Schedule Line Category: ATP check = No
- Requirement Class: ATP check = Yes

Result: No ATP check, because the Schedule Line Category overrides the Requirement Class.

The configuration and material master data described above constitute the core setup required for ATP checks and are applicable to both Classic ATP and Advanced ATP (aATP) in SAP

7. ***Implement Additional Features (Optional)-***
   Configure further aATP features like Backorder Processing (BOP), Product Allocation (PAL), and Alternative-Based Confirmation (ABC) using available Fiori apps and IMG nodes as needed according to business requirements.
8. ***Transport and Testing-***
   Apply all configurations in a sandbox or development system, thoroughly test availability check behavior, and transport the settings to production after successful validation.
9. ***Assign Roles and Authorizations-***
   Ensure relevant users have necessary authorization to use aATP functionalities and related Fiori apps.
10. ***Test Example: Verify Basic ATP Functionality***
    Test Example to verify the Basic ATP working or not-
    Create Finished Goods Material
    a. Assign Checking Group 02
    b. Maintain all relevant ATP fields in the material master
    Check ATP Situation
    c. Use Transaction CO09 to simulate and verify the ATP situation
    d. ATP should behave according to the assigned Checking Group and Checking Rule

9. Functions of aATP and Their Execution Sequence in SAP S/4HANA

Advanced Available-to-Promise (aATP) in SAP S/4HANA strengthens order fulfillment by delivering greater accuracy, fairness, and flexibility. Its different capabilities such as availability checks, product allocation, backorder processing, and supply protection—are executed in a logical sequence throughout the order-processing cycle.

In this overview, we will also examine how aATP works end-to-end and how these functionalities can support your organization in running operations more efficiently, improving customer service, and achieving real, tangible business benefits.

aATP provides multiple key functions, including *Product Allocation (PAL), Backorder Processing (BOP), Alternative-Based Confirmation (ABC), Supply Protection, Supply-Based Confirmation (SBC), and Business Process Scheduling (BPS).* Depending on specific business requirements, companies can implement one, some, or all of these functions. This flexibility allows aATP to be tailored for optimal order fulfillment, efficient stock prioritization, and improved delivery reliability.

Key Functionalities of aATP (Advanced ATP)- aATP consists of the following major capabilities:

#	Functionalities	Details
1	*PAC – Product Availability Check*	The core engine that checks stock, receipts, and requirements to confirm quantities and dates.
2	*PAL – Product Allocation*	Limits how much can be confirmed based on customer, channel, region, or other attributes.
3	*ABC – Alternative-Based Confirmation*	Provides substitution options: Plant substitution, Storage location substitution, Product substitution
4	*Supply Protection*	Reserves supply for priority customers or channels so others cannot consume it.
5	*SBC – Supply Based Confirmation*	Supply-Based Confirmation (SBC) ensures that orders are confirmed based on actual or planned supply.
6	*BPS - Business Process Scheduling*	BPS ensures that ATP confirmations consider all dependent processes, enabling reliable order fulfillment and improved coordination across the supply chain.

7	*BOP – Backorder Processing*	Reallocates supply after the initial confirmation, based on prioritization rules.
8	*Release for Delivery*	Release for Delivery is the process of authorizing a sales order or delivery document for physical shipment.

When creating or changing a sales order, the system performs AATP checks in a specific order, starting with the Product Availability Check (PAC), followed by other aATP functionalities.
In this book, I will explain each aATP functionality in detail and describe how it applies to the sales process.

Execution Sequence When All aATP Functions Are Active - aATP does not execute all functions simultaneously. Instead, they are processed in a defined sequence, aligned with the order-to-cash process, as outlined below:

Confirmation Sequence-

Step	Function	Details
1	*PAC*	• Foundation: Calculates base ATP quantity from stock + receipts - requirements. • When the sales order is created or changed, the system first runs the availability check. • This is the foundation layer. All other checks sit on top of PAC.
2	*PAL*	• Uses time-series allocation objects (e.g., weekly/monthly limits per customer/region/channel). • If PAL is active, AATP checks allocation limits before confirming quantities. • If allocation is exceeded → confirmation is restricted.
3	*Supply Protection*	• Supply Protection is applied after PAC/PAL. It ensures protected quantities stay reserved for priority customers / channels. • If the incoming order would consume protected stock contrary to the rules, confirmation is reduced or denied; higher-priority demand can still be allowed to use lower-priority protection, depending on configuration.
4	*ABC*	ABC tries to find a viable option to satisfy the demand. • Triggered when the original plant/product cannot fulfill the requested quantity/date after the above steps.

		• System re-runs PAC (plus PAL/Supply Protection, if in scope) for alternative plants, locations, or substitute products to find a feasible confirmation.
5	*SBC*	Advanced ABC with supply creation (PP/DS planned orders). • Extends ABC by considering supply creation via PP/DS or embedded planning, where permitted. • Used in advanced scenarios where the system proposes new or adjusted planned orders to cover demand instead of only searching existing stock/receipts.
6	*BPS*	Final: Converts confirmed qty into dates (MAD → delivery). • Converts the confirmed quantities into concrete dates (requested/confirmed, material availability date, loading and goods-issue dates, delivery date). • Applies the relevant scheduling rules and calendars so final confirmations are time-phased correctly for logistics execution.

Note:

BOP (Backorder Processing) - *Executed periodically, not in real time*

BOP (Backorder Processing) runs independently (batch/scheduled) to reprioritize unconfirmed orders across priorities.

After many orders are created, BOP can re-evaluate all confirmations and reallocate supply based on business priorities.

BOP is typically run:

- Daily
- Hourly
- Or during supply disruption

Independent Functionalities-

BOP	Periodic job reprocesses open requirements, consuming new supply per priority segments (delivery priority, sales org, etc.).
Release for Delivery	Post-confirmation step; checks credit/delivery blocks before PGI, independent of ATP sequence.

Execution Flow Example-

Sales Order Entry → PAC proposes initial qty → PAL limits by allocation → ABC substitutes plant/product → Supply Protection reserves → BOP (later) reallocates → Delivery release. Sequence configurable via ATP_IMG (Manage Product Availability Check).

In SAP S/4HANA aATP, not all of the functions you listed run "in a line" during sales order entry; some are online checks (per order) and others are periodic/analytic processes. Conceptually, the runtime sequence around a sales order looks like this:

Online steps during sales order ATP

For a single sales-order item, the typical logical sequence at order save / ATP check is:

1. PAC – Product Availability Check

- Core ATP calculation of dates and quantities based on stock, receipts, and requirements.
- This is the technical heart of every aATP confirmation.

2. PAL – Product Allocation

- Runs as part of the ATP check (if configured) to limit confirmations according to allocation objects and sequences (e.g. by customer, region, channel).
- In practice, PAC checks availability, PAL restricts what portion of that availability can be promised.

3. ABC – Alternative-Based Confirmation

- If the requested product/plant/SL cannot fully confirm within PAL/PAC constraints, ABC searches for alternatives.
- It can perform:
    - Location substitution (alternative plants / storage locations).
    - Product substitution (substitute or successor products).
- ABC then proposes and writes the chosen alternative confirmation back into the sales order (either inline or via subitems).

4. SUP – Supply Protection

- If used, SUP reserves parts of supply for specific "protected" segments (channels, regions, customers) and further restricts what PAC/PAL/ABC are allowed to consume.
- Technically it is another constraint layer around the PAC result; in many designs you think of the online sequence as: Available supply → reduced by SUP → reduced by PAL → confirmed via PAC/ABC.

5. SBC – Supply Based Confirmation

- SBC extends ABC by not just substituting locations/products but actively creating supply when PAC/PAL/Supply Protection cannot confirm:
- Triggers PP/DS to generate planned/production orders.

- Simulates capacity-constrained confirmations (CTP-like).
- Executes after ABC in the sequence if alternatives still insufficient.

6. BPS – Business Process Scheduling / TM-based scheduling

- When integrated with transportation management (TM), BPS/scheduling logic uses the confirmed quantities and locations to derive realistic transportation-based dates.
- From an order-entry viewpoint it is called after or in parallel with ATP result determination to calculate final requested/confirmed dates based on route, transit, and loading times.

Periodic or corrective steps after order entry: These functions are not executed during the initial sales order creation or change. They are performed later through follow-up or periodic processes (for example, back-order processing or rescheduling and Release for Delivery).

1. BOP – Backorder Processing

- Re-allocates existing confirmations across many orders when the supply situation changes or priorities must be rebalanced.
- It can override the original PAC/PAL/ABC result according to strategies (Win, Gain, Redistribute, Fill, Lose).
- Typically scheduled as a background job, not a direct step in the online order-entry sequence.

2. RFD – Release for Delivery

- Works on already confirmed requirements and controls which confirmed quantities are actually released to create deliveries (often rule-based, may use aATP attributes).
- Comes after ATP/BOP in time: first confirm and possibly re-prioritize, then decide what to hand over to Logistics for picking/shipment.

Practical Takeaway-

1. Online during sales order processing:

- Supply Protection (SUP) and Product Allocation (PAL) act as constraints first.
- Then the system performs the Product Availability Check (PAC).
- If the requested source cannot fulfill the demand, ABC (Alternative-Based Confirmation) is triggered, followed by SBC (Supply-Based Confirmation).
- Finally, BPS (Business Process Scheduling) or TM (Transportation Management) enriches and adjusts the delivery dates.

2. After order entry / periodically:

Backorder Processing (BOP) is used to re-allocate confirmations based on business priorities.

RFD (Release for Delivery) controls which confirmed items are allowed to move forward into delivery creation.

10 Product Availability Check (PAC)

In SAP S/4HANA aATP, PAC stands for Product Allocation Check.

The Product Availability Check is a core function within Advanced Available-to-Promise (aATP) that determines on which date and in what quantity a particular customer requirement, such as a sales order item, can be fulfilled.

In short: PAC = ATP check.
PAC (Product Availability Check) is the standard ATP check that determines whether sufficient stock or planned supply is available to confirm a sales order quantity for a given plant and requested delivery date.

Product Availability Check (PAC) is a core ATP function used in both classical ATP and Advanced Available-to-Promise (aATP) in SAP S/4HANA. PAC determines the confirmed date and quantity for fulfilling a requirement such as a sales order item by comparing demand with available stock, open orders, deliveries, purchase orders, and planned receipts. It calculates confirmations by considering current and future availability, replenishment lead times, and shipping timelines.
PAC functions as the central availability engine, generating the foundational confirmation results that other aATP capabilities such as **Product Allocation (PAL), Alternative-Based Confirmation (ABC),** and **Supply Protection (SUP)** build upon. Therefore, PAC forms the essential availability mechanism for both classic ATP and the enhanced aATP framework. **PAC is a basic ATP Function in** Its core functions include:

Reading Requirements-	• Evaluates incoming customer demand (sales orders, forecasts, dependent requirements). • Considers existing confirmed quantities and open requirements.
Checking Receipts-	Verifies available supply: • On-hand inventory (unrestricted, quality, blocked per scope of check) • Planned receipts (purchase orders, production orders, STOs) • Incoming supplies within the check horizon
Providing Delivery Proposal-	Calculations feasible confirmed quantity and delivery date based on: • Available ATP = Receipts - Requirements • Material lead times, transportation, scheduling rules • Returns feasible dates/quantities to the order

This core availability-check functionality exists in both classic ATP and aATP and has long been part of the SAP SD module. However, S/4HANA aATP enhances it with improved performance, better user experience (Fiori), advanced rules, and more intelligent decision-making capabilities. PAC (Product Allocation Check) is part of the foundation for both Classic ATP and Advanced ATP (aATP) in SAP.

Under **Product Availability Check (PAC)**, two configuration nodes are available to maintain the **scope of check**:

1. ***Configure Scope of Availability Check*** *(for classic ATP)*
2. ***Configure Scope of Advanced Availability Check*** *(for aATP)*

Classic ATP Scope	Classic ATP scope is simpler and rule-based, mainly focusing on stock and open orders.
Advanced ATP Scope	aATP scope is modern, flexible, and supports advanced planning and allocation rules, making it suitable for complex supply chain scenarios.

Classic ATP Configuration Node-

SPRO→ IMG→ Cross Application Components → Advanced Available-to-Promise (aATP) → Product Availability Check (PAC) → Configure Scope of Availability Check

Transaction Code – OVZ9

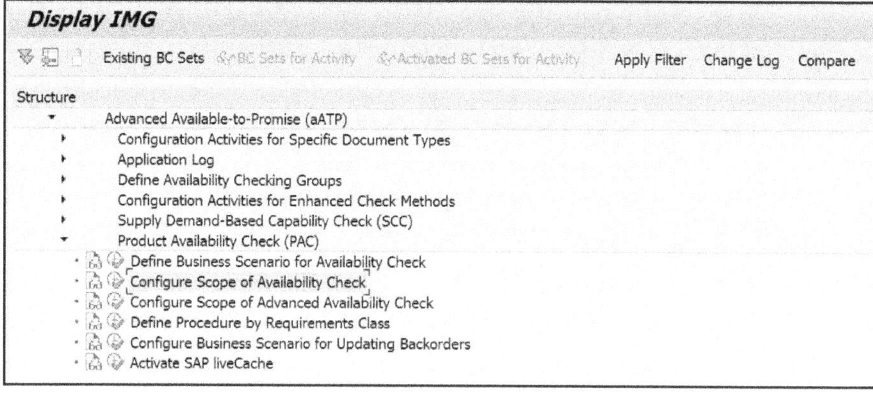

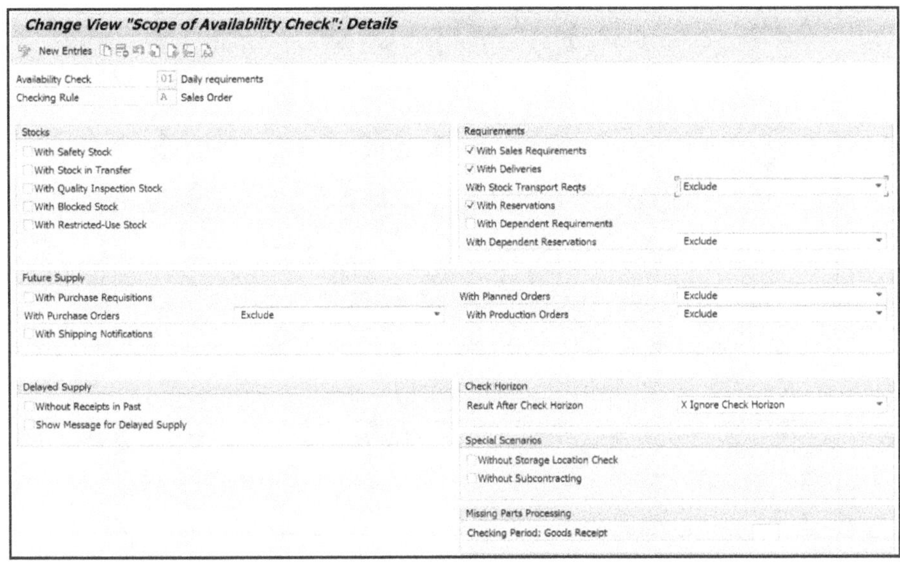

aATP Configuration Node-

SPRO→ IMG→ Cross Application Components → Advanced Available-to-Promise (aATP) → Product Availability Check (PAC) → Configure Scope of Advanced Availability Check

Transaction Code – OVZ9A

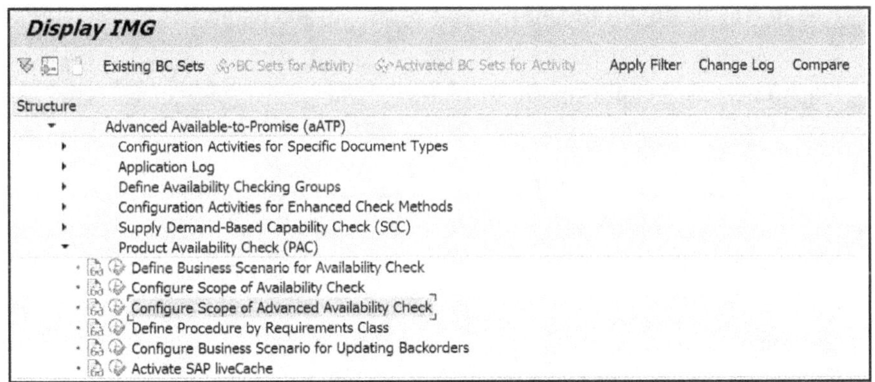

Change View "Scope of Advanced Availability Check": Overview

New Entries

Dialog Structure
- Scope of Advanced Availability Check
  - Stocks
  - Receipts
  - Requirements

Avail	Description	CRI	Checking Rule
02	Individ.requirements	01	Checking rule 01
02	Individ.requirements	03	Goods Issue
02	Individ.requirements	A	Sales Order
02	Individ.requirements	AE	SD order; make-to-order stock
02	Individ.requirements	AQ	SD order; project stock
02	Individ.requirements	AR	Supply Assignment (ARUN)
02	Individ.requirements	AV	SD order; returnable packaging
02	Individ.requirements	AW	SD order; consignment
02	Individ.requirements	B	Delivery

aATP Scope of Check – Header Level Details:

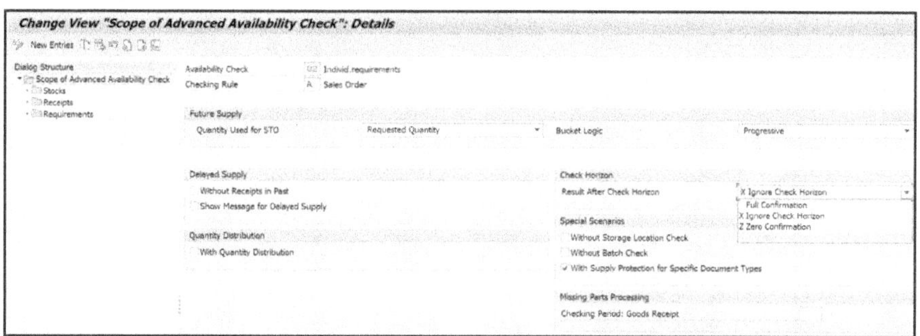

The header level contains multiple fields, several of which play a critical role in ATP calculation. One of the most important parameters is the Check Horizon, which determines how far into the future the system looks for stock and receipts during the availability check.

1. Check Horizon:

Check Horizon is one of the main differences between classical ATP and Advanced ATP (aATP) in SAP S/4HANA is how the "Check Horizon" is handled.

In classical ATP, the check horizon is typically static and based on material lead times maintained in the material master. It defines the fixed period (in working days) within which the system considers supply and demand for confirming orders. Demands beyond this horizon are either ignored or treated with predefined confirmation logics (zero or full confirmation).

In contrast, aATP enhances the check horizon function by providing more flexible and dynamic options. It allows multiple behaviors such as:

- *Full confirmation beyond the horizon,*
- *Zero confirmation beyond the horizon,*

- *Ignoring the check horizon or dynamically adapting it based on supply chain conditions and heuristics.*

This flexibility in aATP helps businesses better manage longer-term planning uncertainties and improve customer service levels by intelligently deciding how to treat demands outside the basic replenishment lead times. Thus, the handling of the check horizon is indeed a key functional distinction between ATP and aATP, with aATP providing more advanced, configurable, and dynamic horizon management to enhance order confirmation accuracy and supply reliability.

There are three types of Check Horizon available:

Full Confirmation	Requirements that fall outside the check horizon are still fully confirmed.
Ignore Check Horizon	The system does not consider Replenishment Lead Time (RLT); confirmations are given purely based on the current stock and receipt situation
Zero Confirmation	Requirements that lie beyond the check horizon are not confirmed at all.

In SAP S/4HANA, including the aATP (Advanced Available-to-Promise) engine, the system uses the RLT value from the Material Master (MRP 3 -> Field: RLT) as the ATP check horizon. RLT (Replenishment Lead Time) = Check Horizon for ATP Checks.

RLT = Check Horizon → Because SAP assumes that replenishment lead time represents the earliest possible time new supply can realistically be available, so the ATP check should not promise based on supply beyond this time horizon.

This value (in days) tells the system how far into the future it should consider incoming supply during an ATP check. And this works as below.

If RLT = 0	• ATP checks only current stock • No incoming receipts are considered
If RLT = 10	• ATP checks stock + all receipts within the next 10 days • Any supply beyond 10 days is ignored for immediate ATP confirmation

2. Special Scenarios in aATP Scope of Check:

1. **Without Storage Location Check** – ATP calculation ignores the storage location and considers stock at the plant level.
2. **Without Batch Check** – ATP ignores batch-specific stock and availability during the check.
3. **With Supply Protection for Specific Document Types** – ATP considers supply protection rules for selected document types to reserve stock accordingly.

3. Quantity Distribution:

1. **With Quantity Distribution** – When a requirement cannot be fully confirmed from a single source, the system splits the quantity across multiple sources (e.g., stock, deliveries, or plants) to fulfill the demand.

aATP Scope of Check Categories in aATP- Stock, Receipts and Requirement

In SAP Advanced Available-to-Promise (aATP), the "Scope of Check" defines the key categories of elements considered during an availability check. The main categories are:

1. **Stocks:** What is physically available right now.
2. **Receipts:** What supply is expected to arrive in the future.
3. **Requirements:** What demand needs to be fulfilled.

Together, these elements enable aATP to make accurate, prioritized, and realistic delivery commitments.

1. **Stocks-**

   This covers the current physical inventory available at the plant, storage location or batch level, including unrestricted stock, quality inspection stock, blocked stock and stock without subcontracting. It represents the inventory physically present and available for fulfilling orders.

   Importance: Provides the foundation for confirming sales orders; without stock, no order can be confirmed unless planned receipts exist.

2. **Receipts-**

   These encompass all incoming supply elements expected to increase stock, such as purchase orders, production orders (planned or confirmed), and Stock transfers from other plants. It may also include shipping notifications depending on the configuration. These receipts account for future inventory expected to arrive within the delivery timeframe.

Importance: Allows the system to **plan future confirmations**, enabling order promising even when stock is not currently available.

3. **Requirements-**

   This category includes all outbound demand elements like sales orders, reservations, stock transport orders, and production demands that consume stock (Planned independent requirements (PIRs). These requirements are taken into account to determine net available quantity.

   Importance: Helps the system **reserve supply** for high-priority orders, manage allocations, and avoid over-committing stock.

Categories Are Important in aATP Because of the Following Advantages

Granularity	By categorizing stock, receipts, and requirements, aATP can perform availability checks at a more detailed level, leading to more accurate and meaningful confirmations.
Priority Management	Categories enable prioritization of demand and supply. For example, orders from VIP or strategic customers can be assigned to higher-priority categories so they receive stock before lower-priority orders.
Flexibility for Advanced aATP Features	Categories act as the foundation for several advanced aATP capabilities, such as: **Product Allocation (PAL)-** Limits confirmations based on category, customer, channel, or allocation group. **Supply Protection (SuP)-** Protects or reserves inventory for high-priority groups to ensure they always receive required supply. **Backorder Processing (BOP)-** Reallocates supply and adjusts confirmations according to category priorities, ensuring that the most important orders remain fulfilled.

Example Scenario-

- **Stock:** 500 Units in Plant P100
- **Receipts:** 300 Units expected from a purchase order next week
- **Requirements:** 3 customer orders:
    - Customer 1001: 200 Units
    - Customer 1002: 400 Units
    - Customer 1003: 100 Units
- **Total Demand**: 700 Units

aATP will check:

- Can we fulfill orders from **stock first**?
- Do we need to include **planned receipts** to confirm orders?
- Are **priority customers** getting allocations first?

aATP Output for Given Scenario (No Priority Rules Assumed)-

PAC Calculation (Step 1)

Available ATP = Stock + Receipts - Existing Requirements

= 500 + 300 = **800 units total supply**

Current confirmed requirements = 0 (no prior orders)

800 units available

Order Confirmation Results (Sequential Entry)

Customer 1001 (200 units): PAC Result: 800 - 200 = 600 remaining
CONFIRMED 200 units (from stock)

Customer 1002 (400 units):
PAC Result: 600 - 400 = 200 remaining
CONFIRMED 400 units (300 stock + 100 from PO)

Customer 1003 (100 units):
PAC Result: 200 - 100 = 100 remaining
CONFIRMED 100 units (from PO)

Final State-

- Total Confirmed: 700 units (all orders fulfilled)
- Remaining ATP: 100 units
- Stock consumed: 500 units
- PO consumed: 200 units (100 remaining)

Key Points-

1. In SAP S/4HANA 2023, it is indeed possible to use both Classic ATP and Advanced ATP (aATP) scope of checks simultaneously within the same system landscape, even for the same material. This determination is controlled via the material master record's availability checking group, which can be set differently per plant.

How it works:

- The availability checking group assigned in the material master (usually in the MRP or Sales views) determines which scope of check is applied during order processing.
- You can configure the system so that for one plant, the material uses the Classic ATP scope of check (basic ATP logic based mainly on stock and planned receipts), while for another plant, the same material uses the Advanced ATP (aATP) scope of check, benefiting from enhanced functionalities like product allocation, backorder processing, and alternative-based confirmation.
- This flexibility allows businesses to transition gradually from classic to advanced ATP without disrupting existing processes or customer orders.
- Configuration is done in Customizing (via transaction ATP_IMG or corresponding Fiori apps) where checking groups, checking rules, and scopes of check are maintained and linked to material master data per plant.
- During availability checks on sales orders or deliveries, the system refers to the checking group of the relevant plant for the material and decides which ATP logic to apply.

This hybrid approach supports diverse operational scenarios, allowing optimized supply fulfillment tailored per plant or business unit while leveraging the strengths of both classic and advanced ATP mechanisms. Material Master record's checking group is key to determining the scope of check used per plant, enabling simultaneous classic ATP in some plants and advanced ATP in others for the same material.

The hybrid ATP determination based on material master checking groups per plant is supported from SAP S/4HANA 2020 release forward through the current versions, including the 2022, 2023 and 2025 latest releases.

2. The scope of check allows fine-tuning which MRP elements—stocks, receipts, and requirements—are included in availability checks. This ensures that ATP calculations realistically reflect actual and planned inventory and consumption, enabling accurate order confirmations in the aATP process.

3. Main Differences Between "Configure Scope of Availability Check" and "Configure Scope of Advanced Availability Check"-

Feature	Configure Scope of Availability Check	Configure Scope of Advanced Availability Check
Purpose	Used for **classic ATP (ECC style)**	Used for **Advanced ATP (AATP) in S/4HANA**
Functionality	Basic availability check based on stock, planned receipts, and open orders	Advanced availability check considering product allocation (PAL), supply protection, rules-based confirmations, backorder processing, and multiple ATP categories
Scope	Limited to standard ATP rules using **Checking Group + Checking Rule**	Flexible, category-based scope including sales order type, customer group, location, product, and allocation rules
Complexity	Relatively simple to configure	More advanced and powerful; supports complex business scenarios like shortages and prioritized customers
Time Horizon	Short-term availability	Short- to mid-term planning with real-time simulations and rules-based confirmations
Key Use Cases	Small-scale ATP checks, stock confirmations	Complex ATP scenarios, priority-based confirmations, stock allocations, supply protection, and advanced order promising

PAC in ECC vs. S/4HANA aATP-

In **ECC**, PAC worked within the SD module, and the availability check was controlled by a combination of **Checking Group + Checking Rule**. However, in **S/4HANA aATP**, the setup is completely different. aATP is **category-based**, where every requirement and receipt element is assigned to a specific **category** (such as stock, planned orders, purchase orders, sales orders, etc.). The confirmation logic is driven by these categories rather than the traditional checking group/checking rule combination.

Additionally, **PAC in aATP supports a checking horizon**, which determines how the system processes requirements:

- **Requirements within the checking horizon** are considered for confirmation based on available allocations.
- **Requirements beyond the checking horizon** may be handled differently, depending on configuration and allocation consumption rules.

This shift from rule-based checking (ECC) to category-based checking (S/4HANA aATP) makes the ATP process more flexible and aligned with real-time supply chain visibility.

How PAC (Product Availability Check) Works in S/4HANA aATP-

PAC is the core engine of SAP Advanced ATP (aATP) in S/4HANA. It determines whether a product can be delivered on time by evaluating current and future supply versus demand. Below is the end-to-end process of how it works.

PAC in aATP integrates stock, receipts, and requirements with advanced features like PAL, Supply Protection, and BOP to provide accurate, prioritized, and realistic delivery commitments. It ensures that demand is confirmed fairly, logically, and in alignment with business priorities.

In the PAC process, the available stock is allocated to the first incoming order. Subsequent orders are confirmed only if additional stock or receipts are available; otherwise, they remain unconfirmed.

This check happens immediately after sales order creation or delivery creation to ensure customers are offered achievable delivery promises. PAC is fundamental to ensuring accurate, transparent, and timely order confirmations by evaluating detailed supply and demand data using defined business rules and configurations.

1. Identify the Check Context

When a sales order, delivery, or other demand element is created, aATP first identifies:

- The checking group (assigned to the material)
- The checking rule (based on the transaction, e.g., sales order vs. delivery)
- The checking horizon (how far ahead in time PAC should look)

These define how the availability check is executed.

2. Gather Key Inputs (Stock/Receipts/Requirements)

aATP collects three main categories of data:

• ***Stock → What is available now***

- Unrestricted-use stock
- Stock in transit (depending on settings)
- Protected stock or allocated quantities (if PAL/Supply Protection is active)

• ***Receipts → What is coming in***

- Purchase orders
- Production orders / planned orders
- Inbound deliveries
- Stock transfers / STO receipts

• ***Requirements → What needs to be fulfilled***

- Sales orders
- Outbound deliveries
- Reservations
- Dependent requirements

PAC evaluates these across time buckets to build a complete availability picture.

3. Apply Category-Based Consumption

aATP uses new ATP categories to classify supply and demand more precisely (sales channels, customer groups, etc.).

Consumption rules determine:

- Which demand can consume which supply
- In what priority order
- Whether some demand is protected or excluded

This enables granular scenarios such as:

- VIP orders consuming supply before standard orders
- E-commerce orders using separate protected stock segments

4. Execute the Allocation Logic (if activated)

aATP supports multiple advanced features that integrate directly into PAC:

• ***PAL (Product Allocation)***

Ensuring demand does not exceed predefined limits per customer/channel/category.

If allocations are exceeded, PAC will not confirm quantities.

• ***Supply Protection***

Reserves inventory segments for specific high-priority groups so others cannot consume it.

• ***Backorder Processing (BOP)***

Reallocates confirmations dynamically based on priority rules.

These features influence how PAC confirms quantities.

• ***ABC for substitutions and multi-location sourcing***

ABC provides substitution rules and multi-location sourcing during PAC, proposing alternatives when primary supply fails.

• **SBC for automatic supply generation**

SBC performs an availability check using PAC or PPAC. PAC determines how much of the requested order can be confirmed immediately, and for unconfirmed quantities, PPAC can trigger planning in PP/DS to propose temporary supply elements like planned orders, purchase requisitions, or stock transfers.

In SAP (especially S/4HANA with PP/DS), PPAC is part of the availability check logic used by SBC (Supply Based Confirmation).

PPAC = Product Availability Check with Planning

(It's often referred to as *Planning-based Product Availability Check.*)

In simple terms-

- **PAC** answers: *"What can I confirm right now with what I already have?"*
- **PPAC** answers: *"If I plan new supply, how much and when can I confirm?"*

5. Perform the Time-Phased Availability Check

For each date in the checking horizon, AATP calculates:

Available Quantity = Stock + Receipts – Requirements

It then determines:

- How much can be confirmed

- On which delivery date
- Whether rescheduling or alternative sourcing is needed

aATP includes advanced simulation-based checking that provides real-time overview of confirmed and unconfirmed quantities.

6. Determine the Confirmation Result (ATP Result)

The system returns:

• Confirmed Quantity

How much of the requested quantity can be delivered.

• Confirmed Delivery Date (CDD)

The earliest date supply is available.

• Alternative Options (if configured)

Such as:

- Plant substitution
- Location substitution
- Alternative delivery proposal

7. Update the Order and the Availability Situation

Once confirmations are made:

- The sales order is updated with confirmed dates/quantities
- AATP reserves the corresponding supply
- The availability situation is recalculated for subsequent checks

Example: Step-by-Step PAC Example for a Sales Order Item in SAP S/4HANA-

1. **Sales Order Creation:**

   A customer placed an order for 500 units of Product ABC with a requested delivery date of 10 days from today.

2. **Trigger PAC:**

   The system performs the Product Availability Check immediately during sales order entry (or when the delivery is created). PAC evaluates the requested quantity and date.

   Trigger PAC (Automatic)- *When you enter the quantity and sales area details in a sales order and then move to the next step or save the document, the system automatically triggers the Product Availability Check (PAC) in real time.*

3. **Check Current Stock:**

   PAC checks the current warehouse stock of Product ABC. Suppose there are 200 units available in unrestricted stock.

4. **Check Inbound Supply:**
PAC examines open purchase orders, production orders, planned orders, and stock in transit. For example, there might be 400 units arriving in 7 days from a confirmed purchase order.

5. **Subtract Existing Commitments:**
PAC considers other confirmed sales orders and outbound deliveries that have allocated some stock, ensuring no double booking.

6. **Apply Scope of Check:**
PAC uses settings defined in checking groups and rules (configured in OVZ9 etc.) to determine which stock types and receipts to include, for example, excluding blocked stock or considering quality inspection stock.

7. **Calculate Confirmation Date:**
Based on stock availability and replenishment lead times, PAC calculates when the complete quantity of 500 units can be delivered.
    - 200 units available today
    - 300 units will arrive by day 7 (part of incoming supply)

    This scenario will generate two schedule lines:
    - *200 units are available today → confirmed on today's date*
    - *300 units become available on day 7 → confirmed on day 7*

    SAP (ATP + PAC) always confirms quantities on the earliest possible dates, so it splits the confirmation by availability date. When stock is available on different dates, SAP creates multiple schedule lines, not a single schedule line with the total quantity. However, if complete delivery is required, the system will create one schedule line with the full quantity confirmed on the later date.

8. **Provide Partial or Full Confirmation:**
PAC may confirm partial delivery of 200 units immediately and the remaining 300 units for day 7 or confirm entire 500 units at day 7 depending on settings like quantity distribution and customer preference.

9. **Result Display:**

The confirmed quantity and date are displayed in the sales order item availability section or in Fiori apps like Review Availability Check Results.

10. **Lock Material-Plant:**
    To avoid overbooking, PAC places a temporary lock or reservation on the material-plant combination during the check until the order is saved.

How does PAC differ from ATP and aATP in S/4HANA-

Here is a clear comparison of PAC, ATP, and aATP in SAP S/4HANA, explaining how they differ.

PAC serves as the foundational availability check, while aATP builds extensively on PAC and basic ATP by adding powerful business-rule driven, scenario-based enhancements to improve order fulfillment accuracy and flexibility in SAP S/4HANA.

1. Product Availability Check (PAC)

Definition:	PAC is essentially the traditional or classical availability check that determines if a specific product is available on a given date and quantity.
Function:	It checks real-time stock, open purchase orders, production orders, and other receipts to confirm availability for a single requirement like a sales order line.
Scope:	PAC focuses on basic availability without advanced allocation or prioritization features.
Example:	When a sales order is entered, PAC verifies if the requested quantity is available from stock or planned receipts and gives a confirmation date.

2. Classical ATP (Available-to-Promise)

Definition:	ATP is the traditional availability check tool in SAP ECC and also present in S/4HANA for backward compatibility.
Function:	ATP confirms product availability based on real-time stock and planned receipts, preventing overselling.
Scope:	ATP covers stock checking aligned with sales requirements but lacks advanced features.
Licensing:	Basic ATP usually does not require additional licensing in S/4HANA.

3. Advanced ATP (aATP)

Definition:	aATP is the enhanced availability check feature in SAP S/4HANA with advanced capabilities.
Function:	Besides checking stock and receipts, aATP includes Product Allocation (PAL), Backorder Processing (BOP), Alternative-Based Confirmation (ABC), and supply protection features.
Scope:	It supports complex supply scenarios with prioritizations, allocations, alternative sourcing, and integration with production and transportation.
Licensing:	aATP requires additional licensing in S/4HANA.
Example:	aATP can prioritize orders for profitable customers, restrict allocation by product allocation rules, suggest alternative products or plants on shortage, and reorder allocations during backorder processing.

Summary Table:

Feature	PAC	ATP	aATP
Check Type	Basic product availability	Stock & planned receipt check	Advanced availability with allocation, substitutions, priorities
Scope	Single requirement	Sales/stock matching	End-to-end supply chain fulfillment
Advanced Features	No	Limited	Yes (PAL, BOP, ABC, supply protection)
Licensing	Included (standard)	Included (standard)	Additional license required
Example Use	Confirm stock for sales order	Basic order confirmation	Prioritized allocation & alternative confirmations

Which master data and sales order fields affect PAC outcomes-

The outcomes of Product Availability Check (PAC) in SAP S/4HANA are influenced by several master data and sales order fields that define how availability is checked and confirmed. Key influencing elements include:

Master Data Affecting PAC-

1. Material Master Data:

Availability Check Group	Availability Check Group: Determines which availability check and scope of check rules apply.
Replenishment Lead Time	Defines how long it takes to procure or produce the material.
Shipping Point/Plant Determination	Impacts stock location for checks.
Batch Management Settings	Batch Management Settings: If batch-managed, PAC considers batch availability.
Stock and Requirements Settings	Including special stock types, quality inspection stock, safety stock.

2. Customer Master Data (Business Partner):

Shipping Conditions	Influence delivery and availability timings.
Customer Priority or Profitability Group	May be used in advanced checks (e.g., aATP prioritization).
Credit Control Area and Limits	Effect on whether orders are blocked or allowed to proceed.

3. Sales Order Fields Influencing PAC-

Sales Document Type	Defines default availability check rules and scope of check applicable for that document.
Schedule Line Category	Controls at the item level when and how availability is checked, whether partial quantities are allowed, etc.
Requested Delivery Date	PAC tries to confirm availability based on this date or propose alternatives if unavailable.
Order Quantity	The quantity requested impacts whether full or partial availability can be confirmed.
Material Number / Batch	Identifies the specific product or batch whose availability is checked.
Shipping Point and Plant	Influence the stock location checked for availability.
Requirement Type	Defines classification of demand element, affecting how the system treats it during checks.

This data collectively governs PAC's ability to estimate realistic delivery quantities and dates by aligning demand, supply, and business priorities.

Test Example to understand the Product Availability Check (PAC) functionality in aATP:

Product Availability Check (PAC) is the core real-time availability check in SAP aATP. It evaluates current stock, planned receipts, and existing requirements to confirm sales order quantities and dates at specific locations, such as plants or storage locations.

1. Material “FG10001” is maintained with Availability Check “02”. Current Stock is available for 500 units. Receipts are 500 units and Issues is Zero.

Availability Overview (CO09)-

CO09 displays a detailed picture of available stock, planned receipts, and existing requirements for material at a specific plant or storage location. It shows how the system calculates ATP (Available-to-Promise) quantities over time.

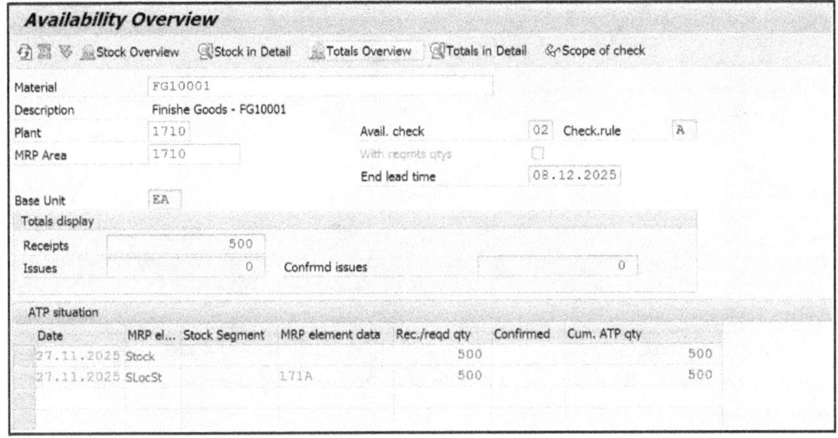

Example Test Scenario-

Consider a material "FG10001" with 500 units in unrestricted stock at Plant 1710 on November 27, 2025. Two sales orders created.

- Sales Order#1 (Customer Standard, 300 units requested for 12/02/2025): PAC checks scope (stock: 500, receipts: none, requirements: none). Confirms 300 units on 12/04/2025, leaving 200 units available.
- Sales Order#2 (Customer Standard, 250 units requested for 12/02/2025): PAC now sees stock reduced to 200 units. Confirms only 200 units on 12/04/2025; remaining 50 units get a later date

12/09/2025 based on future receipts (e.g., incoming PO for 50 units on 12/09/2025, confirming partial on 12/09/2025).

Product Availability Check (PAC) Result-
Order 1 is fully confirmed; Order 2 is partially confirmed (200 confirmed immediately + 50 confirmed for a later date).

Availability Check of the 1s Sales Order-

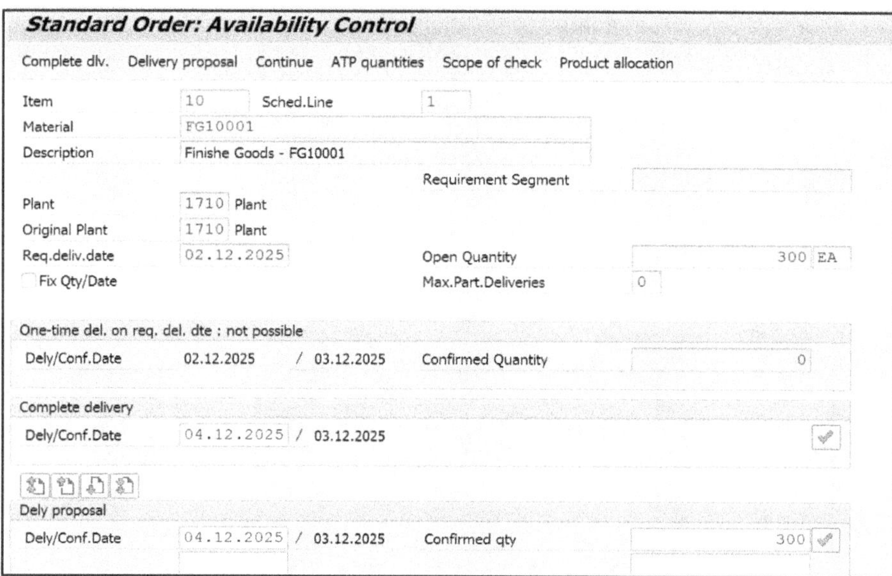

CO09 after creation of the 1st Sales Order-

Availability Check of the 2nd Sales Order-

Standard Order: Availability Control

Complete dlv. Delivery proposal Continue ATP quantities Scope of check Product allocation

Item 10 Sched.Line 1
Material FG10001
Description Finishe Goods - FG10001
Requirement Segment
Plant 1710 Plant
Original Plant 1710 Plant
Req.deliv.date 02.12.2025 Open Quantity 250 EA
End lead time 09.12.2025
Fix Qty/Date Max.Part.Deliveries 0

One-time del. on req. del. dte : not possible
Dely/Conf.Date 02.12.2025 / 03.12.2025 Confirmed Quantity 0

Complete delivery
Dely/Conf.Date 09.12.2025 / 08.12.2025

Dely proposal
Dely/Conf.Date 04.12.2025 / 03.12.2025 Confirmed qty 200
09.12.2025 / 08.12.2025 50

CO09 after creation of the 2nd Sales Order-

Availability Overview

Stock Overview Stock in Detail Totals Overview Totals in Detail Scope of check

Material FG10001
Description Finishe Goods - FG10001
Plant 1710 Avail. check 02 Check.rule A
MRP Area 1710 With reqmts qtys
End lead time 08.12.2025
Base Unit EA

Totals display
Receipts 500
Issues 550 Confrmd issues 500

ATP situation

Date	MRP el...	Stock Segment	MRP element data	Rec./reqd qty	Confirmed	Cum. ATP qty
27.11.2025	Stock			500		0
03.12.2025	CusOrd		Totals record	550-	500	0
08.12.2025	----->		End lead time		0	
08.12.2025	CusOrd		Totals record	0	50	
27.11.2025	SLocSt		171A	500		500

Delivery Proposal Options in aATP at Sales Order - (Availability Control)

In SAP S/4HANA aATP, Delivery Proposal Options at Sales Order level control how the system proposes delivery dates and quantities during ATP - confirmation when a sales order is created or changed.

In SAP aATP, One-Time Delivery, Complete Delivery, and Delivery Proposal are configurable options within the Product Availability Check (PAC). These options determine how partial availability is handled when the full requested quantity is not available on the requested date.

The Availability Control screen appears during the availability check in the sales order.

- In interactive sales order processing (VA01/VA02 or Fiori *Create/Change Sales Orders*), the Availability Control screen is displayed when the system cannot fully meet the requested quantity or date and the availability check is configured to be interactive for that scenario.
- The screen provides proposals such as One-Time Delivery, Delivery Proposal, and Complete Delivery, and displays information such as ATP quantities, scope of check, and alternative plants.

The Availability Control pop-up appears only in interactive sales order processing when ATP cannot fully confirm the requested quantity or date, and Availability Control is configured accordingly. Its behavior is primarily driven by Availability Control customizing, along with availability check settings in the requirements class, schedule line category, sales document type, item category, and the customer/material check groups and scope of check.

In EDI/IDoc-based sales order creation, the Availability Control pop-up is not displayed. Instead, the system applies the configured ATP and delivery proposal logic automatically in the background, without user interaction.

Behavior in EDI/IDoc Processing

- For EDI/IDoc-based sales order creation, processing runs in background (dialog-less) mode, and the interactive Availability Control screen is not shown.
- The system applies the same underlying ATP and Availability Control configuration automatically, using predefined decision

rules (for example, selecting a system-defined delivery proposal or confirming zero), without user input.

A. Manual Sales Order (VA01 / Fiori "Create Sales Order")

When a user creates a sales order, aATP performs the availability check. If the requested quantity or requested delivery date cannot be confirmed, the system displays the **Availability Control** pop-up, allowing the user to choose one of the following:

- One-time delivery
- Complete delivery
- Delivery proposal
- Accept new confirmation date

B. EDI / IDoc Sales Orders (ORDERS05 / API_SALES_ORDER_SRV)

In the case of EDI/IDoc sales orders, the Availability Control pop-up **does not appear**.

aATP executes automatically in the background, and the system applies the delivery rules based entirely on configuration - **no user interaction is possible**.

What happens automatically?

Situation	System Behavior
Partial deliveries allowed	Confirms available quantity
Complete delivery required	Pushes to next date with full supply
Delivery proposal configured	System applies proposal automatically
aATP PAL/Supply Protection	Customer may get reduced/increased confirmation
Missing data	Order may fail › IDoc error queue

All rules work in the background. The same aATP logic is applied as in the manual check, but without any pop-up. The behavior is fully dependent on the configuration setup.

Manual orders require user interaction, so SAP displays the selection screen. EDI orders, however, must process automatically, so SAP selects the option entirely based on configuration.

Thus, the same configuration nodes control both processes, but only manual order creation shows the screen.

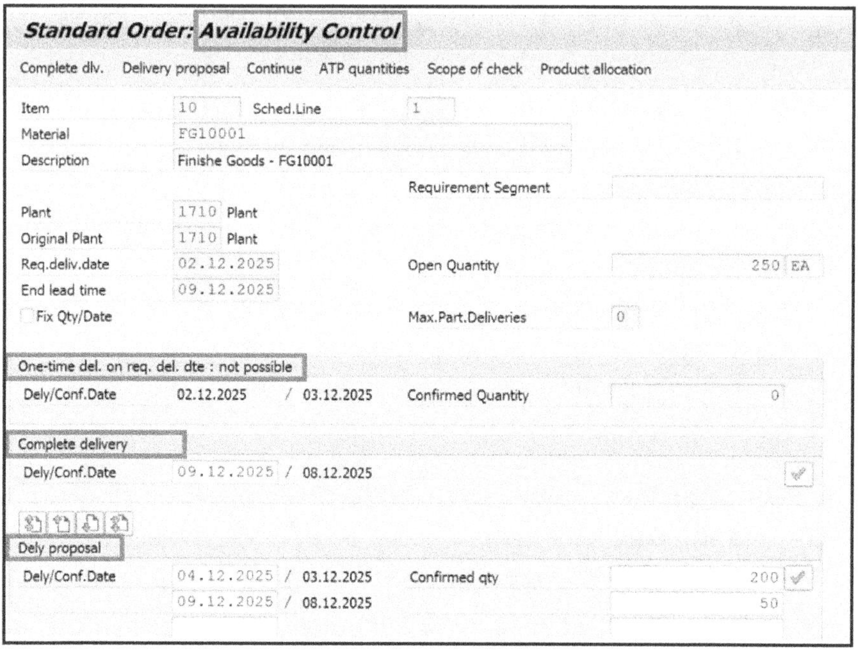

Standard Order: Availability Control

Complete dlv. Delivery proposal Continue ATP quantities Scope of check Product allocation

Item	10	Sched.Line	1
Material	FG10001		
Description	Finishe Goods - FG10001		
		Requirement Segment	
Plant	1710 Plant		
Original Plant	1710 Plant		
Req.deliv.date	02.12.2025	Open Quantity	250 EA
End lead time	09.12.2025		
Fix Qty/Date		Max.Part.Deliveries	0

One-time del. on req. del. dte : not possible

Dely/Conf.Date	02.12.2025 / 03.12.2025	Confirmed Quantity	0

Complete delivery

Dely/Conf.Date	09.12.2025 / 08.12.2025

Dely proposal

Dely/Conf.Date	04.12.2025 / 03.12.2025	Confirmed qty	200
	09.12.2025 / 08.12.2025		50

1. One-Time Delivery (Rule A)-

The system confirms only the quantity available exactly on the requested material availability date, ignoring any later possibilities. All later deliveries are excluded, prioritizing the exact requested timing over full fulfillment. This ensures that the required quantity is delivered on the requested delivery date, based on how much can actually be confirmed.

- The system tries to confirm the entire quantity requested in **one single delivery**.
- If the full quantity cannot be confirmed at once, it will **not split** the delivery unless allowed by other settings.
- Common when customers require a single shipment.

2. Complete Delivery (Rule B)-

The system finds the earliest single date when the entire requested quantity can be delivered in one shipment, even if later than requested. This ensures full quantity in a single delivery but may push out the date. This ensures that the earliest delivery date is selected on which the entire required quantity can be delivered in one shipment.

- Ensures that the **entire order (all items)** is delivered together in one delivery.

- If any item cannot be confirmed, the system may delay confirmation until all items can be delivered at once.
- Typically used when partial shipments are not allowed for the whole order.

3. Delivery Proposal (Rule C)-

The system proposes multiple partial deliveries across different dates until the full quantity is covered, optimizing fulfillment through split shipments. This balances speed and completeness when stock is staggered.

- Allows the system to generate **multiple delivery options** based on stock availability.
- Propose possible delivery splits (e.g., partial delivery now, balance later).
- User can choose the best delivery scenario from the proposed options.
- Ideal when flexibility is needed, and the customer accepts partial or staged deliveries.

Key Point-

In SAP S/4HANA aATP, the Delivery Proposal (the availability control screen that pops up during ATP check) is controlled by configuration under: "Configure Default Settings by Sales Area".

Purpose of "Configure Default Settings by Sales Area"-

This configuration step ensures:

- **Consistent ATP behavior** across the defined sales area.
- **Control over ATP functionalities** that should be active for that sales area.
- **Definition of delivery proposal behavior**, influencing how ATP suggests confirmations.
- **User-friendly and predictable confirmation logic** during sales order processing, improving accuracy and efficiency.

Configuration Path-

SPRO → Cross-Application Components → Advanced Available-to-Promise (aATP) → Configuration Activities for Specific Document Types → Sales Orders and Deliveries → Configure Default Settings by Sales Area (Transaction Code – OVZJ)

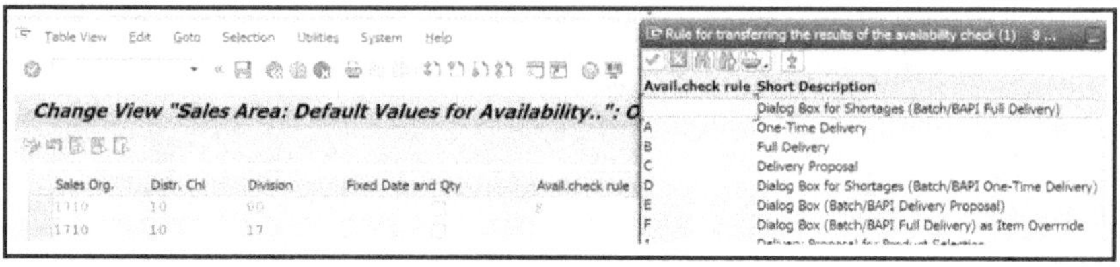

In SAP transaction OVZJ (Default Values for Availability), the Availability Check Rule field (Table: TVTA-REVFP) defines how the system handles ATP results when full confirmation isn't possible. It's sales area-specific and controls dialog behavior and schedule line creation.
Availability Check Rules (A, B, C, D, E, F)-

Rule	Meaning	Schedule Lines Created	Dialog Behavior
A	One-time delivery	1 line (confirmed qty only)	No dialog; partial confirm as single line
B	Complete delivery	0 lines if partial	No dialog; zero confirm if not full qty
C	Delivery proposal	Multiple lines (confirmed + open)	No dialog; proposes all future dates
D	Dialog (one-time delivery)	1 line	Shows Availability Control popup
E	Dialog (delivery proposal)	Multiple lines	Shows Availability Control popup
F	No check	N/A	Skips ATP entirely

Best OVZJ Availability Check Rule: Rule C (Delivery Proposal) is the most widely recommended for general sales scenarios.

Why Rule C is Optimal

- Flexibility: Creates multiple schedule lines showing all future confirmation possibilities (e.g., 50 today + 30 Jan 7 + 20 Jan 10) → Customer sees complete picture.
- Maximizes fulfillment: Unlike A/B (strict one-time/complete delivery), C confirms whatever is available across dates, reducing zero confirmations.
- MRP-friendly: Generates realistic requirements in MD04 on proposed dates, enabling proactive planning.
- Standard practice: Used by 80%+ of SAP customers for MTS/MTO mixed environments.

RACR screen in aATP Fiori-

Use the **"Review Availability Check Results"** Fiori app to view the detailed breakdown, including stock consumption, time-phased scheduling, and the visualization of supply and demand elements.

RACR (Review Availability Check Result) is the Fiori app/screen in SAP aATP that serves as the modern equivalent of the classic Availability Control screen. It provides a clear, interactive overview of aATP results during sales order creation or change when the full quantity is not available on the requested date.

RACR improves transparency for sales representatives by showing exactly why orders are confirmed the way they are and enabling faster, more informed decision-making.

This is the situation after creating the 2nd order – the current stock is **0** (displayed in red).

The 2nd order was created with a quantity of **250 units**, but only **200** units were available, so 200 units were confirmed immediately. For the remaining **50 units**, the system provided a future confirmation date.

RACR Screen-

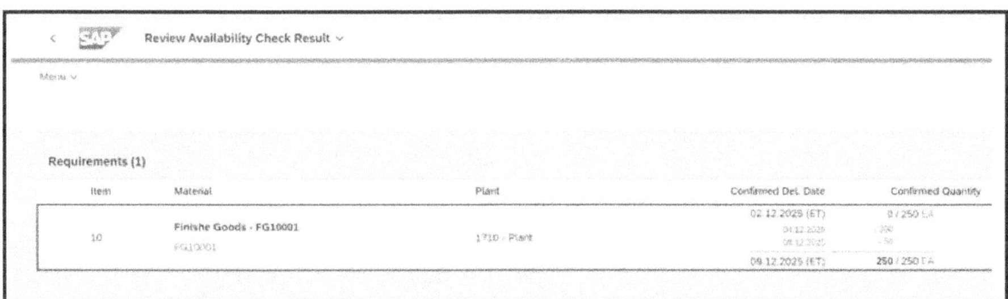

Chosen Confirmation Screen-

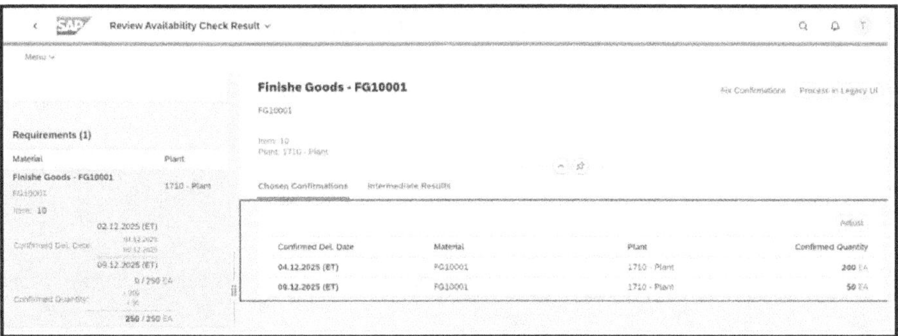

Intermediate Results Screen-

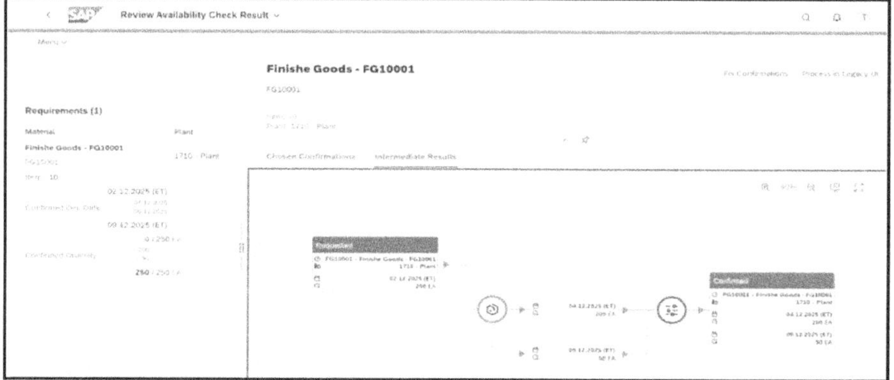

Here is a clear explanation of the purpose and use of RACR in aATP at the sales order level.

Review Availability Check Result (RACR) is a Fiori app/screen in SAP aATP that appears automatically at the sales order level during order creation or change when an item cannot be fully confirmed for the requested date or quantity.

Purpose / Use of RACR in aATP Fiori-

Modern Availability Control	RACR replaces the classic Availability Control screen and provides a modern, interactive view of Product Availability Check (PAC) results.
Transparency for Sales Orders	It shows why a sales order is confirmed fully, partially, or not at all.
Decision Support	Helps sales representatives make quick, informed decisions when requested quantities cannot be fully confirmed on the desired delivery date.
Visualization of Stock and Supply/Demand	Displays current stock, planned receipts, and outstanding requirements in an easy-to-understand, time-phased view.
Partial Confirmations	Shows how the system splits confirmations (immediate vs. future delivery) when full quantity isn't available.

How RACR is Used

Accessing RACR: During sales order creation or change, the app can be launched automatically if a partial or unconfirmed quantity occurs.

Viewing Results:

- Confirmed quantity: Quantity that can be delivered immediately.
- Future quantity: Quantity that can be delivered later (future confirmation date).
- Stock & Demand Overview: See which stock or planned receipts were allocated to this order.

Interactive Analysis: Users can drill down into supply and demand elements, check stock consumption, and understand why the order was confirmed as it was.

Decision Making: Sales reps can decide whether to accept the proposed schedule, split the order, or communicate with the customer about alternate delivery dates.

11 Product Allocation (PAL)

Product Allocation (PAL) in SAP S/4HANA aATP is a functionality designed to control how limited or constrained products are distributed to customers, sales channels, or regions, based on predefined business rules and allocation quantities. PAL ensures that scarce inventory is allocated efficiently and fairly, preventing overselling and ensuring that strategic customers, markets, or regions receive their prioritized share according to established rules and priorities.

In short, it helps prevent overselling and ensures that strategic customers or regions receive their allocated share when supply is limited. It is a feature used to control and fairly distribute limited product quantities during availability checks.

Situation: When demand is high and supply is limited, PAL is used to manage customer demand by strategically distributing available supply. Instead of allocating stock on a first-come-first-served basis, the system applies business-defined strategies (configured rules) to determine which customers receive allocation.

From a supply perspective, you typically have a rough view of available supply coming from production orders, purchase orders, and planned receipts. PAL also enforces capacity constraints (such as monthly production limits) to support fair and controlled distribution decisions across customers or sales channels.

In SAP S/4HANA aATP, Product Allocation (PAL) is not a generic/default feature but must be configured specifically based on business scenarios or requirements like supply shortages, high demand launches, or customer prioritization needs.

It is very important to understand the business purpose of using PAL functionality, because if it is applied indiscriminately, it might disrupt the business cycle.

Practical Examples (Business Scenarios)-

Business Scenarios	Details
1. Short Supply:	• When demand is high but supply is limited, allocation rules are used to balance fulfillment and ensure supply is distributed according to business priorities.
2. Seasonal Requirement:	• For example, during Christmas, orders may be received 2–3 months in advance. We need to manage the demand with available supply, ensuring that no single customer receives all the stock. Supply must be balanced across customers based on business strategies to ensure everyone receives products on time. • PAL is configured to work for this specific period (e.g., Christmas) and not for all times.
3. New Product Launch:	• When a new product is launched, initial supply is often limited. PAL helps prioritize customers or regions, ensuring that key accounts or strategic markets receive stock first. • This prevents overselling and helps manage expectations while ramping up production.
4. Continuity of Business Operations:	• PAL ensures that regular operations continue smoothly, even when demand fluctuations occur, by allocating stock in a controlled and predictable manner.

Note: Planning is done using planning tools, but the results are fed into PAL allocation to control the distribution of available stock.

Usage guidance-

Scenario-driven:	• *Activate PAL only for constrained materials via Product Allocation Objects (PAOs), sequences, and planning data tailored to customer groups/regions.*
Avoid generic use:	• *Applying PAL universally adds unnecessary overhead; use classic ATP for unconstrained products.*

Example-

PAL is particularly useful in scenarios like rationing new mobile phones by region—for example, allocating 20,000 units per month per customer tier—or when combining sales and transport capacity checks. PAL ensures that available supply is distributed according to predefined rules, determining how much each customer or region receives based on the allocation strategy.

Suppose Apple is launching a new iPhone in January 2026. Initially, supply will be limited for the first three months. This represents a rough supply estimate—based on production capacity of how many units can be produced during those three months. PAL itself does not calculate supply; rather, these quantities must be defined upfront based on expected production capacity and planning assumptions.

Month	Supply Quantity Based on Planning
Jan-2026	100,000 Units
Feb-2026	120,000 Units
Mar-2026	80,000 Units

In this situation, the available supply is distributed across regions based on predefined business priorities. This represents the PAL allocation for each month, considering the specific customer regions.

Regions	USA	India	China	Taiwan	Japan
Jan-2026	30,000	20,000	20,000	10,000	20,000
Feb-2026	25,000	25,000	25,000	25,000	20,000
Mar-2026	25,000	25,000	20,000	5,000	5,000

A very important point is that the PAL check is performed after the PAC. The system first runs the Availability Check and calculates the ATP quantity and then applies the maximum allocation for the relevant CVC (Characteristic Value Combination) to determine the final confirmed quantity based on the allocation constraints.

In the context of SAP aATP and PAL, a constraint is a limit or restriction that controls how much of a product can be allocated or confirmed.

CVC (Characteristic Value Combination) in PAL is based on fields such as Sales Organization, Region, Customer, etc., not on system characteristics.

Scenario #1	Results
• Sales order quantity: 10,000 units • Requested delivery date (RDD): within January 2026 • Region: USA • ATP quantity Available: 8,000 units • PAL allocation for USA in Jan-2026: 30,000 units	The system confirms only 8,000 units based on the available ATP quantity, even though the PAL allocation allows a higher quantity.
Scenario #2	**Results**
• Sales order quantity: 30,000 units • Requested delivery date (RDD): within January 2026 • Region: USA • Available ATP quantity: 50,000 units • PAL allocation for USA in Jan-2026: 30,000 units	The system confirms 30,000 units based on the PAL allocation limit, even though additional ATP stock is available.
Scenario #3	**Results**
• Sales order quantity: 50,000 units • Requested delivery date (RDD): within January 2026 • Region: USA • Available ATP quantity: 50,000 units • PAL allocation for USA in Jan-2026: 30,000 units	The system confirms only 30,000 units based on the PAL allocation limit, even though ATP stock is available for the full requested quantity.

In SAP aATP, Product Allocation (PAL) defines the maximum quantity that can be allocated to a customer or to orders based on defined attributes (such as customer, region, or channel). ***The key concept is the maximum allocation limit determined by the CVC (Characteristic Value Combination) combinations.***

Product Allocation (PAL), Backward and Forward Scheduling-

In SAP aATP – Product Allocation (PAL), Backward and Forward Scheduling control from which time bucket (period) the PAL allocation quantity is consumed when confirming a sales order.

PAL allocations are usually maintained per time period (day / week / month).

When a sales order is created, the system first looks at the Requested Delivery Date (RDD) period. If sufficient allocation is not available in that period, backward or forward consumption logic determines where else the system is allowed to consume allocation from.

Backward and forward scheduling in PAL define whether allocation is consumed from earlier or later periods when the requested period does not have enough allocation.

- Backward scheduling allows the system to consume unused allocation from **earlier periods**.
- Forward scheduling allows consumption from **future periods**.

Backward and forward scheduling are not active by default and must be enabled in the PAL configuration to allow allocation consumption from earlier or later periods.

1. Backward Consumption (Earlier Periods)

What it means:	The system is allowed to consume unused PAL allocation from earlier periods.
How it works:	• System checks allocation in the RDD period • If insufficient, it looks backward (previous month/week/day) • Consumes remaining allocation from earlier periods (if allowed)
Use case:	• Demand is delayed, but earlier allocations were not fully used • You want to avoid wasting unused allocation
Example:	• Jan allocation (USA): 30,000 → only 20,000 used (10,000 remaining) • Feb order request: 25,000 • Feb allocation available: 10,000 **Backward consumption allows using the 10,000 unused from Jan, so:** • 10,000 from Feb • 10,000 from Jan (backward) Key point: PAL confirms up to the sum of available allocations across the relevant periods, not more than that. The remaining 5,000 units of the order remain unconfirmed.

Example: Handling Unconfirmed Quantity in PAL & BOP

In this example, 5,000 units remain unconfirmed due to PAL allocation limits. During Backorder Processing (BOP): BOP periodically re-evaluates unconfirmed quantities, checking:

    - New incoming stock (production or purchase orders)
    - Allocation releases or unused quantities in current or future periods
    - Changes in PAL quotas

- If additional stock or allocation becomes available, the unconfirmed 5,000 units can be confirmed.
- The unconfirmed quantity stays open until stock or allocation is available.
- BOP can be run daily, weekly, or manually, depending on business process settings.

Key Points-

- PAL limits are applied at the time of order creation.
- Unconfirmed quantities do not disappear; they remain eligible for confirmation later.
- BOP is the process that attempts to confirm previously unconfirmed orders when supply or allocation changes.

During the BOP run, the system attempts to reallocate (reschedule) the unconfirmed quantity, but it also respects PAL allocation limits and strategies, ensuring that all confirmations comply with the configured allocation rules.

During sales order creation, the system first performs the PAC (Product Availability Check), then checks PAL (Product Allocation). Later, Backorder Processing (BOP) runs according to its schedule to confirm unfulfilled quantities. PAL can become complex, especially when orders from other regions do not arrive as expected planners need to adjust allocations to ensure stock is utilized effectively.

2. Forward Consumption (Later Periods)

What it means:	The system is allowed to consume PAL allocation from future periods.
How it works:	• System checks allocation in the RDD period • If insufficient, it looks forward (next month/week/day) • Consumes future allocation early (if allowed)
Use case:	• Customer demand comes earlier than planned • Business allows pulling future supply forward
Example:	• January allocation (USA): 20,000 • Order request in January: 30,000 • February allocation (USA): 15,000 System confirms: • 20,000 units from January allocation • 10,000 units from February allocation (forward consumption) Total confirmed: 30,000 units Key point: Forward consumption allows the system to use allocation from later periods when the current period does not have enough allocation, but only up to the defined allocation limits.

Difference Between PAL and BOP-

Here's a detailed explanation of the difference between PAL (Product Allocation) and BOP (Backorder Processing) in SAP S/4HANA aATP:

Feature	PAL (Product Allocation)	BOP (Backorder Processing)
Purpose in SAP S/4HANA AATP - Main Purpose	Pre-allocate stock to customers or regions based on allocation rules, periods, and priorities. Ensures fair distribution of limited stock.	Re-assign stock to unfulfilled or partially confirmed sales orders when supply becomes available.
Timing of Execution - When it executes	During order entry or availability check, before confirming the order.	After an order is partially or completely unfulfilled, when new stock arrives or becomes available.
Mechanism in SAP S/4HANA AATP - How it works	Uses allocation objects, sequence, sequence groups, and constraints to allocate stock proactively. Can consider backward/forward consumption and multiple schedule lines.	Uses backorder processing runs to reschedule and prioritize open orders. Orders are sorted by priority, rules, or customer categories and stock is reassigned automatically.
Outcome / Result - Result	Ensures planned allocation of available stock across customers or regions. Prevents over-allocation and ensures supply meets business rules.	Ensures pending or partially confirmed orders are fulfilled as soon as stock is available, according to priority and rules.
Key Difference	PAL: Proactive stock control before order confirmation. Prevents conflicts by controlling allocation across customers/orders.	BOP: Reactive stock management after orders cannot be fulfilled. Resolves backorders when new stock arrives.

Product Allocation (PAL) in ECC vs. S/4HANA aATP

Product Allocation (PAL) is also available in ECC; however, setting it up in ECC requires nearly 30 configuration steps. In contrast, S/4HANA aATP simplifies the entire setup, allowing you to configure PAL end-to-end using just five Fiori apps, with no SPRO configuration required except for activating PAL—everything else is done solely through the Fiori apps.

The technical complexity of PAL has been significantly reduced in S/4HANA aATP, making it much easier to adopt compared to ECC PAL. However, the business scenarios themselves remain complex and cannot be simplified or eliminated.

While the technical architecture is now streamlined and user-friendly, PAL still requires careful design due to the complexity of real-world business requirements. Additionally, after go-live, organizations typically need at least 6–8 months to fine-tune the setup, especially adjustments related to CVC values and allocation strategies, to achieve optimal results.

In ECC, PAL was adopted by very few clients due to its complexity. Analysis and monitoring tools were not available, making it difficult to track allocation consumption, under-consumption, or over-consumption. Planners had to manually adjust allocations month by month to ensure correct utilization of available stock. Correct allocation was critical; for example, if a region like Japan had allocated stock but orders were not received in that month, planners needed to manually reassign quantities to ensure proper utilization.

PAL in S/4HANA aATP: Simplified end-to-end setup using **just five Fiori apps**. Monitoring dashboards (available from 2020 onward) allow planners to:

- Track which CVC combinations are being consumed
- Identify under-consumption or over-consumption
- Make timely adjustments to ensure allocation aligns with actual orders
    - Allocation is dynamic and based on supply projections, which are not 100% fixed.
    - Planners can adjust PAL allocations according to updated supply projections to optimize stock usage across months.
- S/4HANA aATP makes PAL more user-friendly and manageable.
- Real-time monitoring and adjustment reduce stock wastage and improve allocation efficiency.
- Dynamic allocation ensures supply is distributed based on business priorities while adapting to changing demand.
- Allocation quantities can be monitored and change directly into production system.

PAL significantly improves **order promising**, ensuring that limited or constrained products are allocated fairly and efficiently according to business priorities.

PAL works as an additional constraint on top of the standard aATP check: even if physical stock is available, the confirmed quantity cannot exceed the allocated quantity in PAL for the relevant period and characteristics. During sales order creation or change, aATP reads the relevant product allocation object and consumption data to decide how much can be confirmed and on which dates.

Core purpose	• PAL ensures that scarce or high-demand materials are allocated based on predefined business rules (such as by customer, region, or sales channel), **rather than on a *first-come, first-served* basis.** • It prevents a single large order from consuming all available stock, thereby protecting supply for priority customers or markets.
Typical business use cases-	• Launches of new or high-demand products where production capacity is constrained, and stock must be reserved for strategic customers and markets. • Situations with supply shortages (for example component or raw-material constraints) where fair and rule-based distribution of limited stock is required.
PAL Related aATP features	• PAL can be combined with Backorder Processing (BOP) to reallocate inventory in line with allocation rules when priorities change. • It also works with supply protection to reserve quantities for specific segments while still enforcing overall allocation limits.

In SAP S/4HANA aATP, PAL (Product Allocation) is the main allocation mechanism that checks quotas against customer/product characteristics after the basic PAC (Product Availability Check), imposing hard limits on confirmations even when stock exists.

Priority sequence-

PAL always takes precedence. The system first performs the PAC check (to verify if stock is available) and then immediately checks the relevant PAL object. If the PAL quota is *fully consumed*, the confirmed quantity is reduced even if ATP stock is available."

When multiple PAL sequences exist, they are checked in the defined priority order, with the highest-priority sequence evaluated first.

Key differences-

PAC	PAL
Standard ATP check against physical stock/ATP quantities, first-come-first-served without allocation rules.	Additional constraint layer that consumes from predefined time-bucket quotas by characteristics (customer group/region), overriding PAC results.

Here's a simple business example to explain how PAL takes precedence over PAC:

Example#1 - **PAL vs PAC in SAP aATP-**

The order is reduced or blocked because **PAL limits are reached**, not because of stock shortage.

Example	System Behaviors (Results)
• Available stock (PAC): 100 units (Product ABC1001). • Product Allocation (PAL) quota for Customer 10001: 50 units. • Customer 10001 already consumed: 50 units. • New sales order quantity: 30 units.	1. SAP aATP first checks PAC → stock is available (100 units). 2. System then checks PAL → allocation quota is fully consumed. 3. Result: Confirmed quantity = 0 units, even though stock is available.

Example#2 **Partial Confirmation due to PAL.**

This is a **partial confirmation caused by PAL**, not by stock shortage.

Example	**Results**
• Available stock: 100 units (Product ABC1001). • Product Allocation (PAL) quota for Customer 10001: 60 units. • Customer 10001 already consumed: 40 units. • Order quantity: 30 units.	• Remaining PAL quota = 20 units • Confirmed quantity = 20 units • Remaining 10 units are not confirmed, despite available stock.

PAL Activation Settings (SPRO and Fiori Apps)-

1. Activate Product Allocation through SPRO Configuration

SAP determines allocation priority between customers in Product Allocation (PAL) primarily through configurable Product Allocation Objects (PAOs) and allocation sequences defined in aATP, where businesses assign higher quotas or checking precedence to priority customer groups, regions, or hierarchies during availability checks.

Product Allocation (PAL) in aATP only affects confirmations if it is explicitly activated in customizing; without activation, allocations are defined but not considered in the check.

Product Allocation (PAL) in aATP allows you to restrict sales order confirmations based on predefined allocation quantities. But crucially:

- PAL only affects ATP checks if it is explicitly activated in customizing.
- If not activated, PAL definitions (allocation objects/sequences) exist but are ignored during order confirmation checks.

In short, in SAP aATP, after the PAC (Product Availability Check), the system evaluates Product Allocation (PAL) *if it is activated and relevant*. If PAL is not activated or not applicable, the system skips it and continues with the next activated aATP functionality in the checking sequence.

- PAL is checked only when it is configured and relevant for the material, plant, and sales area.
- Even if stock is available in the PAC, PAL can still reduce or block the confirmed quantity if the allocation quota is exhausted.
- If PAL is not active or not applicable, the system proceeds to the next aATP step (such as ABC classification, Backorder Processing relevance, Supply Protection, BSP, etc.), depending on the configuration.

SPRO Path-Activate Product Allocation

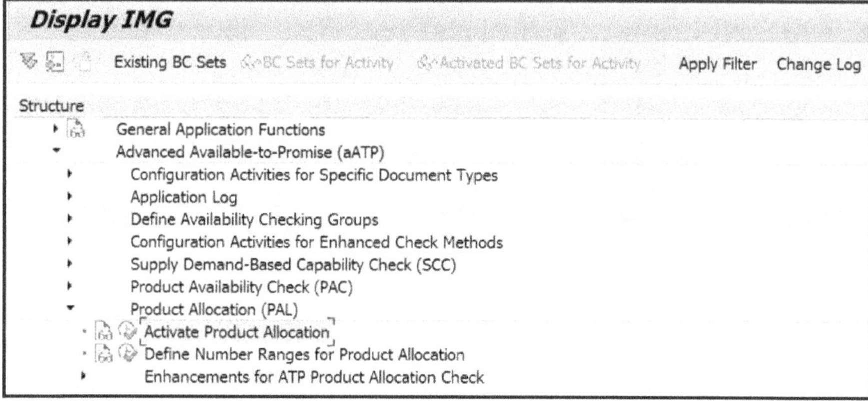

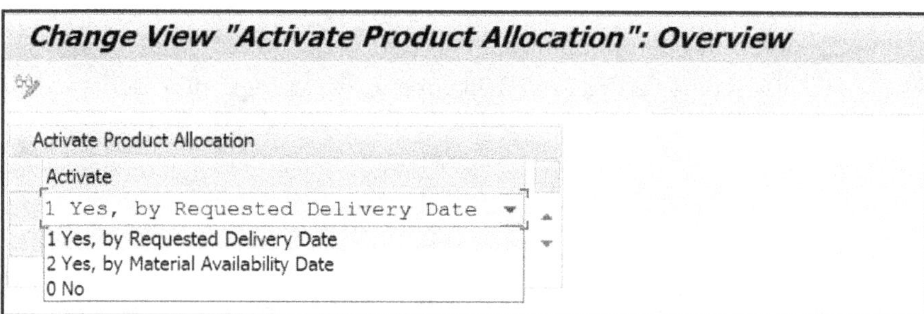

In the Activate Product Allocation customizing, you choose how the system finds the relevant allocation sequence by deciding which date it uses RDD or MAD.

The system identifies the allocation requirement based on the Requested Delivery Date (RDD) or the Material Availability Date (MAD), depending on the business scenario and configuration. In our example, this determines which allocation period (January, February, or March) is considered for the allocation.

Activate Value	Details
1	Activate at Requested Delivery Date: The system searches for allocation sequences using the requested delivery date from the order item. PAL consumption and available allocated quantities are evaluated along that date axis.
2	Activate at Material Availability Date: The system searches allocation sequences based on the material availability date (MAD), that is, the internal logistics date when the material must be available for picking/loading.
0	No Activation: Product allocation is not active in aATP; the ATP check ignores PAL, regardless of existing allocation objects and sequences.

2. Apart from this, some generic configurations must be completed at the Requirement Class and Schedule Line Category levels:

1. Activate Product Allocation at the Requirement Class level (enable the checkbox).
2. Activate Product Allocation at the Schedule Line Category level (enable the checkbox).

Activating these settings ensures that the system enforces the defined PAL rules during sales order creation, changes to sales orders, and BOP runs.

3. Additionally, the required key configurations for PAL must be completed using the Fiori apps to ensure proper functionality.

1. Fiori Apps
2. Configure Product Allocation Object
3. Configure Planning Data
4. Maintain Product Allocation Sequence

Note - *This key Fiori configuration is maintained without a Transport Request and must be set up individually in each system. Therefore, for the Production system, the configuration must be performed directly in Production.*

The **"Manage Characteristics Catalog"** Fiori app in SAP S/4HANA Product Allocation (PAL) is used to define and manage the characteristics that determine how allocations are controlled.

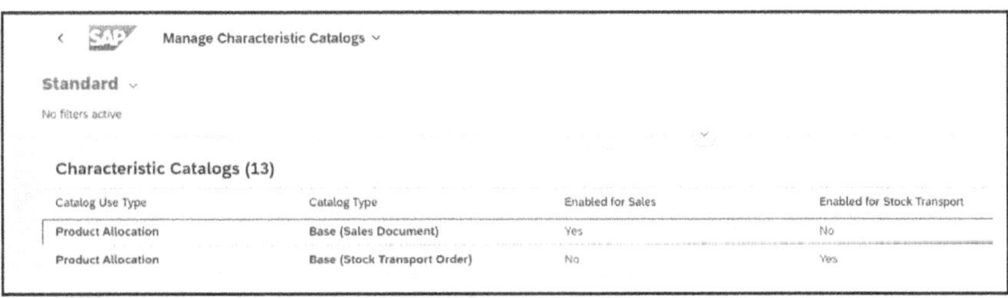

This app is where you define the building blocks (characteristics) that PAL uses to control and monitor allocations.

Purpose:	• It allows you to create, maintain, and assign characteristics that describe your allocation objects, e.g., customer, region, sales organization, product, etc. • These characteristics are used to define Characteristic Value Combinations (CVCs), which are the basis for PAL allocations.

Use in PAL:	•	When you define a PAL object, the system uses the CVCs to track allocation consumption.
	•	Characteristics help segment allocations (e.g., region-wise, customer-tier-wise, or channel-wise).
	•	They allow flexible and granular allocation rules.
Example:	•	Characteristics: Region, Customer Group, Material.
	•	CVC: USA | Premium Customers | Product A
	•	PAL allocation can be applied specifically to this combination.

Here's a **clear, simple explanation** of each **Product Allocation (PAL) Fiori app** and its purpose in **SAP S/4HANA aATP**:

PAL Fiori Applications-

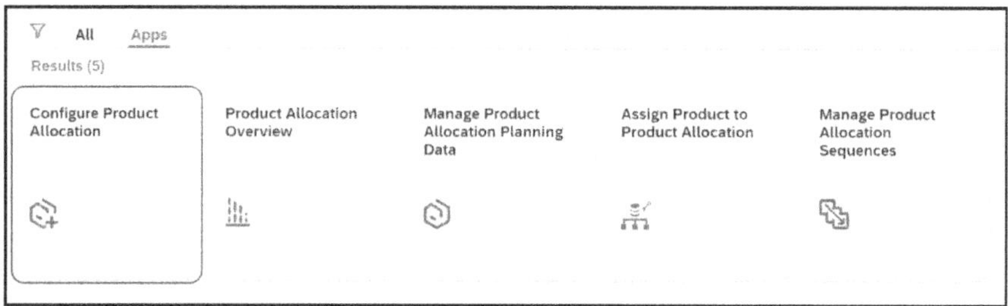

#	PAL Fiori Apps	Details
1	*Configure Product Allocation*	Used to define the overall PAL setup. You configure: • Allocation strategy • Time granularity (day/week/month) • Backward and forward consumption rules • Consumption logic This app defines how PAL works.
2	*Manage Product Allocation Planning Data*	Used to maintain allocation quantities for each period. You enter: • Monthly/weekly/daily allocation quantities • Allocation limits per CVC This app defines how much can be allocated.
3	*Manage Product Allocation Sequences*	• Used when multiple PAL strategies exist. • Defines the priority and sequence in which allocation strategies are checked. This app controls which allocation strategy is applied first.
4	*Assign Product to Product Allocation*	• Used to link materials/products to a PAL strategy. • Without this assignment, PAL will not apply to the product. This app defines which products are controlled by PAL.

5	*Product Allocation Overview*	• Provides a monitoring dashboard for PAL.
		• Helps planners:
		  • Track allocation consumption
		  • Identify under-consumption or over-consumption
		  • Analyze allocation usage by CVC (customer/region/channel)
		This app helps you monitor and analyze PAL usage.

Fiori key Configuration Steps.

To configure PAL allocation, the following five Fiori apps must be executed sequentially for key settings.

Step#1 – Configure Product Allocation (Define Product Allocation Object with CVC)

In this step, you define the **Allocation Object** and the characteristics (CVC fields) that the system uses to control and monitor allocation consumption.

Here, we have maintained the following fields:

- Allocation Object – "ZSSP_PAL1"
- Allocation Object Description
- Quantity Unit
- Period Type – Day/Week/Month
- Check Date and Time – Goods Issue Date, Material Availability Date, Requested Delivery Date
- Purpose – Use Sales Document/Stock Transport
- Collective – There are three types of collective allocation: Collective Allocation Managed by System, Collective Allocation Managed Manually, No Collective Allocation
- Period Time Zone
- Factory Calendar
- Characteristics – Here we have created 4 characteristics (Sales Organization, Distribution Channel, Division, Customer)

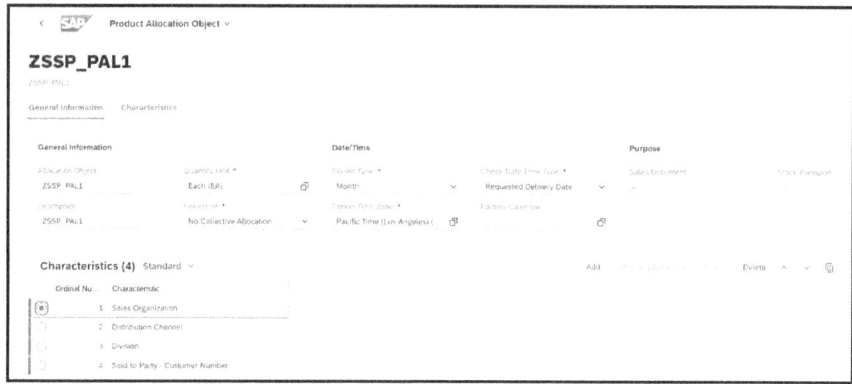

Details of Important Fields-

This section describes some of the key fields used in SAP S/4HANA aATP Product Allocation (PAL) and explains how they influence allocation behavior during sales order processing.

1. Field "Check Date Time Type" - In SAP S/4HANA aATP PAL, Check Date Time Type determines which date from the sales order schedule line is used to identify the relevant time bucket for allocation consumption.

Available options:

i. Material Availability Date

- Default. Uses the calculated material staging date (after backward/forward scheduling).
- Example:
    - Requested delivery date: May 15
    - Material availability date: May 8
    - System checks the PAL bucket containing May 8.

ii. Requested Delivery Date

- Uses the customer-requested delivery date directly and ignores scheduling lead times.
- Example:
    - Customer requested delivery date: May 15
    - System checks the PAL bucket containing May 15, regardless of the actual material availability date.

iii. Goods Issue Date

- Use the calculated goods issue date after transportation and lead-time calculation.

- Example:
  - Delivery date: May 15
  - Goods issue date: May 13
  - System checks the PAL bucket containing May 13.

Note: Wrong date type shifts consumption to different time buckets, affecting confirmation quantities and dates.

2. Field "Collective" - In aATP PAL, collective allocation is a way to reserve one shared bucket of quantity for "generic" characteristic values (like a group of customers) instead of maintaining separate allocations for each specific combination. The three options behave like this:

- **Collective Allocation Managed by System** - The system automatically creates collective allocations (for example, using a placeholder such as # for all low-priority customers not explicitly listed), and uses that shared bucket when an order's characteristics are not covered by a specific CVC.

- **Collective Allocation Managed Manually** - You manually define the collective CVCs (for example, Sales Org = US, Dist. Channel = 10, Sold-to = #), and assign quantities to them; orders matching these generic patterns consume from those manually maintained buckets.

- **No Collective Allocation** -
  PAL checks only the exact characteristic value combinations you planned; any customer or combination that is not explicitly in the allocation object gets no allocation from a collective bucket and will not be confirmed via PAL.

Step#2 – Manage Product Allocation Planning Data (Define Planning Data)

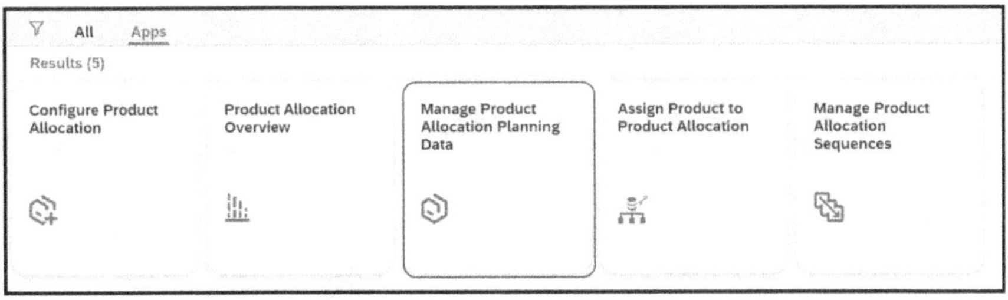

All Apps
Results (5)
Configure Product Allocation
Product Allocation Overview
Manage Product Allocation Planning Data
Assign Product to Product Allocation
Manage Product Allocation Sequences

Create planning data using the allocation object "ZSSP_PAL1".

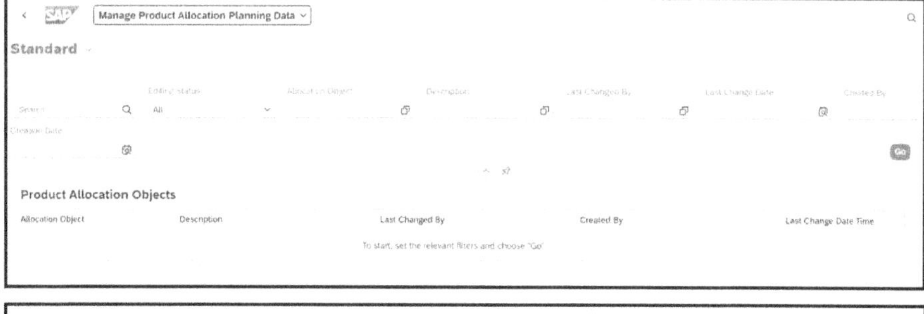

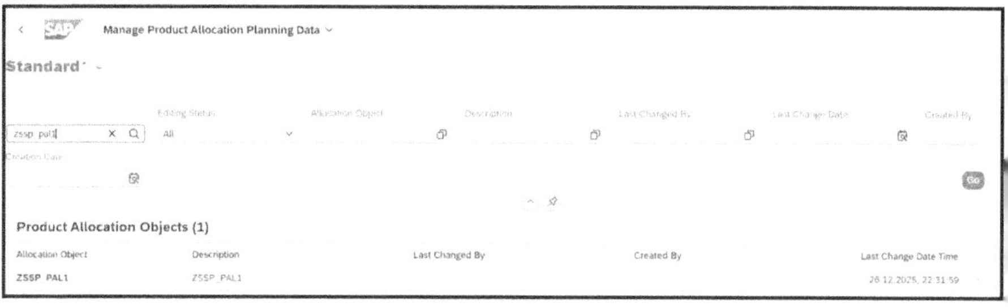

Here, you can see the Product Allocation planning data. Based on the allocation object fields, values have been updated for Sales Organization, Distribution Channel, Division, and Customer according to the respective months.

Key Fields-

1. Constraint Status in Product Allocation Objects (PAOs) controls whether/how allocations are enforced during availability checks. The three main types are:

Configuration: Set in Manage Product Allocation Objects app; Sequence level overrides PAO level for granular control.

i. Restricted Availability

- PAL active with quotas enforced – Orders consume from defined time bucket allocations (customer/region limits apply).

- Use when: Supply constrained, need fair distribution.

ii. Unrestricted Availability

- PAL active but unlimited quotas – No hard limits; acts like classic ATP but tracks consumption for reporting.
- Use when: Monitoring demand patterns without blocking orders.

iii. As in Sequence Constraint

- Inherits status from the specific Sequence – Each sequence can override PAO-level status independently.
- Use when: Different sequences need different behaviors (e.g., Sequence 1 restricted, Sequence 2 unrestricted).

2. Show Consumption - provides a graphical diagram showing how much quantity has already been confirmed and what quantity is still available.

Planned Quantity	Planned Quantity - The total quantity planned or allocated for a specific product, period, and characteristic combination (Sales Org, Distribution Channel, Division, Customer, etc.).
Available Quantity	The remaining quantity that is still available for confirmation. (Calculated as: Planned Quantity – Consumed Quantity)
Negative Available Quantity	It indicates that the allocated quantity has been exceeded.
Consumed Quantity	The quantity that has already been used or confirmed by sales documents or stock transport orders.

In this chart, January 2025 initially shows a planned quantity of 100 units and an available quantity of 100 units. Once a sales order for January is created with a quantity of 20 units and the allocation is consumed, the chart updates to display all three values: a planned quantity of 100 units, an available quantity of 80 units, and a consumed quantity of 20 units.

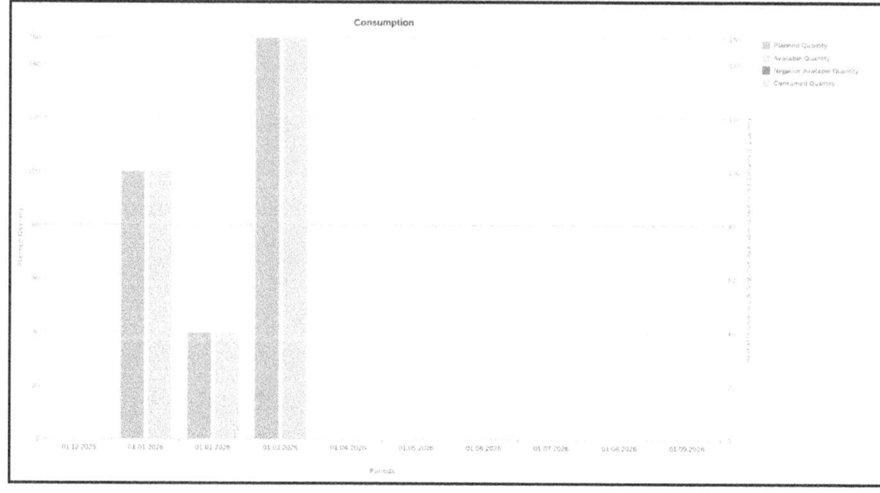

Key Points-

1. The download and upload functionality is available to perform mass updates of planning data.
2. Once allocation has been consumed, the planning data cannot be deleted. Planning data becomes non-deletable after consumption.

Step#3 - Manage Product Allocation Sequences

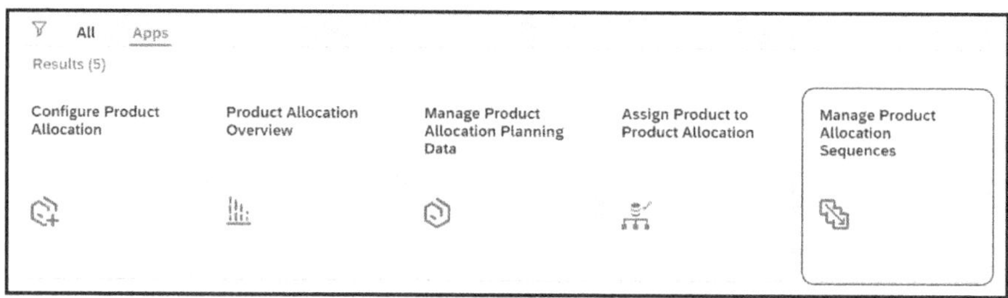

Allocation Sequence-

Creating a New Sequence-

To manage product allocation sequences, follow these steps:

In PAL, the allocation sequence has two levels: Sequence Group and Constraint.

- **Sequence Group:** The higher-level grouping that organizes multiple constraints under a logical structure.
- **Constraint:** Defines specific rules or limits for allocation and is linked to an allocation object.

Steps-

1. **Sequence (Seq)** – The top-level entity in product allocation.
    - A sequence can have multiple Sequence Groups (Seq Groups).

2. **Sequence Group (Seq Group)** – Groups within a sequence.
    - Each sequence group can contain multiple constraints.
    - There are two types of sequence groups in PAL:

1.Sales Sequence Groups	Define sales-related constraints, such as customer, region, or order priority.
2.Capacity Sequence Groups	Define warehouse or capacity-related constraints, such as storage location limits or production capacity.

Example Scenario: Warehouse Capacity Constraint (Daily 5,000 Units)

- For the respective customers, we want to deliver 5,000 units per day in the sales order schedule line based on allocation.
- However, the warehouse capacity allows dispatching only 4,500 units per day.
- In this case, the system will not confirm the full 5,000 units. Instead, it will confirm only 4,500 units, respecting the warehouse capacity constraint.
- The schedule line will therefore be confirmed for 4,500 units.
    - Order Quantity: 5,000 units
    - ATP Quantity: 5,000 units
    - PAL Allocation: 5,000 units
    - Sales Constraint: 5,000 units
    - Warehouse Capacity: 4,500 units
- The system first checks the sales constraints (5,000 units) and then the warehouse capacity (4,500 units).
- Since the warehouse can only dispatch 4,500 units, the sales order schedule line is confirmed for 4,500 units.
- The system always considers the most restrictive constraint when confirming schedule lines.
- Warehouse capacity constraints can be maintained **daily or weekly**, depending on business requirements.

3. **Constraints**– Defines the allocation for specific customers, sales orgs, or other characteristics.
    - Constraint is assigned to an Allocation Object.

Hierarchy Summary-

- **Sequence → Sequence Groups → Constraint → Allocation Object**
- One sequence can have multiple sequence groups.
- One sequence group can have multiple constraints.
- Each constraint is linked to a specific allocation object.

This combination of sequence, sequence groups, constraints, and allocation objects, together with sequence control, determines how PAL allocations are executed, including the order of month checks, backward/forward consumption, and schedule line creation.

Backward and Forward Consumption in Product Allocation-

In SAP S/4HANA aATP PAL, backward and forward consumption periods can be maintained at three levels: Sequence Constraint, Sequence Group, and Sequence. Most implementations set them specifically at Sequence and Sequence Group levels for granular control, while Sequence Constraint provides the base/fallback values.

Hierarchy (most specific overrides)-

- Sequence (highest priority) → Individual availability check rules.
- Sequence Group → Groups of sequences sharing strategy.
- Sequence Constraint (base level) → PAO-specific defaults.

Backward and forward consumption help the system determine availability across periods. Setting these values ensures that allocations are consumed efficiently and according to business priorities.

1. Backward Consumption

- You can specify a number of months for backward consumption.
- The system will check prior periods for available quantity up to the defined number of months.

Note: This refers to allocation quantity, not ATP quantity.

2. Forward Consumption

- You can specify a number of months for forward consumption.
- The system will check future periods for available quantity up to the defined number of months.

3. Consumption Direction

Here are simple numeric examples you can use to illustrate each consumption direction code.

01 – Backward then Forward

- First, consume from previous periods (backward) within the allowed backward window.
- If it is still not enough, consume forward into later periods.
- Example (monthly buckets, backward 1 month, forward 1 month).

02 – Backward earliest first, then Forward

- When going backward, start from the earliest period in the backward range, then move forward up to the requirement period.
- After that, if still not enough, go forward into later periods.

- Example (monthly buckets, backward 2 months, forward 1 month)

4. Past Period Allowed - is a setting in the allocation sequence that controls whether *unconsumed allocation in past time buckets* can still be used for confirmations on or after today.

- If Past Period Allowed = ON, the system could consume remaining quota from earlier periods (for example, unused January allocation for a February order), subject to your backward consumption settings.
- If Past Period Allowed = OFF, any unused allocation in past buckets is "frozen" and cannot be used for current/future confirmations; only current and future periods are considered.

5. This example demonstrates how allocations are processed using backward and forward consumption.

These are the **PAL-allocated quantities** for the respective months.

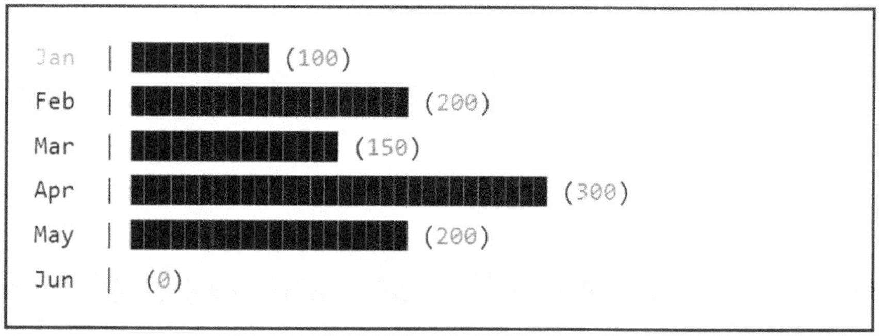

Case	Backward Consumption	Forward Consumption
#1	0	0
#2	1	0
#3	0	1

Case #1 – We received a Sales Order for March with a quantity of 200 units and ATP quantity of 300 units. Backward and forward consumption are not allowed (0 months). In this case, the system's PAL allocation will be only 150 units, because March's allocation is 150 units.

Note: Backward and forward consumption applies to PAL allocation, not ATP.

Case #2 – We received a Sales Order for March with a quantity of 200 units and ATP quantity of 300 units. Backward consumption is 1 month,

and forward consumption is 0 months. In this case, the system's PAL allocation will be the full 200 units (150 from March and 50 from February). Since backward consumption is enabled, the system will check the previous month (February) for available allocation and consumption from it. ATP must be available and greater than 200 units.
In this case, the system will generate only **1 schedule line** with a total quantity of 200 units.

Backward consumption direction-

1. **Backward then forward** → system will go 1 month back (February).
2. **Backward with earliest first then forward** → system will start from January, then February.

Case #3 – We received a Sales Order for March with a quantity of 200 units and ATP quantity of 300 units. Backward consumption is 0 months, and forward consumption is 1 month. In this case, the system's PAL allocation will be the full 200 units (150 from March and 50 from April). Since forward consumption is enabled, the system will check the next month (April) for available allocation and consumption from it. ATP must be available and greater than 200 units.
In this case, the system will generate **2 schedule lines**: 150 units in March and 50 units in April.

This behavior is based on PAL's design to allocate stock based on months, so delivery to the customer may span multiple months (e.g., March and April) with multiple schedule lines.

Note-
Min/Max quantities work together with **backward/forward consumption**. For example, even if backward consumption is allowed, the system will **never allocate more than MaxQty** in total across all time buckets.

- PAL is not global: It applies only to sales orders that match the CVC (Customer-Product-Plant) combination.
- Orders outside PAL scope: Sales orders that do not match the CVC bypass PAL and consume only ATP.
- Acts as a maximum cap: PAL defines the maximum quantity that can be allocated, even if more ATP is available.

- Confirmation logic: Final order confirmation always considers ATP first, then PAL allocation.
- Final Confirmed Quantity = min(Available ATP, Remaining PAL Allocation)

Step#4 - Assign Product to Product Allocation

To enable PAL allocation, each product must be assigned to an allocation object, which determines how stock is distributed according to business rules.

- Only products assigned to an allocation object are controlled by PAL.
- Products not assigned to PAL will bypass allocation and consume only standard ATP.

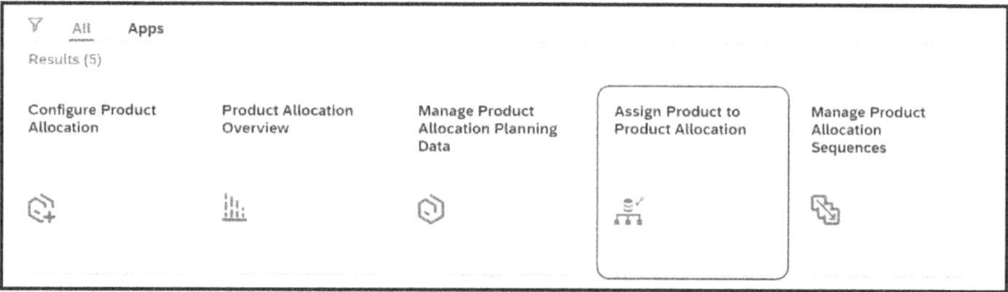

Select Allocation Sequence-

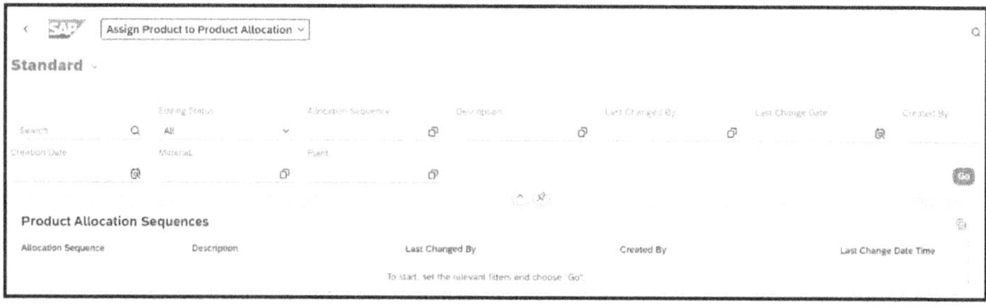

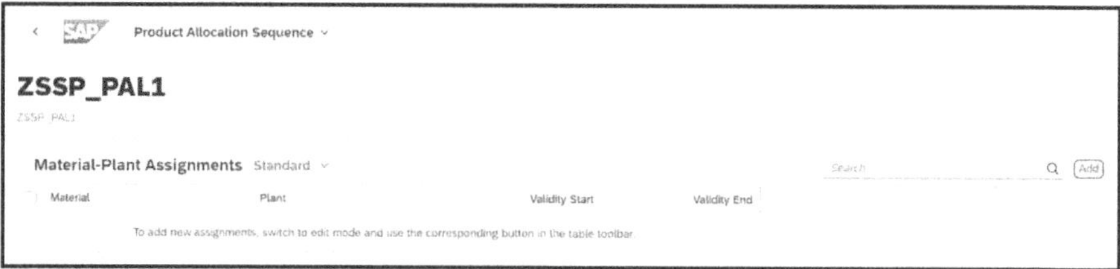

Select the Product Allocation Sequence "ZSSP_PAL1" and assign it to the relevant material/plant combination.

- This is a key setup step to enable PAL functionality for the material in the system.
- Without assigning the sequence to the material/plant, PAL allocation will not be executed during sales order processing.
- There is also a mass upload functionality available to download data and upload data.
- Material – "FG10002" and Plant – "1710".

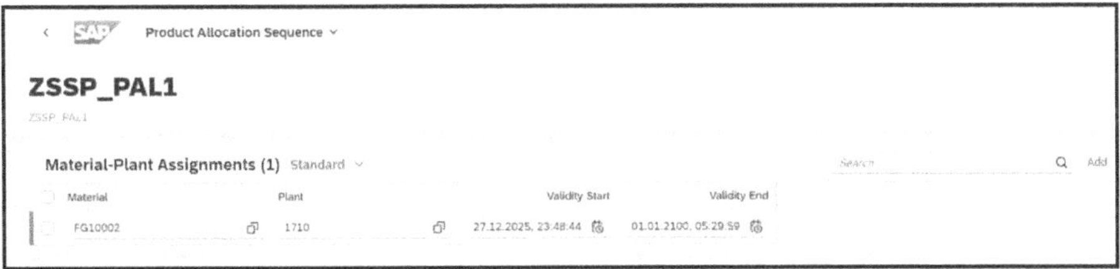

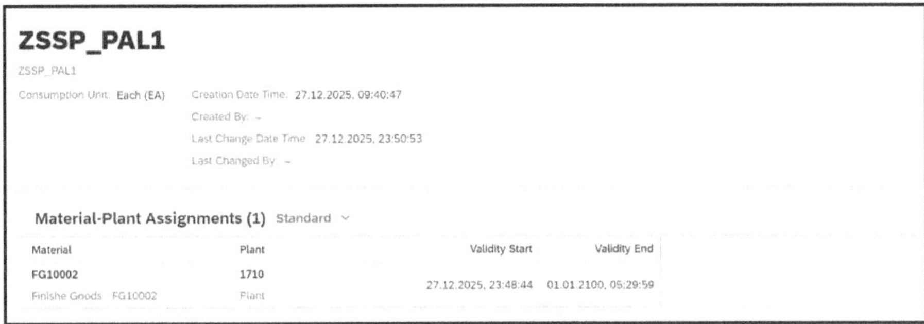

Step#5 – Product Allocation Overview

Product Allocation Overview is used to see, on one screen, how much allocation exists, how much is already consumed, and how much is still available for your Product Allocation Objects, periods, and characteristic combinations.

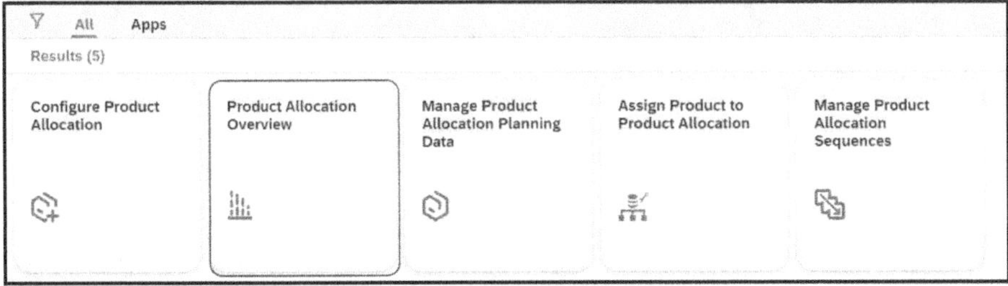

Main uses

- Monitor shortages and bottlenecks early by checking remaining allocation versus incoming demand per period (week/month) and per customer/region.
- Support operational decisions (for example, whether to confirm, reallocate, or increase supply) by giving planners a cockpit to review the current PAL situation before and after order entry.

This gives planners a traffic light dashboard to spot bottlenecks, unused capacity, and take action (increase quota, reallocate, expedite supply). The Product Allocation Overview app displays:

- Overloaded periods: Consumption > 100% (quota exceeded, orders rejected).
- Underloaded periods: Consumption < threshold (e.g., <50%, unused quota).
- High loaded periods: Consumption near 100% (e.g., 80-99%, warning zone).
- Product Allocation Order Items: Drill-down to specific sales orders consuming the allocation.

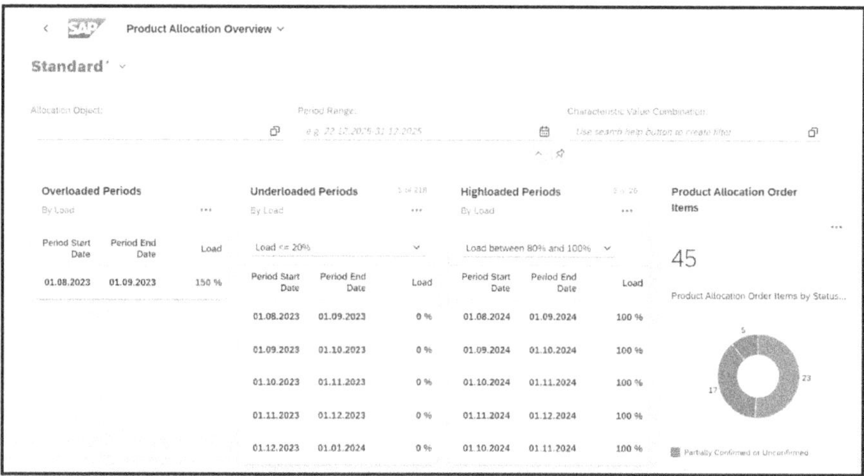

1. Example Without Product Allocation (PAL) Setup-

We have product FG10002 in location (**Plant 1710**) with a current available stock quantity of **500 units.**

Sales Order	Customer	Required Qty	Confirmed Qty	Stock Available
2000100	10001	500	500	500
				0
2000101	10002	500	0	0
2000102	10003	500	0	0

Based on the above example, when we create the **first sales order (2000100)** for a quantity of **500 units**, the system will confirm the full quantity and provide a delivery date because sufficient stock is available.

For the **second order (2000101)**, the system will not confirm any quantity, as there is no stock remaining after the first order is confirmed. The order will stay unconfirmed until new stock becomes available.

The same behavior will occur for the **third order (2000102)**—no confirmation will be provided.
In this situation, **Customer 10002 and Customer 10003 are high-priority customers**, but they will not receive any confirmed quantity because the available stock was entirely consumed by Customer 10001, whose order was created earlier.

To avoid such business risks, we can use Product Allocation (PAL) functionality. PAL helps control and limit the maximum confirmation quantities for customers, ensuring that strategic or priority customers always receive their fair share of limited supply.

Suppose we configure customer-level allocation as follows: (Example allocations would be listed here.)
In this example, allocations are shown at the **customer level**, but PAL can also be configured based on multiple other attributes such as **Order Type, Sales Organization, Division, Customer Group, Region, Channel**, and many more.

2. Example with Product Allocation (PAL) Setup-
The allocation has been set up based on the customers as shown below:

Customer	Allocation Qty – Maximum Customer Confirmation
10001	100 Units
10002	200 Units
10003	200 Units

Sales Order	Customer	Required Qty	Confirmed Qty	Stock Available
2000100	10001	500	100	500
				400
2000101	10002	500	200	200
2000102	10003	500	200	0

a. In this case, for the **first sales order (2000100)**, the system will confirm only **100 units**, even though the available stock is **500 units,** and the requested quantity is **500 units**.
This is because the **customer allocation limit is 100 units**. After this confirmation, the remaining stock is **400 units**.

b. For the **second order (2000101)**, the system will confirm only **200 units**, even though **400 units** is available, and the requested quantity is **500 units**.
This is due to the **customer allocation limit of 200 units**. After this confirmation, the remaining stock is **200 units**.

c. For the **third order (2000102)**, the system will again confirm only **200 units**, even though **200 units** are available, and the requested quantity is **500 units**, because the **customer allocation limit is 200 units.**
After this confirmation, the remaining stock becomes **0 unit**.

Key Points-

1. In SAP S/4HANA aATP, Product Allocation (PAL) can be configured for both sales orders and Stock Transport Orders (STOs). This capability was not available in ECC classic ATP and was introduced with S/4HANA aATP, with further enhancements delivered in later releases.
PAL applies to both sales and capacity allocations, including allocation at the transportation capacity level.

2. During a BOP run, ATP is executed again, and the system follows a defined sequence between PAC, PAL, and BOP. The system re-runs the Product Availability Check (PAC/ATP) for open and unconfirmed quantities because

- Stock may have changed
- New receipts may exist
- Allocations may have been adjusted

Sequence During Sales Order Creation	**When a sales order is created (VA01):** 1. PAC (ATP Check) • Checks available stock and planned receipts at plant level 2. PAL (Product Allocation) • Applies allocation limits based on CVC (customer, region, etc.) • May reduce the confirmed quantity even if ATP is available 3. Result- • Order is partially confirmed or unconfirmed • Remaining quantity becomes a backorder
Sequence During BOP Run	**When BOP is executed:** 1. PAC (ATP Check) runs again • Recalculates available supply based on current stock and receipts 2. PAL is applied again • Ensures confirmations respect allocation strategies and limits 3. BOP Logic (Prioritization & Rescheduling) • Orders are sorted by priority (e.g., customer priority, requested date) • Available supply is redistributed fairly 4. Final Confirmation Update • Previously unconfirmed quantities may now be confirmed (fully or partially)

During BOP, the system re-runs ATP (PAC) first, then reapplies PAL, and finally redistributes supply based on BOP priorities. Simple flow is - **VA01 → PAC (ATP) → PAL → BOP**

- ATP (PAC) runs during sales order creation and again during BOP.
- PAL is always applied after ATP, including during BOP runs.
- BOP does not bypass PAL rules; allocation strategies are always respected.
- Final confirmed quantity = whichever is lower: ATP or PAL

Example#1 • *ATP available: 50,000 Units* • *PAL allocation: 30,000 Units*	Confirmed quantity = 30,000 (PAL restricts it)
Example#2 • *ATP available: 8,000 Units* • *PAL allocation: 30,000 Units*	Confirmed quantity = 8,000 (ATP restricts it)

Note: BOP may change *which orders* receive confirmation based on priority, but it will never exceed PAL limits.

12 ABC - Alternative Based Confirmation

Alternative Based Confirmation (ABC) in SAP S/4HANA 2023 is a functionality within the Advanced Available-to-Promise (AATP) framework that allows sales order requirements to be fulfilled by calculating possible alternatives when the initially requested product, plant, or storage location cannot meet the demand for the desired delivery date or quantity. ABC automatically computes alternative options such as substituting a product, selecting a different plant or storage location, or adjusting quantities and delivery dates, using a heuristic approach to find the best possible substitutes. This functionality enhances confirmation flexibility by enabling the system to suggest the most advantageous alternatives during sales order processing, thus improving order fulfillment when the original availability is insufficient.

In SAP S/4HANA aATP, Alternative-Based Confirmation (ABC) enables the system to automatically propose and confirm alternative plants, products, or storage locations (when ATP checks are performed at the storage location level) to ensure customer demand is fulfilled.

For a sales order requirement (product, quantity, plant, date), if the requested plant or storage location has insufficient availability, ABC can replace it with an alternative plant, an alternative storage location (if ATP is performed at the storage-location level), or even an alternative product, based on the configured substitution strategy and master data.

Alternative-Based Confirmation (ABC) enables substitutions in SAP aATP. There are 3 types: Plant, Product, and Storage Location substitution.

ABC substitution supports both 1:1 and 1:n scenarios across plant, product, and location substitutions.

1:n Product Substitution with Percentage Distribution

ABC also supports 1:n substitution with percentage-based distribution. For example, if a primary product FG10101 can be substituted by multiple alternative products—FG10105, FG10106, and FG10107—the confirmed quantity can be automatically distributed based on predefined percentages:

- *FG10105 – 50%*
- *FG10106 – 30%*
- *FG10107 – 20%*

This allows the system to intelligently split confirmations across multiple substitute products while still fulfilling the customer's overall demand.

An important point to note is that aATP supports two types of substitution: actual substitution and availability substitution, each serving different business scenarios.

Types of Substitution in aATP-

Actual Substitution	In actual substitution, the original material is replaced with an alternative material. The sales order and delivery are updated to reflect the substitute product, and the customer receives the alternative item instead of the originally requested one.
Availability Substitution	In availability substitution, the original material remains on the sales order, but the availability check considers alternative materials or locations only for the purpose of confirming quantities and dates. The substitution is used to determine availability, while the original product reference can still be maintained from a business or reporting perspective.

The substitution strategy defines the sequence in which the system should try the alternatives such as alternative storage locations first, then alternative plants, and finally alternative products. aATP then optimizes the combination of these alternatives to fulfill as much of the customer demand as possible.

Location substitution and product substitution were not available in ECC or in Basic ATP. These capabilities are newly introduced and supported in S/4HANA aATP. aATP provides advanced substitution logic through **Alternative-Based Confirmation (ABC)**, enabling:

- **Plant Substitution** (alternative plants)
- **Location Substitution** (alternative storage locations)
- **Product Substitution** (alternatives products)

Alternative-Based Confirmation (ABC) activates during sales order creation/modification (VA01/VA02) and Batch Backorder Processing (BOP) runs.

ABC in Sales Order Creation/Modification

- ABC activates automatically when the availability check runs in VA01/VA02 (or Fiori sales apps) if the requested material-plant-date combination lacks sufficient supply.

- It evaluates substitution rules and alternative controls in real-time, proposing confirmations (e.g., alternative plants or materials) via the Review Availability Check Results (RACR) screen.
- Confirmed alternatives update schedule lines immediately, integrating with PAC/PAL checks.

ABC in Backorder Processing (BOP)

- BOP (Fiori app "Manage Rescheduling") re-processes open sales orders in bulk (e.g., nightly jobs) to re-allocate based on current supply.
- ABC runs as part of BOP if configured in the rescheduling horizon/strategy, searching alternatives for unconfirmed or partially confirmed lines.
- Results can split lines, shift dates, or reassign plants/materials without manual intervention, respecting priorities like win/lose rules.

This dual application ensures consistent promising logic across interactive and mass processing, reducing backlogs while honoring ABC rules.

ABC Activation Setting-

The ABC functionality only works if it has been activated at the **availability checking group** level. In the screenshot below, you can see that **Availability Checking Group "02"** has ABC activated and is therefore active for alternative-based confirmations.

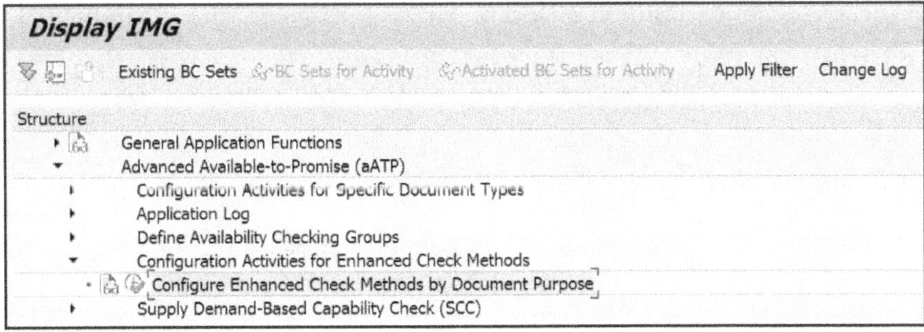

1. Plant Substitution in ABC (Alternative-Based Confirmation)

Plant substitution in SAP S/4HANA Advanced Available-to-Promise (aATP) is a key feature that allows the system to propose alternative delivery plants when the originally requested plant has insufficient stock to fulfill a sales order. This enables improved order fulfillment rates and more reliable delivery promises.

This means that instead of waiting for supply at the original/base plant, the system proactively proposes another plant to fulfill the order. This process is called **Plant substitution**. By using plant substitution:

- Order fulfillment rate increases.
- Customer delivery promises to improve.
- Delays due to stock shortages are reduced.

In S/4HANA aATP, ABC (Alternative-Based Confirmation) plant substitution works the same way for both plants and storage locations.

- The system checks availability at the original requested plant/storage location first.
- If stock is not available, it searches for predefined alternative plants or storage locations.
- The order is confirmed from the first alternative that has available stock.
- The logic and rules for substitution are consistent for both plant-level and storage location-level substitution.

This ensures flexibility in sourcing while improving order fulfillment and delivery reliability.

Thus, plant substitution is an integral part of aATP that enhances customer order promising by dynamically using alternative supply points in the network when the primary plant is unable to fulfill the order.

Example - We have the following situation regarding the product, the plants, and the current stock levels.

There are multiple ways to handle this scenario. If we create a sales order with plant P100 and there is no stock available, the system will not confirm the schedule line. We could run MRP to generate stock from the manufacturing plant, or we could transfer stock from plant P101 by creating an STO to move inventory from P101 to P100 before delivery.

However, to make this process automatic, we can set up plant substitution with rules for a group of plants. In this case, if a sales order is created with plant P100 and no stock is available, the system can automatically propose confirmation from the next available plant P101.

Product	Location (Plant)	Stock
FG10101	P100	0 Units
FG10101	P101	100 Units
FG10101	P102	50 Units
FG10101	P103	20 Units

In this case, when a sales order is created for product 'FG10101' with a requirement of 100 units at plant P100 and there is no available stock at P100, the system will propose an alternative location, P101, to fulfill the demand, provided the substitution strategy and master data are set up accordingly.

For ABC to propose an alternative plant, that plant must be included in the Location Substitution Rule within aATP. This means that if you want ABC to check all four plants—P100, P101, P102, and P103 - then all of these locations must be listed in the Substitution Strategy / Substitution Rule.

If a plant is not included, ABC will not propose it - even if stock is available and even if it is technically feasible for supply.

aATP does not automatically search all plants. It only searches for the plants defined in the Location Substitution Rule that you assign to the Check Instruction.

ABC supports both full and partial confirmations - it does not require the alternative plant to satisfy the entire demand.

a. Full Confirmation- If the alternative location can supply 100% of the requested quantity, ABC will confirm the entire order quantity from that location.

b. Partial Confirmation- If no single plant/location can fulfill the entire quantity, ABC can still:

- Confirm partially from the best alternative location
- Perform multi-source confirmations (split confirmation from multiple plants), if configured

This behavior depends on what in S/4 HANA aATP -

- The substitution strategy
- Rule-based ATP (RB-ATP) configuration
- Whether partial delivery is allowed in the sales order
- Whether multi-level substitution is enabled

The system follows the sequence below to execute and obtain an ABC confirmation:

- You created a sales order for product FG10101 with a requirement of 100 units at delivering plant P100. aATP first checks availability in P100 and finds no stock or insufficient supply.
- Because an ABC location substitution strategy is defined for product FG10101 and plant P100, the system then checks the alternative plants (such as P101) in the sequence maintained in the substitution group/control.
- If P101 has sufficient available quantity for the requested dates and meets any additional substitution rules (such as priority, distance, or building logic), aATP confirms the requirement from P101. The system then proposes P101 in the sales order schedule line or as a substituted item—depending on the substitution item type (inline vs. sub-item) configured in aATP.
- This substitution can happen inline (replacing the plant on the sales order line) or by generating sub-items per plant to fulfill partial quantities.
- The substitution logic follows rules defined in the substitution strategy configuration, such as preferring the plant that can deliver the full quantity earliest or the one that minimizes overall delivery time.
- This feature helps reduce delivery delays and increases customer satisfaction by leveraging inventory distributed across multiple plants.

Example Key configuration Setup-

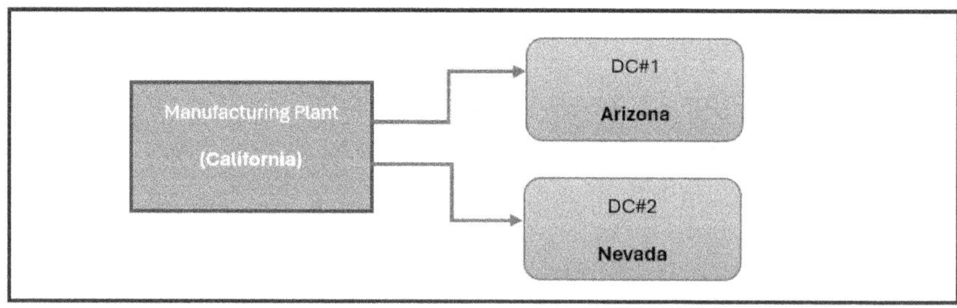

In this example, California is the Manufacturing Plant, and we have two Distribution Centres (DCs): one in Arizona and one in Nevada. The current stock situation is as follows:

- DC#1 (Arizona) has no stock available.
- DC#2 (Nevada) has 5000 units of stock available.

A sales order with a quantity of 1000 units for Product FG10109 is created against DC#1 (Arizona) to fulfil the customer's order.

In this case, we have multiple options to fulfil the customer's requirements:

Option 1: Stock Transfer Requirement (STR)

Since there is no stock available in DC#1, no schedule line will be confirmed. Instead, an STR (Stock Transfer Requirement) will be generated based on the MRP run to produce stock in the California Manufacturing Plant and transfer it to Arizona DC for delivery to the customer.

Option 2: Manual Plant Change

We could manually change the plant from DC#1 to DC#2 (Nevada), where stock is available, in order to fulfil the customer's demand.

Option 3: Automated Plant Replacement

However, we are looking for this process to happen automatically. In this case, DC#1 is the original plant designated to provide the customer delivery. During the availability check, if the system finds that there is no stock in DC#1, it should automatically check the next DC (DC#2) and, if stock is available, replace the plant in the sales order line item (from DC#1 to DC#2).

This plant replacement should happen automatically during sales order creation, order changes, and the BOP (Backorder Processing) run, where the system will replace the plant automatically based on availability from the respective plant, and the schedule line will be confirmed for delivery to the customer.

Expected Outcome:

- The system will automatically select the appropriate DC (Nevada) when stock is not available at the original DC (Arizona), ensuring the customer demand is fulfilled without manual intervention.
- The sales order line item will reflect DC#2 (Nevada) for fulfilment, and the corresponding schedule line will be confirmed for delivery to the customer.

This automatic plant substitution should be achieved using Alternative-Based Confirmation (ABC) configuration in aATP, where:

- *The system is configured to substitute plants automatically when the original plant (DC#1) does not have stock.*
- *We can also control how many plants the system checks during Sales Order creation/change. For example, if there are 10 plants in total, but you want the system to check only 4 plants for availability.*
- *This process should run seamlessly during sales order creation, updates, and BOP to ensure real-time availability checks.*

Setup-

ABC (Alternative-Based Confirmation) includes the following three essential apps, which are crucial for maintaining and applying the 3 objects: Plant, Location, and Product Substitution:

1. Configure Alternative Control
2. Configure Substitution Strategy
3. Configure Alternative Determination

Here's a breakdown of the purpose and use of each of the three apps in **Alternative-Based Confirmation (ABC)**:

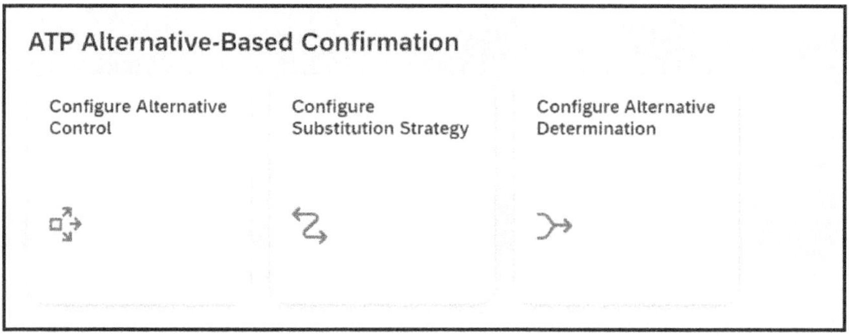

ATP Alternative-Based Confirmation

Configure Alternative Control

Configure Substitution Strategy

Configure Alternative Determination

Substitution Strategy defines the logic, Alternative Determination defines the rules, and Alternative Control activates when and where ABC runs.

- **Substitution Strategy** defines the overall ABC framework and contains the building rules that control how substitution works.
- **Alternative Determination** defines the custom-building rules used to identify valid alternatives.
- **Alternative Control** activates ABC and controls when it is applied, through substitution master data and characteristic-based activation.

In short: Strategy = logic, Determination = rules, Control = activation.

1	*Configure Alternative Control*	• Purpose: This app is used to define and configure the rules and sequence of how the system should handle alternatives during the availability check.
		• Use:
		    • You define alternative control strategies such as plant substitution, material substitution, or delivery date shifts.
		    • The app helps set the priority and sequence of checking different alternatives (e.g., first check for available stock in a primary plant, then in alternative plants).
		    • It defines whether the system should look at other substitutes or simply shift the delivery date when the requested product is unavailable.
		• Example: If stock is not available in the requested plant, it will check the next plant in the sequence, ensuring the order is fulfilled without manual intervention.

2	*Configure Substitution Strategy*	• Purpose: This app allows you to define the substitution strategies that determine which substitutes can be used for a given product and under what conditions.
		• Use:
		    • It specifies whether a product can be substituted with exact replacements or whether there are multiple substitute options (1:1, 1:n).
		    • You can configure different substitution rules based on specific conditions like sales organization, material group, or customer group.
		    • It also controls the substitution hierarchy (e.g., whether material A can be substituted with B, then C if B isn't available).
		• Example: If Product A is unavailable, it defines which other products (e.g., Product B or C) can replace it based on availability and other criteria.

3	*Configure Alternative Determination*	• Purpose: This app is used to define how the system should determine alternatives for products during the availability check.
		• Use:
		    • It sets the conditions and criteria under which the system should search for alternatives.
		    • You can define specific access sequences and the order in which the system should look for alternatives (e.g.,

		substitute materials, alternative plants, alternative delivery dates). • This app integrates with both substitution strategy and alternative control to ensure the correct alternative is determined automatically based on stock availability. • Example: The system will first check if a substitute is available in the same plant, then check other plants or locations, or even shift the delivery date.

Let's dive deeper into the "Configure Substitution Strategy" app and its role in Alternative-Based Confirmation (ABC) in SAP S/4HANA.
App: Configure Substitution Strategy

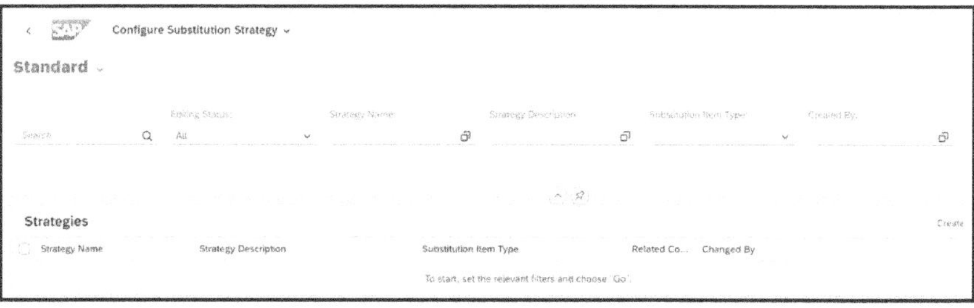

When creating a new strategy, you will see three areas: General Information, Substitution Methods, and Building Rule.
This strategy will be applicable to all three objects: **Plant**, **Location**, and **Product Substitution**. You will find this option under **"Substitution Methods"**, where you can add multiple substitution methods by using the **"Add"** button.
When creating a new substitution strategy, you do not need to explicitly mention whether it's for Plant, Location, or Product in the strategy itself. Instead, the strategy can be created as a single, overarching strategy that is applicable to all three objects—Plant, Location, and Product—depending on how you configure the substitution methods within the strategy.

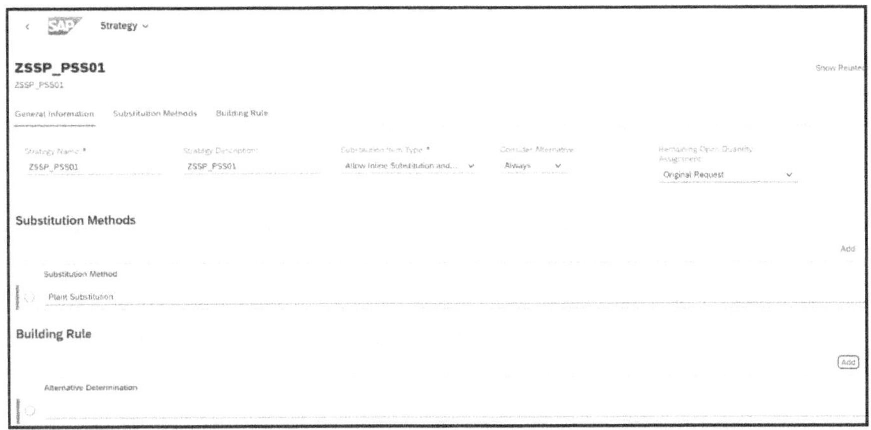

In **General Information**, we will define the Strategy Name and Description, along with other settings such as Substitution Item Type, Consider Alternative, and Remaining Open Quantity Assignment.

1. Substitution Item Type

Here, we have three options: in Substitution Item Type

- Allow Inline Substitution and Subitem Generation
- Force Inline Substitution
- Force Subitem Generation

a. Allow Inline Substitution and Subitem Generation

This option allows the system to perform both inline substitution and subitem generation when a product substitution occurs. It enables the system to substitute products directly in the sales order line and create additional subitems when necessary.

- The system decides dynamically whether to use inline substitution or subitems.
- If a single plant can fully cover the requirement, it does inline substitution: original item stays, but the delivering plant is replaced (no subitems).

- If multiple plants or product/storage-location substitution are needed, it keeps the original item and creates subitems for each confirming plant / storage location / product.
- Use this if you want the system to pick the "simplest" solution automatically and avoid subitems when no substitution is actually needed.

Assumptions (same in all three):

- Material: FG10101
- Plant: P100 (requested plant)
- Customer orders 100 pcs
- Only 40 pcs are available in P100
- An alternative plant P200 can supply the full 100 pcs
- ABC is active and triggers plant substitution

Example A – one alternative plant can cover everything	Example B – quantity needs multiple plants
1. You create the order: Item 10: FG10101, 1000 units, Plant P100 (requested) 2. ATP with ABC: • Plant P100: 400 units available • Plant P200: 1000 units available (one plant can cover the full 1000 units) 3. System behaviour: • Uses inline substitution, because one alternative plant can cover the full quantity. • No subitems created. 4. Result in the order: • Item 10: FG10101, 1000 units, Plant P200 (original item, plant changed) • No item 11, no subitems.	1. Same order: Item 10: FG10101, 1000 units, Plant P100 2. Availability: • Plant P100: 400 units • Plant P200: 400 units • Plant P300: 200 units 3. System behaviour: • Cannot cover 1000 units from a single alternative plant. • Uses subitem generation. 4. Result in the order: • Item 10: FG10101, 1000 units (requirement item) • Item 11: FG10101, 400 units from P100 • Item 12: FG10101, 400 units from P200 • Item 13: FG10101, 200 units from P300

b. Force Inline Substitution

This option forces the system to perform inline substitution and directly replace the out-of-stock product with the substitute product in the same line item without generating any new subitems.

- The system will automatically replace the unavailable product with the substitute product, but there will be no additional line items created for tracking purposes.
- The system must use *inline* substitution (no subitems).
- The original requirement item remains as one line; its delivering plant is simply replaced by the alternative plant.
- Only plant substitution is possible with this option, and only if a *single* alternative plant can fully confirm the requested quantity.

Here the system is *not allowed* to create subitems; it must use inline plant substitution, and only if a single plant can cover the full requested quantity.

Example – one plant can cover everything	Example – quantity split required (what happens)
1. You create the order: Item 10: FG10101, 1000 units, plant P100 2. Availability: • Plant P100: 0 units • Plant P200: 1000 units 3. System behaviour: • Forces inline substitution. • Entire requirement is moved to plant P200; no subitems allowed. 4. Result: • Item 10: FG10101, 1000 units, Plant P200 • No subitems.	1. You create the order: Item 10: FG10101, 1000 units, plant P100 2. Availability: • Plant P100: 400 units • Plant P200: 600 units (no single plant with full 1000 units) 3. Because "Force Inline Substitution" does not allow subitems: System can either: • Confirm only up to what a single plant can supply in line (e.g., 600 units from plant P200, 400 remain open), or • Depending on release/logic, not use substitution for the split case at all. In any case, it will not create subitems. Result (typical): Item 10: FG10101, 600 units confirmed in plant P200 (inline), 400 units remain open/unconfirmed on item 10.

c. Force Subitem Generation

This option forces the system to generate a subitem for the substituted product, but the original product remains in the sales order line. The substitute product will be treated as a separate line item (subitem).

- This is useful when you need to track both the original product and the substitute product separately within the same order for additional processing, such as pricing, shipping, or inventory management.
- The system must use *subitem* generation whenever a substitution happens.
- The original item remains as a "header" requirement; one or more subitems are created to represent the actual confirmed quantities from substitute plants/products/storage locations.
- This is useful when you want full transparency of which plant or substitute product fulfilled which quantity, or when multiple substitutions might be combined in one order.

Here the system is *not allowed* to do inline substitution; any substitution creates subitems, even if one plant could cover everything.

1. You create the order:
    - Item 10: FG10101, 1000 units, Plant P100
2. Availability:
    - Plant P100: 400 units
    - Plant P200: 1000 units
3. System behavior:
    - Must use subitem generation.
    - Original item 10 stays as requirement/reference.
    - Confirmation is written to subitems.
4. Result in the order:
    - Item 10: FG10101, 1000 units (requirement; often no own confirmation)
    - Item 11: FG10101, 400 units from plant P100
    - Item 12: FG10101, 600 units from plant P200

Even though plant P200 could cover the full 1000 units, the system still uses subitems because the configuration says "Force Subitem Generation".

When to Use Each Option:

- **Allow Inline Substitution and Subitem Generation**: Use this when you want to replace a product and also generate a new subitem to track any separate processing or attributes of the substitution.

- **Force Inline Substitution**: Use this when you only need to replace the product with the substitute directly in the same line item, without any need for additional tracking via subitems.

- **Force Subitem Generation**: Use this when you need to **keep the original product** visible in the sales order while creating a separate subitem for the substitute. This is helpful for cases where you want to track and process the substitute as a distinct item.

In Short-

- **Inline Substitution** replaces the original product with the substitute within the same line item.
- **Subitem Generation** creates a new line item for the substitute, allowing it to be tracked separately.
- The combination of these options provides flexibility in how the system handles product substitutions, allowing businesses to control whether to treat substitutions as a seamless replacement or as a separate entity for further processing.

2. Consider Alternative

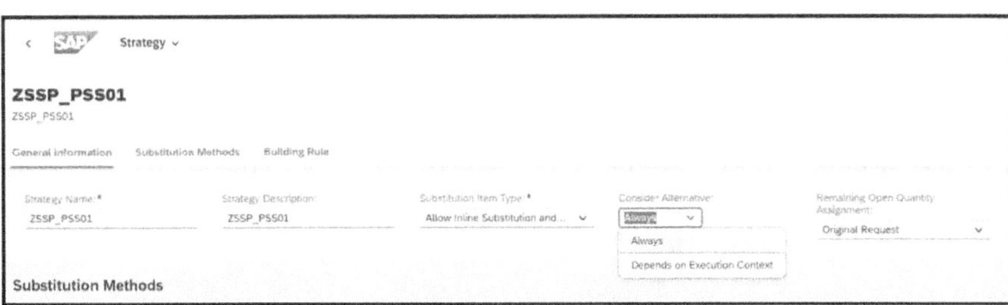

Here, we have two options: in Consider Alternative

- Always
- Depends on Execution Context

These two options define **when ABC is allowed to look for alternatives during the ATP check** when using this strategy.

- If "Always" is selected, the strategy is applied during both sales order creation and change.
- If "Depends on Execution Context" is selected, the strategy can be applied selectively, for example only during creation or only during change, depending on the execution context.

In aATP, the **"Change" execution context primarily refers to Backorder Processing (BOP)**. Selecting **"Depends on Execution Context"** allows control over whether ABC is executed during order creation, during BOP, or both, preventing unintended substitutions during re-confirmation runs.

- When you select "Depends on Execution Context", the system displays a popup window where you can choose "Consider When". There are two options:
  - New – ABC is executed during sales order creation.
  - Posted – ABC is executed during sales order changes, including Backorder Processing (BOP).

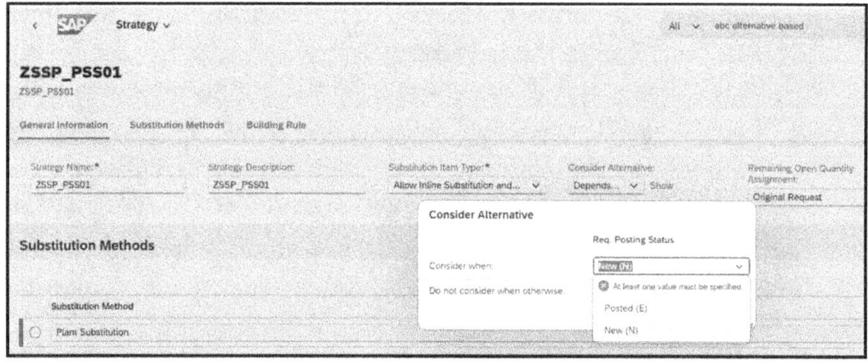

Example-

Assumptions for all examples

- Material: FG10101
- Plant: P100
- Customer orders 1000 units
- Only 400 units are available for FG10101 in P100.
- ABC is configured and there is a valid substitution (for example, FG10102 or another plant).

Option#1 – Always

Details - The system considers alternatives in all supported execution contexts where this substitution strategy is used (for example, during sales order creation and change, and in background runs such as BOP, provided other prerequisites are met).

Example –

Case 1: Consider Alternative = Always

You create a new sales order (VA01):

1. You enter: Item 10: FG10101, 1000 units, Plant P100.
2. You trigger ATP:
   - System sees: 400 units available for FG10101 in P100.
   - ABC is active and "Always" allows alternatives in this context.
   - System looks for alternatives and finds a valid substitution (for example, FG10102 in P200, or same product in another plant).
3. System confirmation might look like:
   - Item 10: confirmed 400 units with original settings (depending on your substitution item type).
   - A subitem is created for the substitute (if you use subitem generation), for example:
   - Item 11: FG10102, 600 units from substitute plant.

Result:

On initial order creation, the system already considers alternatives and proposes them because "Always" is set.

Option#2 - Depends on Execution Context

Details - The system only considers alternatives in specific contexts, such as only when the item is already posted (change mode like VA02 or BOP) and not and not during initial order creation (VA01), depending on how the execution context is defined in your release.

Example-

Same business situation, but "Depends on Execution Context" is configured so that alternatives are not considered during initial creation, only in a later context (for example, BOP run or change mode).

Step 1 – Initial creation (VA01)

1. You enter: Item 10: FG10101, 1000 units, Plant P100.
2. You trigger ATP in VA01:
   - System confirms only what is available for the original item (400 units).

- Because of the execution context rule, it does not search alternatives in this step.

3. After saving, the order looks like:
    - Item 10: FG10101, ordered 1000 units, confirmed 400 units, 60 units remain open/unconfirmed.

No subitems, no alternative yet.

Step 2 – Later processing (e.g., VA02 or BOP)

Later, you:

- Run a backorder processing (BOP) job, or
- Open the order in VA02 where the context is allowed by your execution rules.

Now:

- ABC is allowed to consider alternatives in this context.
- System checks the 600 units open on item 10 and looks for alternatives.
- It finds a suitable alternative (FG10102 or other plant) and confirms, for example, 600 units on a substitute subitem.

Result after this step:

- Item 10: FG10101, 1000 units, confirmed 400 units (original).
- Item 11: FG10102 (or other alternative), 600 units confirmed as subitem.

So, with Depends on Execution Context:

- At *creation time*, the system may not look for alternatives.

Later, in the "allowed" context (change, BOP, specific process), ABC steps in and assigns the remaining open quantity using alternatives.

3. Remaining Open Quantity Assignment

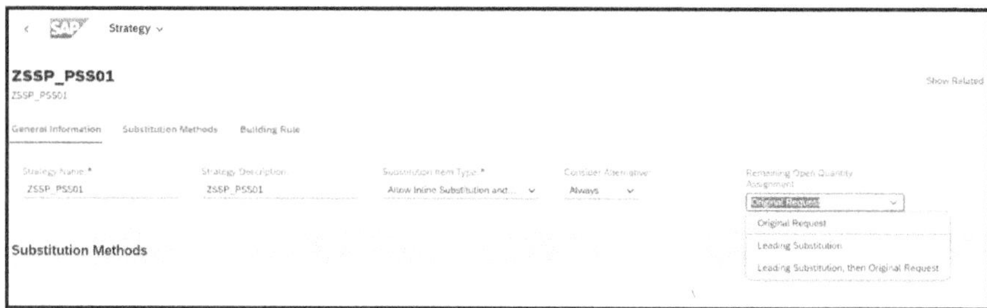

Here, we have two options: in Remaining Open Quantity Assignment

- Original Request
- Leading Substitution
- Leading Substitution, then Original Request

These three options controls *where* any remaining unconfirmed quantity stays when subitems are created in a substitution scenario.

Meanings of the options-

Option#1- Original Request

Details - Any remaining open (unconfirmed) quantity is kept on the original main item. The substitute subitems contain only the confirmed quantities; the "shortage" stays with the original request.

Example-
Behaviour: Remaining open quantity stays on the original item.
You create the sales order:

- Item 10: FG10101, 1000 units, Plant P01 (requested)

During ATP with ABC:

- 600 units are confirmed for FG10101 in plant P01.
- 400 units are confirmed for substitute FG10102 in P02 as a subitem.
- The open (unconfirmed) quantity remains on item 10.

Resulting items:

- Item 10: FG10101, 1000 units
- 600 units confirmed
- 400 units still open (or later to be backordered/cancelled etc.)

- Item 11 (subitem): FG10102, 400 units, plant P02
- 400 units confirmed

Use case: You want the original order line to show the "shortfall" and handle it centrally there.

Option#2 - Leading Substitution

Details - The remaining open quantity is assigned to the subitem that is flagged as the leading substitute. The original item and other subitems only hold their confirmed portions, while the lead-substitute subitem carries the unconfirmed remainder.

Example-
Behaviour: Remaining open quantity is assigned to the leading substitution subitem.
You define one substitute as the "leading" substitution (for example, FG10102 in plant P02).
You create the sales order:

- Item 10: FG10101, 1000 units, plant P01

During ATP with ABC:

- 600 units confirmed on FG10101 in plant P01.
- 400 units confirmed on FG10102 in plant P02 (leading substitution).

- Any remaining open quantity (if there were still some unconfirmed) is moved to the leading subitem, not kept on item 10.

Resulting items (in this example, all 1000 units are covered, so no open qty remains):

- Item 10: FG10101, 1000 units
    - 600 units confirmed
    - 400 units "moved" to subitems, no remaining open qty here.
- Item 11 (leading subitem): FG10102, 400 units, plant P02
    - 400 units confirmed, and if there were any leftover unconfirmed, they would sit on this item.

Use case: You want any leftover "problem" quantity to sit with the leading substitute instead of the original item.

Option#3- Leading Substitution, then Original Request

Details - The system first tries to assign the remaining open quantity to the leading substitute. If no unique leading substitute can be determined (for example, none or more than one is marked as leading), the remaining quantity is assigned back to the original request item instead.

Example-

Behaviour: Try to assign remaining open quantity to the leading substitution; if that is not possible, keep it on the original item.

You again have a leading substitution (FG10102 in plant P02).

You create the sales order:

- Item 10: FG10101, 1000 units, plant P01

Scenario A: normal case (leading substitute is clear)

Same as "Leading Substitution":

- 600 units on FG10101 in plant P01.
- 400 units on FG10102 in plant P02.
- Any leftover would be assigned to item 11 (leading subitem).

Scenario B: leading substitute cannot be determined (e.g., two alternatives marked equally, or configuration issue)

- 600 units confirmed on FG10101 in plant P01.
- 400 units confirmed on FG10102 in plant P02 (but not clearly leading), or possibly multiple subitems.
- Remaining open quantity is then left on the original item 10, because the system cannot decide a unique leading subitem.

Use case: You prefer to move open quantity to the leading substitute but want a safe fallback so that, if there is no clear leading subitem, the open quantity remains on the original line instead of being spread ambiguously.

4. Substitution Method

Here, we have three options: in Substitution Method

- Plant Substitution
- Product Substitution
- Storage Location Substitution

These three settings define *what* ABC is allowed to substitute during the ATP check for a given strategy.

Substitution Method	Details
Plant Substitution	Allows ABC to replace the delivering plant with an alternative plant that can better fulfil the requirement.
Product Substitution	Allows ABC to replace the requested product with one or more alternative products from the product-substitution master data.
Storage Location Substitution	Allows ABC to replace the requested storage location with another storage location in the same plant according to storage-location substitution master data.

a. Plant Substitution

- The system replaces the delivering plant of the requested item with an alternative plant that has better availability (more quantity, earlier date, or both).
- Product and storage location stay the same; only the plant can change.
- Key Point for Plant Substitution –
    - *For plant substitution to work, the material must be extended to all plants that are intended to participate in the substitution process.*
    - *ABC plant substitution only considers plants with available stock, ensuring the system selects a deliverable alternative.*

- Example:
    - Sales Order: FG10101, 1000 units, Plant P100.
    - Plant P100 has 0 units; Plant P200 has 1000 units.
    - With plant substitution, the confirmation is moved from plant P100 to plant P200 (inline or via subitems, depending on your Substitution Item Type).

b. Product Substitution

- The system replaces the requested product with one or more alternative products defined in product-substitution master data.
- Plant can be the same or also be substituted if you combine methods in different strategies, but this method primarily focuses on changing the material.
- Example:
    - Sales Order: FG10101, 1000 units.
    - No stock for FG10101, but FG10102 is defined as a valid successor/alternative with stock.
    - ABC confirms FG10102 instead of FG10101 (inline only for plant, subitems when products change, depending on your configuration).

c. Storage Location Substitution

- The system replaces the storage location within the same plant by another storage location that has available stock.
- Product and plant usually remain the same; only the storage location changes in line with the storage-location substitution graph.
- Example:
    - Sales Order: FG10101, 1000 units, Plant P100, Storage Location 0001.
    - Storage Location 0001 has no stock, Storage Location 0002 in the same plant has 1000 units.
    - ABC confirms from Storage Location 0002 as the substitute storage location.

Substitution Method Sequence-

In the Substitution Method, you can define the sequence (order of determination) in which ABC tries different substitution methods during the ATP check.

You can add multiple substitution methods to a single strategy—such as Plant, Storage Location, and Product Substitution—and they are processed top-to-bottom in the order you arrange them.

Example Sequence-

1. Plant Substitution (first priority)
2. Storage Location Substitution
3. Product Substitution (last resort)

During the ATP check, the system follows this logic:

1. The system first tries Plant Substitution
    - Searches for alternative plants for the original product.
2. If Plant Substitution cannot fully confirm the quantity, the system then tries Storage Location Substitution
    - Searches within the original plant and any already-substituted plants.
3. If Storage Location Substitution is still insufficient, the system finally tries Product Substitution
    - Searches for substitute products as a fallback option.

This sequencing gives you full control over the substitution hierarchy, ensuring that the system follows business priorities (for example, changing plants before changing products).

5. Substitution Method Sources

Each Substitution Method—Plant, Storage Location, and Product—has its own Substitution Method Sources.

To view and maintain these sources, select the respective Substitution Method and then use the navigation option to drill down into the detailed configuration.

Substitution Method Sources-

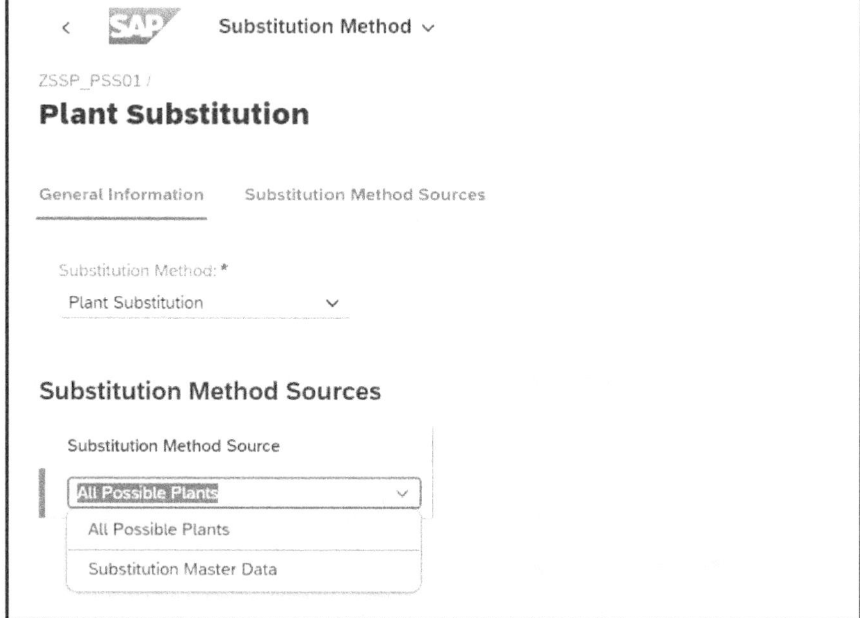

When you choose Plant Substitution, the Substitution Method Source tells the system where to look for alternative plants.

For Plant Substitution, there are two Substitution Method Sources:

- All Possible Plants - (Check everywhere)
- Substitution Master Data – (Check only what I allow, in the order I define)

All Possible Plants	• System considers all plants where the requested material is maintained/extended. • No master data maintenance required. • ABC ranks all eligible plants by availability (stock + incoming supply) and confirms from the best option(s). Use case: You want maximum flexibility and don't want to maintain plant hierarchies.

Substitution Master Data	• System only considers alternative plants defined in plant substitution master data (separate configuration). • You define exact plant relationships, sequences, priorities, and validity dates. • ABC follows your defined plant substitution chain/graph exactly. Use case: You want precise control over which plants can substitute which others (business rules, geography, capacity limits).

Key Points-

1. For "Product Substitution", the only available source is "Substitution Master Data", because:
   - Unlike plants, you cannot assume that any product can replace another product.
   - There could be thousands of products, and it is not feasible to automatically determine all possible substitution relationships.
   - Substitutions must be explicitly defined because:
     - Products may have different specifications, features, or pricing.
     - Only certain products are approved to fulfil the demand of another product.
   - Therefore, ABC relies on the pre-maintained product substitution master data to determine which product(s) can replace the original product.

Example

- Original Product: FG10106
- Substitution Master Data defines:
  - FG10107 → 50%
  - FG10108 → 30%
  - FG10109 → 20%

During ATP check, ABC uses this master data to propose substitutes in the defined sequence and percentages.

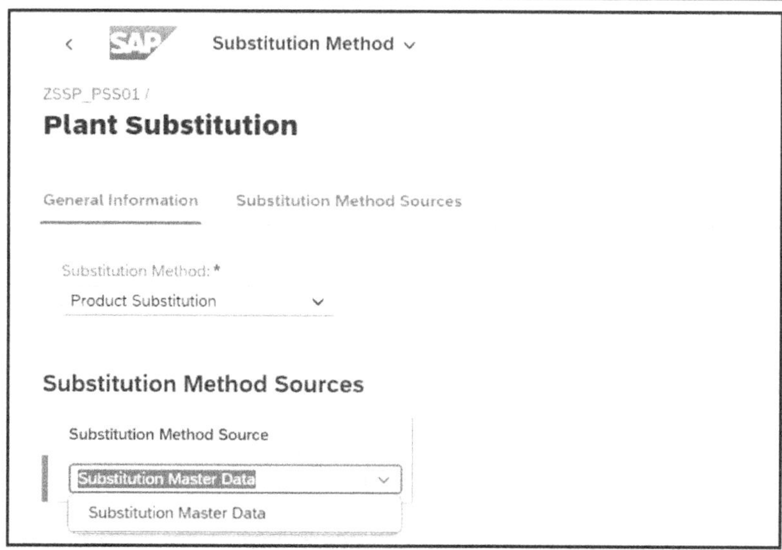

2. For "Location Substitution", also the only available source is "Substitution Master Data", because:

- Locations (storage locations or delivery locations) are specific and predefined, and you cannot assume any location can automatically replace another.
- Substitutions must be explicitly defined to ensure accurate stock allocation and correct fulfilment.
- This gives full control over which locations can act as alternatives, avoiding unintended deliveries or stock errors.

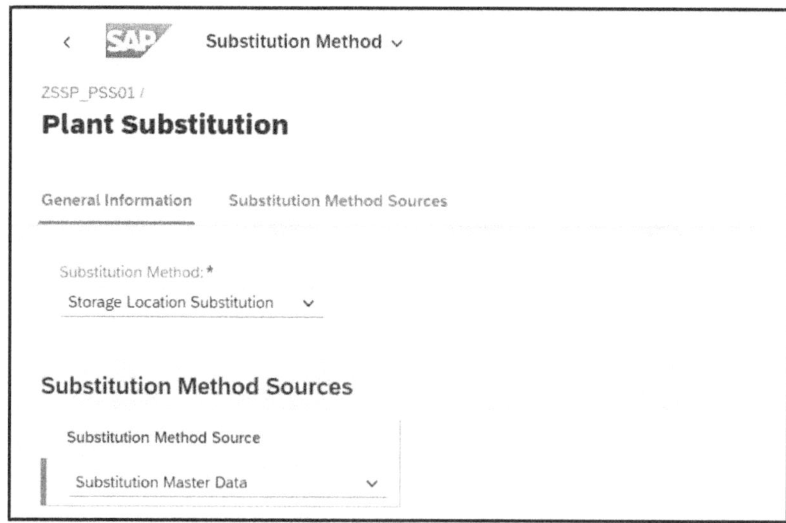

6. Building Rule

In the Building Rule configuration of ABC substitution strategies, SAP provides these three standard options that control *how* the system selects and ranks alternatives during ATP checks.

Building Rule (how ABC selects alternatives)-

- Full Confirmation – picks alternative that confirms 100% quantity (ignores date).
- Max Earlier Confirmation – picks alternative with maximum quantity earliest.
- On Time Confirmation – picks maximum quantity on/before requested date.

Alternative Determination

Search

Determination Name	Determination Description
FULL_CONFIRMATION	Full Confirmation
MAX_EARLIER_CONFIRMATION	Maximum Confirmation Earlier
ON_TIME_CONFIRMATION	On-Time Confirmation

These rules determine which plant/product gets chosen when multiple alternatives are technically feasible.

Building Rule Options-

1. Full Confirmation-

Full Confirmation	• Selects the alternative (plant/product/SLoc) that can confirm the complete requested quantity, regardless of delivery date. • Prioritizes 100% quantity fulfilment over delivery timing.

Full Confirmation prioritizes 100% quantity fulfilment over delivery timing.

Sales Order Example-Scenario:

- Customer orders: 1000 units of FG10101 from Plant P100
- Requested delivery: Jan 28, 2026
- ABC Building Rule: Full Confirmation

Availability across plants:

Plant	Available Qty	Earliest Date
P100	400 Units	Jan 28
P200	800 Units	Jan 28
P300	1000 Units	Feb 05
P400	900 Units	Jan 30

What happens during ATP:

1. ABC evaluates all alternative plants
2. P300 wins because it offers full 1000 units (ignores the later date)
3. Other plants lose despite earlier dates because none can do 100%

Result in sales order:

Item 10: FG10101, 1000 units confirmed from Plant P300 on Feb 5, 2026

Business impact: Customer gets complete order (no partial shipments), but delivery is delayed 8 days vs. what partial confirmations could achieve.

Use case: Products where customers refuse partial deliveries (e.g., complete machine kits, full project deliveries).

2. Max Earlier Confirmation-

Max Earlier Confirmation	• Selects the alternative that confirms the maximum quantity earliest (considers both quantity and delivery date). • Balances quantity with faster delivery.

Max Earlier Confirmation balances maximum quantity with the earliest possible delivery date.

Sales Order Example

Scenario:

- Customer orders: 1000 units of FG10101 from Plant P100
- Requested delivery: Jan 28, 2026
- ABC Building Rule: Max Earlier Confirmation

Availability across plants:

Plant	Available Qty	Earliest Date
P100	400 Units	Jan 28
P200	800 Units	Jan 28
P300	1000 Units	Feb 05
P400	900 Units	Jan 30

What happens during ATP:

1. ABC scores alternatives by quantity × earliness
2. P200 wins with 800 units on Jan 28 (best combination of quantity + date)
3. P300 loses despite full quantity (too late)
4. P400 loses (less quantity than P200, later date)

Result in sales order:

Item 10: FG10101, 800 units confirmed from Plant P200 on Jan 28, 2026
Item 10: 200 units remain open (or handled per your Remaining Open Qty Assignment)
Business impact: Customer gets maximum quantity (800/1000) on-time vs. waiting for full 1000 units later.
Use case: Products where partial but fast delivery is better than complete but delayed delivery.

3. On Time Confirmation-

On Time Confirmation	• Selects the alternative that confirms the maximum quantity on or before the requested delivery date. • Strictly respects the requested date; no late deliveries.

On Time Confirmation strictly prioritizes delivery on or before the requested date.

Sales Order Example

Scenario:

- Customer orders: 1000 units of FG10101 from Plant P100
- Requested delivery: Jan 28, 2026
- ABC Building Rule: On Time Confirmation

Availability across plants:

Plant	Available Qty	Earliest Date
P100	400 Units	Jan 28
P200	800 Units	Jan 28
P300	1000 Units	Feb 05
P400	900 Units	Jan 30

What happens during ATP:

1. ABC only considers plants that can deliver on or before Jan 28
2. Plant P200 wins with 800 units on Jan 28 (maximum quantity on-time)
3. Plant P300 eliminated (too late - Feb 5)
4. Plant P400 eliminated (too late - Jan 30)
5. Plant P100 offers only 400 units (loses to Plant P200's 800 units)

Result in sales order:

Item 10: FG10101, 800 units confirmed from Plant P200 on Jan 28, 2026
Item 10: 200 units remain open (or handled per your Remaining Open Qty Assignment)

Business impact: Customer gets maximum possible quantity on-time (800/1000) vs. waiting for full quantity later.
Use case: Products where on-time delivery is critical (e.g., event merchandise, seasonal goods, just-in-time manufacturing).

4. Custom Building Rules-

In addition to SAP's three standard Building Rules, you can define custom building rules to implement your specific business prioritization logic.
Building Rule

- Standard options: Full Confirmation, Max Earlier Confirmation, On Time Confirmation
- Custom rules: Define your own ranking logic (distance, cost, capacity, business priorities, etc.) for complex scenarios.

This gives you full flexibility to match your exact business requirements beyond SAP's standard logic.

Custom Building Rules-
To create your own building rules, you can use the Fiori app "**Configure Alternative Determination**." –

You can create custom rules that go beyond the standard Full/Max Earlier/On Time options. These let you define complex ranking criteria like:

- Distance/cost between plants
- Preferred customer/supplier relationships
- Capacity utilization limits
- Custom scoring (e.g., 40% quantity + 30% date + 20% transport cost + 10% plant preference)
- Tie-breaker logic when multiple alternatives score equally

You can create a new building rule in the Fiori app "Configure Alternative Determination" by copying from an existing rule using one of the following options:

- Create with Business Add-In (BAdI) – Add custom logic via a BAdI implementation.
- Create with Attribute and Constraint – Use specific attributes and constraints to define the rule.
- Create from Template – Use a predefined template as the starting point.

You can review existing rules to understand the current logic and determine what is needed in the new rule. While reviewing, you can check:

- Rating Attributes – Define how alternatives are rated or prioritized.
- Hard Constraints – Define strict conditions that must be met for confirmation.

Example:

- Rating Attribute – Could be used to prioritize one alternative over another.
- Hard Constraint – A rule for full confirmation may require a confirmation ratio = 100%.
- In this case, if a sales order requests 1,000 units, but only 999 units are available, the system will not consider this stock, because the confirmation ratio requirement is not met.

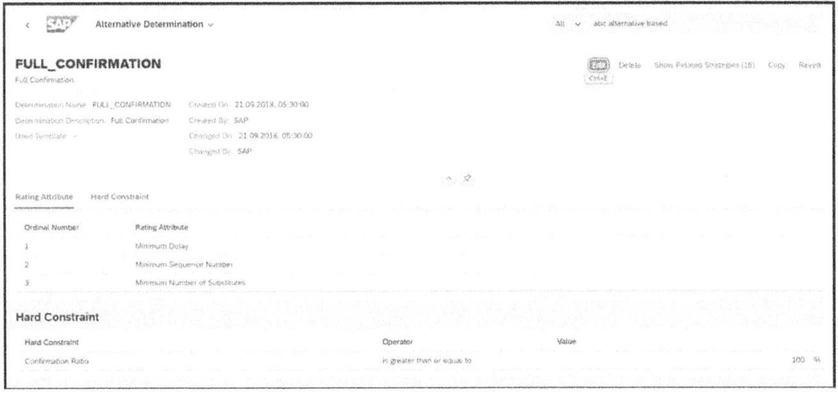

You can create a new custom rule by copying from the existing "Full Confirmation" rule using the option "Create with Attribute and Constraint."

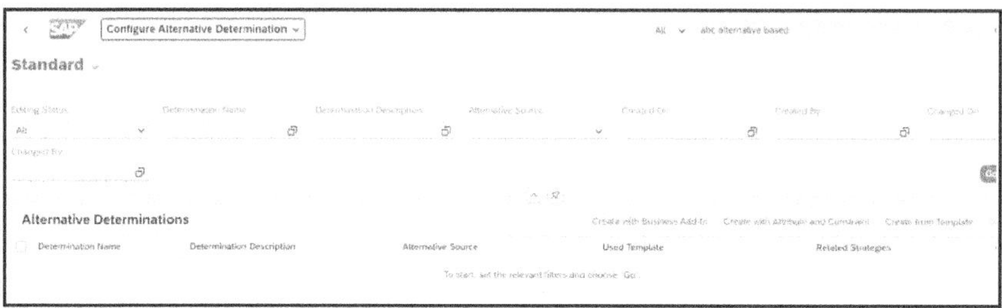

Copied from Full Confirmation-

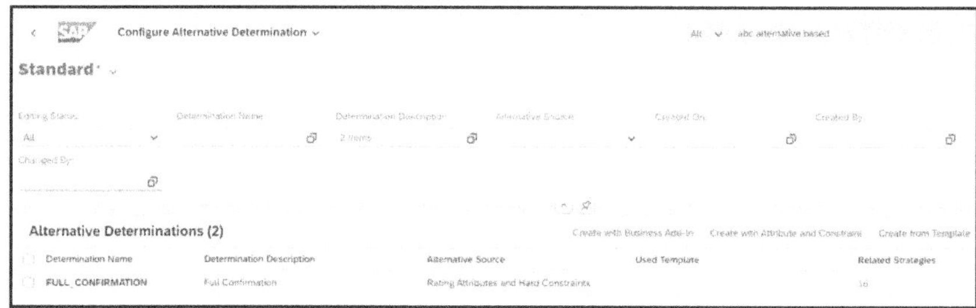

A new custom rule is created with a determination name and description, along with its rating attributes and hard constraints.

Example:

- The confirmation ratio is set to 90%, meaning the system will consider available stock if at least 90% of the requested quantity can be confirmed.

Key Points for the consideration-

1. Limit the number of plants per sales order

- Based on business feasibility, a sales order should not be split across too many plants.

- Example: If a sales order requests 1,000 units, and the system proposes delivery from 5 different plants, this may not be feasible for the business.
- Therefore, it is important to set rules or constraints to limit the number of plants used for fulfilling a single order (e.g., maximum 2 plants per order).

2. Create with Business Add-In (BAdI)

- This option allows custom logic to be added via a BAdI implementation.
- Example: Before substituting a plant, the system can check the distance and transportation cost based on predefined rates.
- Based on this evaluation, the system can decide whether to consider a substitute plant or not, adding more business intelligence to the ABC logic.

5. Tie Breaking Logic-

Tie Breaking Logic kicks in *after* standard/custom building rules when multiple alternatives score equally (tie situation).
Tie-breaking logic is SAP's hard-coded, sequential rule set used to select a single valid record when multiple records qualify equally.
How Tie Breaking Works

Building Rule → Finds alternatives with SAME score ↓ **Tie Breaker → Applies 4 sequential rules (in order) to break the tie**

The 4 Tie Breaker Rules (in priority sequence)

1. Ship-to Party and Plant from Same Country	Priority #1: Prefer plant in same country as ship-to party Example: US customer → US plant over Mexico plant (same score)
2. Ship-to Party and Plant from Same Region	Priority #2: If countries match → prefer same region/state Example: California customer → California plant over Texas plant
3. Shorter Transit Time (Route-based)	Priority #3: Prefer plant with shorter transit time from route master Example: If no route → transit time = 0 (equal)

	Plant A: 2 days route → Plant B: 3 days route → Plant A wins
4. Shorter Delivery Time (Plant + Shipping Point)	Priority #4: Total time = Plant loading + Pick/Pack + Shipping point processing Example: Plant with faster processing wins

Sales Order Example

Order: 1000 units FG10101, Ship-to: California, USA

Requested: Jan 28, 2026

Building Rule "Max Earlier Confirmation" → 3 plants tie at 1000 units Jan 28

Priority	Rule	P100	P200	P300	Winner
1	Same Country as Ship-to	USA	USA	Mexico	P100/P200 Tie
2	Same Region as Ship-to	California	Texas	-	P100 Win
3	Shorter Transit Time	1 Day	2 Days	0 Days	P100 Win
4	Shorter Delivery Time	2 Days	3 Days	4 Days	P100 Win

Result: Processing stops at Priority #2 – Plant P100 is the clear winner.

3 plants tie after "Max Earlier Confirmation"
↓
Priority 1: Country check → P300 eliminated (Mexico)
↓ *[P100 vs P200 remain]*
*Priority 2: Region check → **P100 wins** (California = California)*
↓ *[End - P100 selected]*

Key Point: Sequential Priority

- Tie-breaker rules are evaluated top to bottom.
- The first rule that resolves the tie determines the winner; remaining rules are not evaluated.
- In this case, plant P100 did not require rules #3 or #4 because the tie was already resolved at rule #2.
- This behaviour is standard SAP, hard-coded logic. It is not configurable and is applied automatically during the tie-breaking process.

Parent Plant in SO Creation-

Parent Plant is the initial plant determined during sales order creation, used as a baseline for intelligent plant substitution. Tie-breaking logic ensures SAP consistently selects the best plant when multiple options qualify.
Parent Plant is the initial / baseline plant that SAP determines when a sales order is created (VA01). It is the starting point used for further plant substitution or optimization logic (like ABC).
"SAP's first choice plant before any intelligent substitution happens."

During SO creation, SAP determines a plant in this priority order:

1. **Customer-Material Info Record (CMIR)** - Customer prefers a specific plant for that material
2. **Customer preference** - Explicit customer-level plant preference
3. **Material Master** - Default delivering plant maintained in material data
4. **Manual Entry** - User manually selects the plant in the SO

The plant derived from any of the above becomes the Parent Plant.
The Parent Plant is not always the final delivering plant.
It is used as: A baseline and ABC then substitute from this "parent" plant baseline. Meaning:

- SAP first determines a valid plant
- Then ABC logic evaluates alternate plants
- Substitution is done relative to the parent plant

Parent Plant is the starting reference plant that ABC uses to intelligently search, compare, and substitute the best possible delivering plant.

Configure Alternative Control-

Configure Alternative Control determines when and under which business conditions ABC is executed, acting as the activation and filtering mechanism for substitution logic.

Configure Alternative Control decides WHEN and WHERE ABC should run.
Think of it as an on/off switch with conditions for ABC.

- Substitution Strategy = *how substitution works*
- Alternative Determination = *which alternatives exist*
- Alternative Control = when substitution is allowed to happen

Why This App Is Important-

Without Alternative Control:	With Alternative Control:
• ABC would run everywhere, which may: o Hurt performance o Cause unwanted substitutions o Break business rules	• ABC runs only where it makes business sense

In this app, you control whether ABC is triggered or skipped based on business conditions. Specifically, you control:

1	*Activation of ABC*	• Turn ABC ON or OFF • Without this app, ABC will not execute, even if strategy and rules exist • Decide whether ABC is enabled or disabled for a given scenario.
2	*Scope of Application*	Control where ABC applies, such as by: • Sales organization • Distribution channel • Division • Plant • Document type
3	*Condition-Based Execution (Characteristics)*	ABC runs only when certain characteristics are met, such as: • Product type • Customer group • Order reason • Priority / urgency • ATP situation Example: Run ABC only for urgent orders Do NOT run ABC for normal orders

4	*Multiple Controls for Different Scenarios*	Maintain multiple control records to activate ABC differently for: • Domestic vs. export orders • Normal vs. rush orders • Specific customers or products

How Does Alternative Control Work (Step-by-Step)?

1. Sales order is created (VA01)
2. Parent Plant is determined
3. SAP checks Alternative Control
4. If conditions match:
   - ABC is activated
   - Substitution Strategy is executed
5. If conditions do NOT match:
   - ABC is skipped
   - Parent plant remains final

Simple Example

Business Requirement

- Run ABC only for:
  - Customer group = Retail
  - Order priority = High

Configuration

- In Configure Alternative Control:
  - Maintain characteristics:
    - Customer Group = Retail
    - Priority = High
  - Activate ABC

Result

- Retail + High priority → ABC runs
- Anything else → ABC does NOT run

2. Storage Location Substitution in ABC (Alternative-Based Confirmation)

ABC at the storage-location level means using Alternative-Based Confirmation to automatically substitute one storage location with another (and optionally plants or products) during the aATP check when the originally requested storage location cannot fully supply the demand.

- It uses storage location substitution master data as the source for ABC. The system starts with the requested storage location and follows a predefined sequence or graph of substitute storage locations for the same plant and product.
- The substitution master data defines:
    - Which storage locations can replace others
    - The order/priority of substitution
    - Validity periods
    - Any locations that are excluded or inactive
- During the aATP check, if there is insufficient stock in the requested storage location, the ABC strategy with the method Storage Location Substitution searches the substitute storage locations from the defined sequence/graph.
- It filters by validity and relevance for aATP, and checks availability in the substitute locations.
- The system confirms the requirement from the first suitable substitute storage location(s) according to the strategy rules.
- This can also be combined with plant substitution and product substitution in one strategy, so the sequence can be: ***requested storage location → other storage locations in the same plant → alternative plant/storage location → alternative product.***

The scope-of-check indicator "Without Storage Location Check" – actually disables storage-location-level checking; to consider storage locations in ATP, this flag must not be set.

This setting is available in the Scope of check (Transaction - OZV9A) under special scenarios – "Without Storage Location Check".

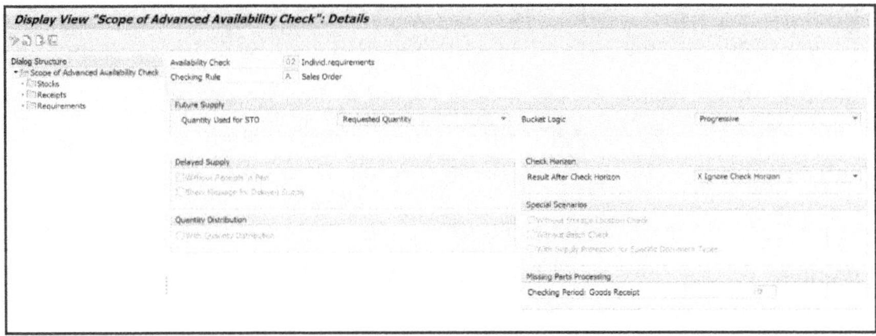

What does the flag mean-

1. When the flag is active (box ticked)

- The system does not perform a storage-location-level ATP check.
- Stock and requirements are aggregated at plant (or MRP area) level, even if a storage location is entered on the sales order or delivery; the check ignores that storage-location detail.
- Storage-location-driven logic (for example, strict check against one storage location or storage-location substitution in ABC) cannot work as a true location constraint, because the check is not restricted to that storage location.

2. When the flag is not active (box empty)

- The system does consider storage locations in the ATP check.
- If a storage location is specified on the document, ATP is executed against that storage location's stock and requirements; if not, the system can still differentiate by storage location according to other settings (MRP/ATP indicators, substitution).
- This is the prerequisite for using storage-location-level ATP and for features like storage location substitution within ABC to behave as intended.

Example - We have the following situation for the product, the locations, Storage Location and the current stock levels.

Product	Location (Plant)	Storage Location	Stock
FG10101	P100	S100	0
FG10101	P100	S101	100
FG10101	P101	S101	50
FG10101	P102	S102	20

In this case, when a sales order is created for product 'FG10101' with a requirement of 100 units at plant P100 and storage location S100, and there is no available stock at that location, the system will propose an alternative storage location, S101, to fulfill the demand, provided that the substitution strategy and master data are set up accordingly.

For ABC to propose an alternative storage location, that location must be included in the Storage Location Substitution Rule within aATP. This means that if you want ABC to check both storage locations S100 and S101 - then both must be listed in the Substitution Strategy / Substitution Rule.

Configuration:

- Location substitution is configured using SAP Fiori apps such as "Configure Substitution Strategy" where substitution groups, reasons, and rules are maintained.
- Substitute plants must be set up and assigned to substitution groups; these are then used by the availability check logic during order creation.
- The system respects legal, tax, and business constraints during substitution, ensuring compliant deliveries.

3. Product Substitution in ABC (Alternative-Based Confirmation)

Product substitution in Alternative-Based Confirmation (ABC) within SAP S/4HANA and aATP is a functionality that allows the system to automatically or manually replace an originally requested product with substitute products during the availability check process, when the ordered product is unavailable or insufficient in quantity. This is designed to improve order fulfillment and customer satisfaction by offering equivalent or alternative products. Substitution can occur under various scenarios such as:

- Replacement by an upgraded or newer product version.
- Substitution due to phasing out or discontinuation of the original product.
- Partial substitution where part of the order quantity is confirmed with the original product stock, and the remainder with substitute products.
- Substitution by multiple products especially in cases where sales units differ (e.g., cases instead of pieces).

ABC manages product substitution by defining substitution reasons and substitution strategies in master data, which include hierarchical product relationships and directed product graphs to facilitate one-to-one or one-to-many substitution cases.

During sales order processing, the system either replaces the product directly or generates subitems with substitute products, depending on configuration choices such as inline substitution or subitem generation. This allows flexible and granular confirmation handling while keeping traceability of the original ordered products.

Product substitution in ABC enables SAP S/4HANA aATP to confirm sales orders by intelligently substituting products based on availability and business rules, supporting automated substitution, upgraded products, partial and multiple product substitutions, and maintaining transparency in the order confirmation process.

Product	Location (Plant)	Stock
FG10101	P100	0
FG10101	P101	0
FG10101	P102	0
FG10101	P103	0
FG10102	P100	120

In this case if we create a sales order for Product "FG10101" with required Qty 100 units with plant P100 but as there is no stock available in any Location P100, P101, P102 and P103 then system will propose alternative Product "FG10102" form Location P100 to fulfill the customer demand.

Example Use Case: Product Substitution with Different Sales Units.
The system handles different sales units (10 lb → 5 lb bags) via unit conversion in substitution master data.
Normally, rice is sold in 10 lb bags. However, in the event of a stock shortage, the product can be substituted with 5 lb bags. For example, if a customer orders 10 units of 10 lb bags, the system can automatically propose a substitution and confirm delivery of 20 units of 5 lb bags, ensuring the total ordered quantity is fulfilled.

Example Use Case: Product Discontinuation with Replacement
If we discontinue product FG10101 and replace it with FG10109, once no availability exists for FG10101, the system will automatically substitute with FG10109 during ATP check.

Configuration Setup for ABC-
aATP (Advanced Available-to-Promise) includes ABC (Alternative-Based Confirmation) to confirm shortages using substitutes (plants, materials, or storage locations) when the original item or plant cannot be confirmed.
ABC can handle different substitution strategies and rules, and is usually configured alongside the other aATP components (PAC, PAL, BOP) in S/4HANA.

- aATP activation, SD integration, up-to-date master data (materials, plants, routes, shipping points), and access to relevant Fiori apps for configuring ABC and substitution rules.

Below are the configuration details for SAP S/4HANA aATP – ABC.

1. Prerequisites and Basic Setup
In aATP 2023, the following prerequisites must be in place:

- aATP is activated in S/4HANA, and the relevant SD integration is completed.
- Master data is up to date, including materials, plants, MRP settings, stock, routes, and shipping points.
- Access to the required Fiori apps for aATP configuration is assigned, in particular:

- Manage Substitution Products
- Manage Substitution Reason Products
- Manage Substitution Groups Products
- Manage Substitution Controls Products

2. Fiori Apps: The following two apps are mandatory for ABC configuration and must be used:

- Manage Substitution Products
- Manage Substitution Reason Products

The following apps are optional and are typically used for one-time configuration activities. They help simplify maintenance and provide additional control over ABC logic but are not mandatory for standard operation. These two apps are used to maintain technical hierarchies.

- Manage Substitution Groups Products
- Manage Substitution Controls Products

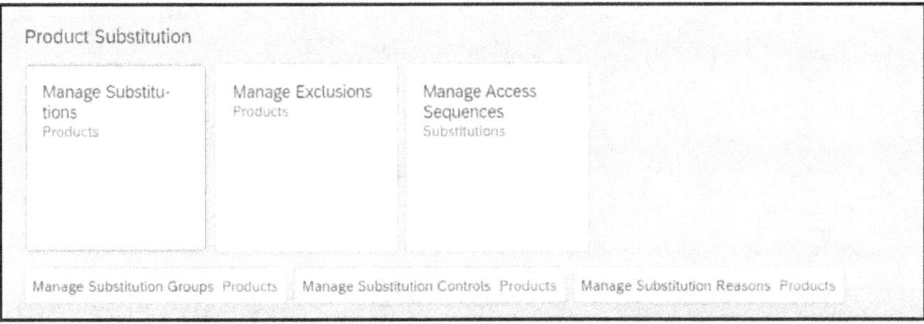

Fiori Apps	Details
1. Manage Substitution Products	• Purpose: Defines the alternative materials that can be used when the requested material cannot be confirmed. Key Features: • Create, edit, or delete alternative materials for a given product • Define priority or sequence of alternatives • Support 1:1 or 1:n substitutions • Supports percentage-based distribution among substitutes Use in ABC: During ATP check or BOP, the system automatically refers to these entries to determine which substitute material(s) can fulfill the order.

2. Manage Substitution Reason Products	• Purpose: Defines substitution reasons that explain why a material is being substituted. Key Features: • Configure reasons for substitution (e.g., stock shortage, plant unavailability, quality issues) • Link reasons to substitution logic to control which alternatives are proposed • Optionally influence confirmation rules, prioritization, or reporting Use in ABC: The system uses the substitution reason to justify changes in the sales order and update schedule lines accordingly.
3. Manage Substitution Groups Products	Purpose: Allows grouping of substitution products to simplify maintenance and reduce redundancy. Key Features: • Assign multiple alternative products to a substitution group • Maintain group-wide substitution rules instead of individual product entries • Streamlines setup when multiple materials share the same alternatives
4. Manage Substitution Controls Products	Purpose: Defines control parameters for substitution behavior. Key Features: • Set global or material-specific substitution rules • Define priorities, allowed substitution types, and limits • Helps control system behavior during ABC checks and BOP runs

3. Conceptual Objects in ABC

For configuration, focus on three main objects:

- ABC Strategy / Alternative Control
  Defines *what kind of alternative* is allowed and in *which order* (e.g., "Plant then Material", or "Material only", or "Plant → Date shift").
- Substitution Rules
  Hold the *business rules* for which alternatives are valid:
    - plant sequence,
    - material substitution hierarchy,

    - date shift limits,
    - validity periods and conditions.
- Assignment to Business Context
  Connects the above to the "sales order" context, so that the system knows when to run which ABC logic.

4. Step-by-step configuration flow (high level)

Use these steps as functional details for system setup.

Step	Scenario	Details
1	*Design your ABC scenarios*	Clarify business rules first: • Do you require plant substitution, location substitution, and/or product substitution? • Are substitutes 1:1 functionally equivalent, 1:n equivalents, or do they require customer-specific approval? • Should later delivery dates be proposed automatically? • Are different rules required per sales organization, distribution channel, or customer group? This design drives how you configure characteristics and alternative controls.

2	*Define Alternative Controls (ABC strategy)*	Via the ABC configuration app (often called **"Configure Alternative Control"** or **"Configure Alternative-Based Confirmation Strategy"** in Fiori):
		• Create an **Alternative Control** for **Document Class = Sales Order**.
		• Define the **ordered characteristics** that the system evaluates, for example:
		  o Sales organization
		  o Distribution channel
		  o Customer group
		  o Material
		  o Plant
		• Assign a **substitution strategy type**, such as:
		  o Plant substitution
		  o Material substitution
		  o Plant and material substitution

		○ Delivery date shift (or *"try alternatives, then move date"*) Here, you define the **sequence of checks**, for example: 1. Check availability in the requested plant. 2. If not available, check alternative plants in the defined sequence. 3. If still insufficient, check substitute materials. 4. If confirmation is still not possible, shift the delivery date within the defined tolerance.
3	*Maintain Substitution Rules*	Using the **"Manage Substitution Rules"** Fiori app (app names may vary slightly by release): • Create **substitution rules per scenario** (for example, *"US Domestic Plant Substitution"* or *"EU Material Substitution"*). For each rule, maintain: • **Validity dates** (from/to). • **Condition criteria** (for example: Sales Organization = 1710, Distribution Channel = 10, Customer Group = Retail). • **Plant determination sequence** (for example: requested plant 1710 → US20 → 1010). • **Material substitution chains** (ordered material → substitute 1 → substitute 2, and so on). • **Delivery date shift limits** (maximum backward and/or forward shift, if used as part of ABC). The substitution rules are **linked to the relevant Alternative Control**, enabling the system to determine which rule set to apply during confirmation.
4	*Material Assignment and Business Context*	• ABC (Alternative-Based Confirmation) is configured via Fiori apps like Configure Alternative Control and Manage Substitution Rules, not directly in the Material Master. • The Material Master → aATP view contains standard ATP settings (e.g., ATP check group, MRP elements, stock/plant info) that contribute to the availability check, but it does not have a field to assign ABC directly.

		• During an ATP check, ABC is triggered automatically when the sales order characteristics match an Alternative Control configuration. • PAC (standard ATP) still runs first, and ABC works on top of PAC to propose alternatives if PAC cannot fully confirm the order. Key point: ABC activation is rule-based via Alternative Control, not a material master switch.
5	*Integration with Product Availability Check (PAC) and Product Allocation (PAL)*	ABC sits on top of PAC and can work together with PAL: ABC in relation to PAC and PAL • PAC still controls the basic ATP logic (receipt/issue elements, horizon, safety stock). • PAL (Product Allocation) can restrict how much quantity can be confirmed per period or per customer. • ABC is invoked to search for alternatives if PAC + PAL cannot fully confirm the requested quantity for the original combination. Typical execution flow in many implementations: 1. Run PAC (and PAL, if active) for the requested material/plant/date. 2. Trigger ABC when a shortfall exists. 3. ABC applies the Alternative Control and Substitution Rules to find alternative plants, materials, or delivery dates.
6	*Testing and fine-tuning*	Recommended Test Approach for ABC: 1. Prepare Test Data • Create a simple test material and 2–3 plants with controlled stock levels. 2. Assign Configuration • Assign the new Alternative Control and Substitution Rules to the test material. 3. Create Test Sales Order • Enter a sales order in VA01 for a plant with insufficient stock. 4. Observe System Behavior • Check how the RACR / aATP result screen proposes plant or material substitutions.

		• Review how the final schedule lines are split by plant, material, and delivery date. 5. Adjust and Validate • Fine-tune rule sequencing, date tolerances, or substitution lists until the system behaviour aligns with business expectations.

5. Typical ABC design patterns

Common patterns in real projects:

Network plant substitution:	Ship from the customer's "home plant" if possible, otherwise automatically pick the nearest plant (shortest transit or total delivery time) in the same country, else fallback to global plant.
Good–better–best product substitution:	Confirm the requested product first, then an equivalent substitute, then an upgraded product if the customer allows it.
ABC + PAL fairness:	Use PAL to cap quantities per customer/region and ABC to find another plant or material so high-priority customers still get confirmed without breaking allocation rules.

Set-up for the Product Substitution-

1. First, we define the substitution reason using the Fiori app **"Manage Substitution Reason Products."**

Manage Substitution Reasons Products

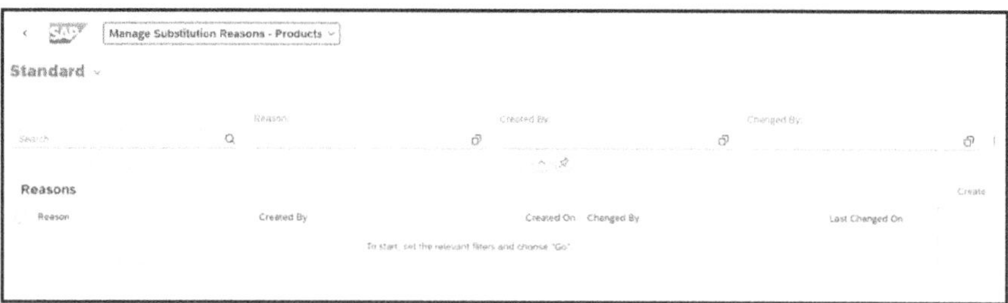

Create a Reason - ZSSP_PS01

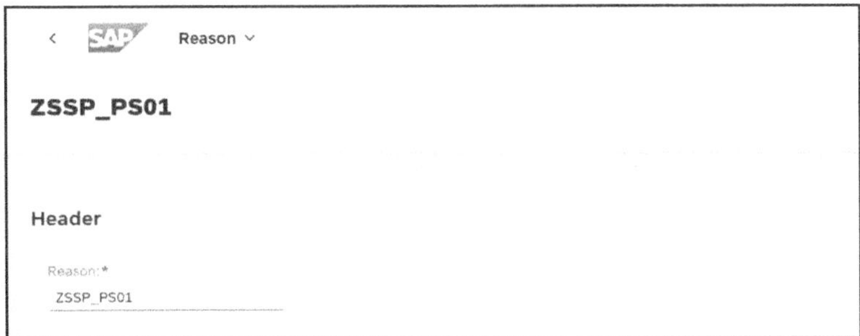

Similarly, we can define **Groups Products** and **Control Products** using the Fiori apps below. These apps are **optional** and not mandatory for standard operation.

In this step, you can create substitution Groups and link them to the corresponding Control Products for ABC.

- Manage Substitution Groups Products
- Manage Substitution Controls Products

These are just objects (Substitution Reason, Groups Products, Control Products) and do not require configuration. They exist to support ABC functionality but are not mandatory for standard operation.

2. Now we define the **Substitution Products** using the Fiori app **"Manage Substitution Products."** - In this step, we specify **which product can replace which product**.

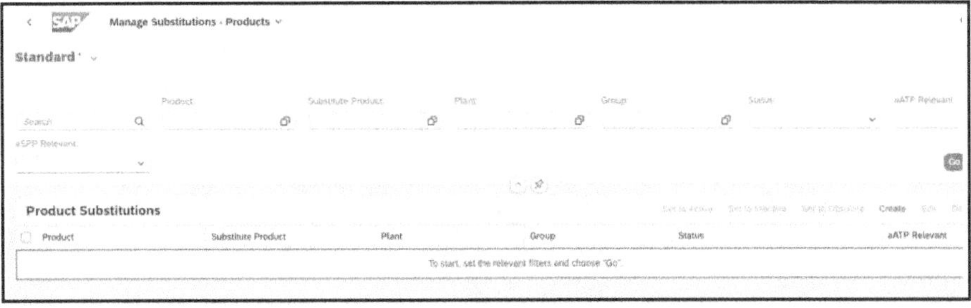

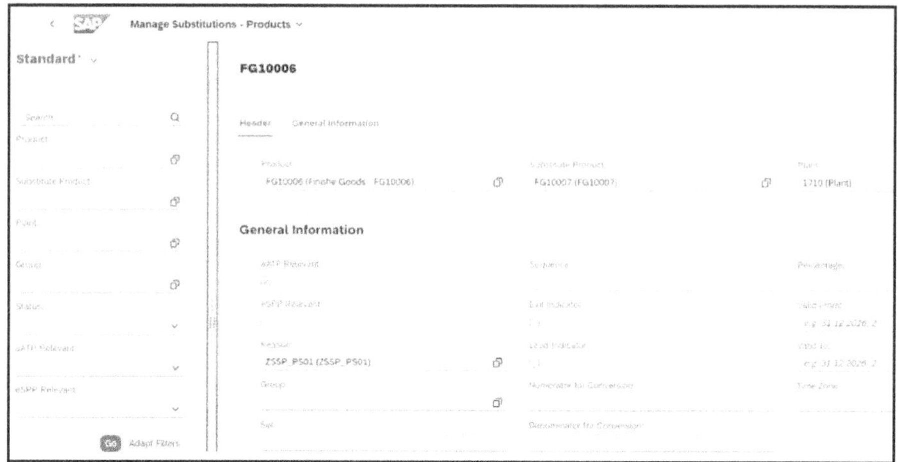

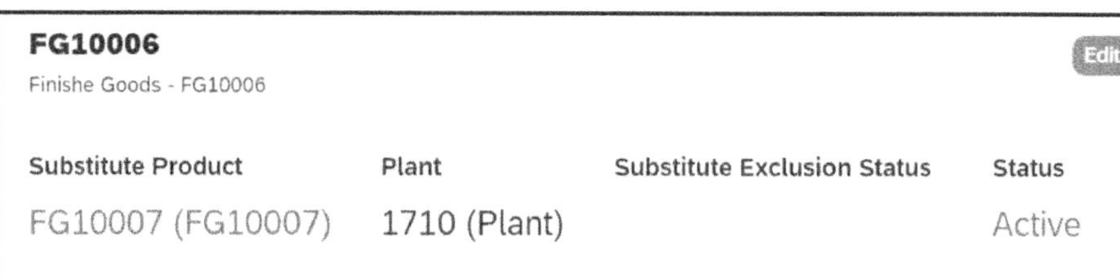

The following fields can be maintained here: in manage substitution product.

#	Field	Details
1	*Product*	Product that will be used as the replacement (e.g. FG10006).
2	*Substitute Product*	Product that will be replaced (e.g. FG10007).
3	*Plant*	Plant for which the substitution is valid.
4	*aATP Relevant*	Indicates whether the substitution is relevant for aATP.
5	*eSPP Relevant*	Indicates whether the substitution is relevant for eSPP. **eSPP (Extended Service Parts Planning)** originated in SAP APO and is now embedded in SAP S/4HANA as the successor to the APO-based solution. The eSPP relevance indicator enables reuse of substitution data for service parts planning scenarios. eSPP is optional for aATP / ABC, requires an additional SAP license, and is not included in the standard aATP scope.
6	*Reason*	Substitution reason.
7	*Group*	Substitution group assignment.
8	*Set*	Substitution set.

9	*Percentage*	Percentage of the quantity that can be substituted.
10	*Sequence*	Sequence in which the substitute is considered.
11	*Exit Indicator*	Stops further substitution search if this entry is used.
12	*Lead Indicator*	Marks the leading substitution product.
13	*Numerator for Conversion*	Conversion factor numerator.
14	*Denominator for Conversion*	Conversion factor denominator.

Key Points-

1. The Lead Indicator and Exit Indicator are used to define the start and end points of a substitution chain.

For example, if Product A has substitute Product B, Product B has substitute Product C, and Product C has substitute Product D, these indicators control how the system navigates the chain:

- The Lead Indicator marks the starting product of the substitution chain.
- The Exit Indicator marks the end point of the substitution chain, where the system stops searching for further substitutes.

2. **Set and Percentage** - The Set groups substitution products that are considered together, while the Percentage defines the maximum portion of the requested quantity that can be fulfilled using products from that set.

Example:

Part A can be substituted with Parts B, C, and D. Using the Set and Percentage fields:

- Part B -> 20% of the requested quantity
- Part C -> 30% of the requested quantity
- Part D -> 50% of the requested quantity

This ensures that when Part A is short, the system splits the order quantity among B, C, and D according to these percentages.

Once all the above fields are maintained to create a product substitution, you can change the status to "Active". You can also view the substitution graph, which displays the product and its corresponding substitute products.

Status – Active

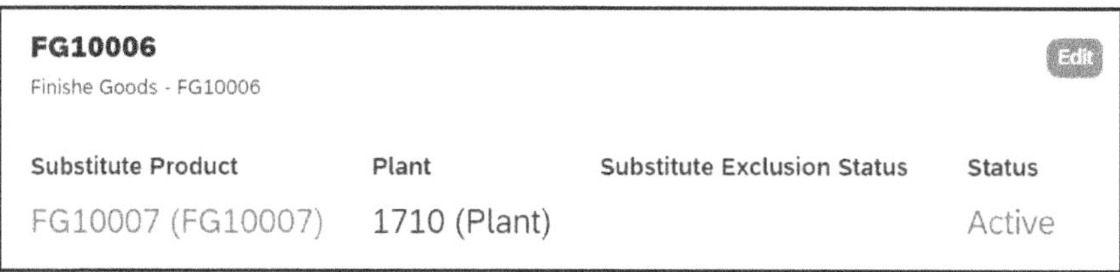

Product Network Graph-

The Product Network Graph in SAP is a tool used to visualize and manage the relationships between products and their substitutes within the Alternative-Based Confirmation (ABC) process in SAP S/4HANA aATP. Key Points about Product Network Graph:

- The graph provides a visual representation of product substitution chains, allowing you to see how different products (materials) are related to each other.
- It helps in identifying substitutes and the sequence in which the system will consider them during the availability check.

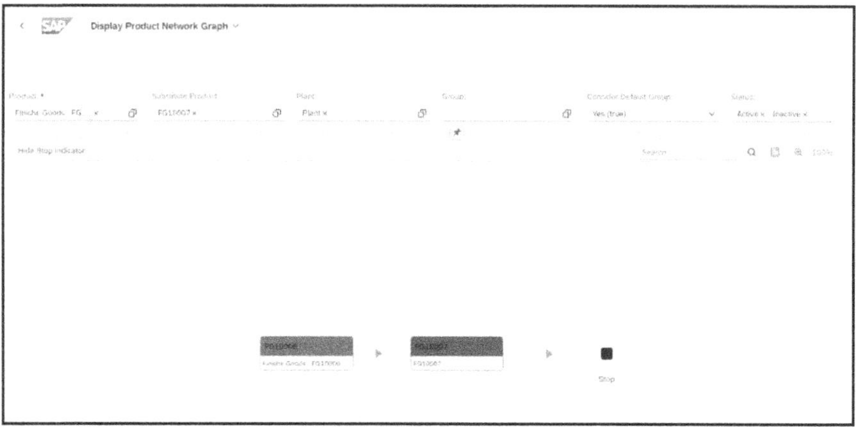

3. Next, we define the **Access Sequence** using the Fiori app **"Manage Access Sequences Substitutions."** This step tells the system the order in which to search for alternative products or plants during the availability check.

Why define Access Sequence in ABC - it tells the system where to look, in what order, and when to stop.

Example:
When defining product substitutions, you need to tell the system how to search for substitutes by creating an Access Sequence.
For instance, an Access Sequence could be:

1. Sales Organization / Plant / Material – Check for a substitute that matches all three.
2. Plant / Material – If not found, check for a substitute by plant and material only.
3. Material – If still not found, check for a substitute by material only.

This sequence ensures the system searches for alternatives in the correct order, following your business rules.

1	*Determine search priority*	• When a material is unavailable, ABC can check multiple alternatives (plants, substitute materials, dates). • The Access Sequence defines which alternatives are tried first, second, etc.
2	*Control substitution logic*	• For example, you may want the system to: 1. Try **alternative plants first** 2. Then try **substitute materials** 3. Finally, **shift delivery date**
3	*Avoid unnecessary checks*	• A well-defined access sequence prevents the system from checking irrelevant combinations, improving performance and ensuring business rules are followed.
4	*Link to rules*	• Each Alternative Control / Substitution Rule can reference an Access Sequence, so the system knows exactly how to traverse the substitution chain.

4. Next, we define the **Exclusion Products** using the Fiori app **"Manage Exclusion Products."** This step tells the system **which products should never be considered as substitutes** during the availability check, even if they meet other substitution criteria.

Example:

Suppose Product A can be substituted with Products B, C, and D. However, Product C is not allowed for certain customers or regions.

- In the "Manage Exclusion Products" app, you can mark Product C as an excluded product for those scenarios.
- During the ABC / aATP check, the system will skip Product C and only consider Products B and D as valid substitutes.

5. Tested Scenario – Sales Order with Product Substitution

1	*Materials used:*	• FG10006 – No stock available • FG10007 – Sufficient stock available
2	*Test Steps:*	• Created a sales order for FG10006 with order qty 10 units
3	*System Behaviour:*	• During the aATP / ABC check, the system identified that FG10006 had no available stock. • The system automatically substituted FG10006 with FG10007 based on the configured substitution rules.
4	*Result:*	• The sales order was successfully confirmed using the substitute product FG10007.

In the sales order, you will see **two lines**:

Original line (Item 10): Product "FG10006" & Substitute product line: Product "FG10007".

The Item Category of the main line is TAPA, and the substitute line is TAN.

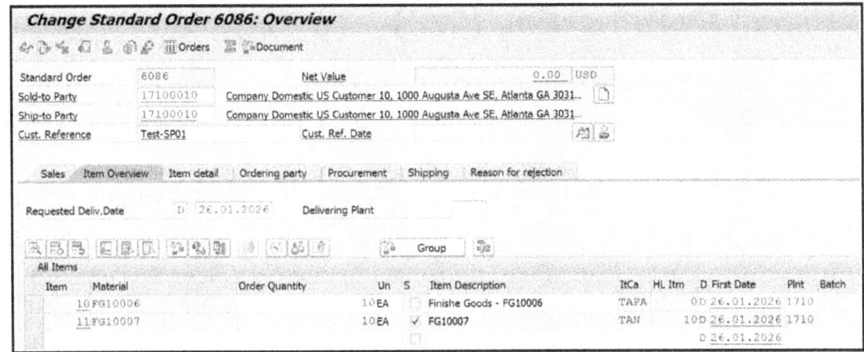

In the substitute line, you will see the material entered as "FG10006".

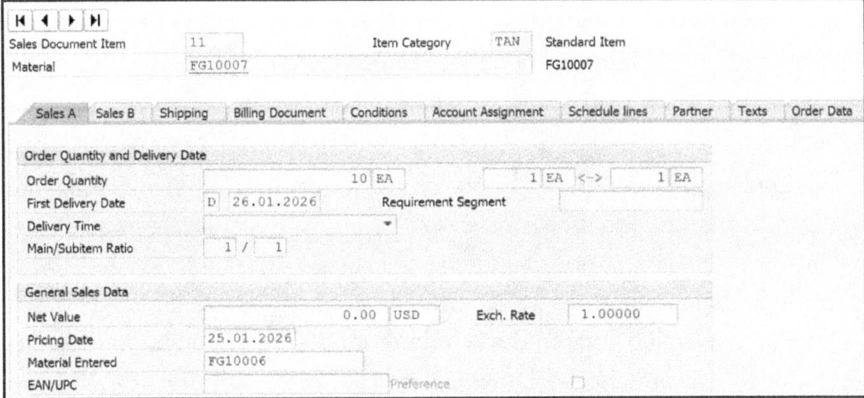

13 Supply Protection (SuP)

In SAP S/4HANA, supply protection is an Advanced ATP (aATP) function that reserves portions of available stock for specific customer groups, channels, or markets so that critical demand is always prioritized and cannot be consumed by lower-priority orders. It is configured per material/plant and characteristics (such as country, customer group, or distribution channel) and then applied automatically during ATP checks in sales and stock transport orders.

Supply Protection (SUP) is a mechanism used to reserve or protect specific quantities of a material at a given plant from being consumed by other demands. The protected quantities can be time-dependent, meaning they apply to defined periods. Supply Protection ensures that prioritized requirements are fulfilled even when overall demand exceeds available supply.

In short supply scenarios, supply protection in S/4HANA aATP ensures that a minimum quantity is reserved for eligible customers or segments. This protected quantity cannot be consumed by other demand, while additional quantities may be confirmed based on the available ATP stock. Supply Protection guarantees a minimum reserved quantity (protected quantity) for priority customers, while allowing additional fulfillment based on available ATP stock.

Supply protection defines "protected quantities" for combinations like material + plant + channel/region/customer attributes, which act as virtual reservations during order confirmation.

The main objective is to safeguard limited inventory for high-priority segments—such as e-commerce or strategic customers—so that other demand can only access stock not reserved for them. This ensures that critical channels or customers have guaranteed supply and prevents lower-priority demand from depleting limited stock.

Supply protection is a powerful feature used to prioritize and reserve limited stock for key business needs. The best use cases are scenarios where inventory is constrained and you need to ensure that critical demand is fulfilled before lower-priority demand. Here's a structured explanation: Supply protection is ideal whenever demand exceeds supply and certain customers, channels, or time-sensitive needs must be guaranteed inventory. It ensures strategic fulfillment, revenue protection, and improved customer satisfaction.

Best Business Use Cases for Supply Protection in SAP S/4HANA AATP-

#	Business Use Cases	Details
1	*Strategic Customers / Key Accounts*	• **Scenario:** You have VIP or strategic customers who must receive products even during stock shortages. • **Use:** Protect stock specifically for these accounts so general demand cannot consume it. • **Benefit:** Ensures customer satisfaction and strengthens strategic relationships.
2	*E-Commerce / High-Priority Channels*	• **Scenario:** E-commerce sales are time-sensitive or revenue-critical. • **Use:** Reserve stock for online orders while allowing physical retail or less urgent channels to use remaining stock. • **Benefit:** Maximizes revenue from priority channels without stockouts.
3	*Seasonal / Promotional Products*	• **Scenario:** Limited edition or seasonal products (e.g., Black Friday, Christmas) where demand exceeds supply. • **Use:** Allocate stock based on forecasted demand or promotion schedule. • **Benefit:** Prevents early sell-outs and ensures availability during peak periods.
4	*Product Launches / New Product Allocation*	• **Scenario:** New product launch with limited initial stock. • **Use:** Protect stock for launch campaigns, important geographies, or key distributors. • **Benefit:** Smooth market introduction and avoids customer disappointment.
5	*Multi-Channel / Multi-Location Fulfillment*	• **Scenario:** Multiple distribution centers or sales channels competing for the same inventory. • **Use:** Protect stock at specific locations for priority channels or regions. • **Benefit:** Optimizes supply allocation across the network.

Key Features in SAP S/4HANA AATP for Supply Protection-

#	Key Features	Details
1	*Stock Protection Groups*	• Categorize stock to be reserved for specific segments or channels.
2	*ATP Checks with Supply Protection*	• During order creation, only non-protected stock is available to general demand.
3	*Dynamic Adjustments*	• Stock protection levels can be adjusted based on sales forecasts or live demand.
4	*Integration with Prioritization Rules*	• Combine with rules like customer priority, product priority, or sales channel priority.

Supply protection is used when you want to "reserve" limited stock for certain markets, channels, or customers and enforce that priority already at the ATP check; product allocation is used when you want to cap confirmation quantities against a planned allocation (often forecast-based) per period and characteristic combination. In practice, supply protection protects a *portion* of stock for priority segments, while product allocation controls the *maximum* that can be confirmed for any segment in a given time bucket.

In short - Supply protection guarantees minimum supply for priorities, while product allocation restricts maximum demand.

Supply Protection	Supply Protection guarantees a minimum quantity of a product for priority customers or segments, ensuring they receive at least this reserved amount even in short supply situations.
Product Allocation	Product Allocation is an SAP functionality that restricts the maximum quantity of a product that can be sold or confirmed to a customer, customer group, or channel during a defined period. It ensures fair distribution of limited stock when demand exceeds supply.

Note -

Supply Protection (SUP) is relatively new compared to older capabilities like PAL (Product Allocation), BOL (Backorder Processing), and ABC (Availability-Based Customer Prioritization). While it may seem simple conceptually - "reserve stock for high-priority segments" - its implementation is more complex because it directly affects order fulfillment and inventory availability across multiple channels.

Supply Protection looks simple but is technically sophisticated. The priority is ensuring correctness and reliability in the system before making stock commitments to customers. Once validated, it becomes a powerful tool to manage limited supply strategically, protecting key segments and channels.

Key differences – Product Allocation vs. Supply Protection

1. Product allocation in aATP checks requested quantities against predefined allocation quantities per product, characteristic combination, and period (for example, "Customer group A can get 1000 units per week").
2. It is mainly a limiting mechanism: once the allocation for a bucket is consumed for that combination, further orders are restricted or confirmed later, even if physical stock still exists.
3. Supply protection "protects" stock for priority demand segments; product allocation "limits" confirmations to planned caps for all segments, typically forecast-driven.
4. Supply protection is priority-oriented (who should win in a shortage), while product allocation is usually fairness- or plan-oriented (nobody exceeds their allocated share, regardless of priority).

When to Combine Product Allocation (PAL) and Supply Protection (SUP)-

1. Product Allocation (PAL)

- Purpose: Limits the quantity of product that can be sold to certain customers, customer groups, or sales channels.
- Typical Use Cases:
    - Seasonal promotions
    - Fair distribution of scarce products across regions
    - Limiting high-demand products per customer

2. Supply Protection (SUP)

- Purpose: Reserves stock for specific channels, segments, or strategic customers so other demand cannot consume it.
- Typical Use Cases:

  - Protecting inventory for key e-commerce channels
  - Reserving stock for VIP or strategic customers
  - Allocating limited stock over a season or campaign

3. When to Combine

- Scenario: You have limited stock, want to reserve it for key customers (SUP), and also want to limit how much each customer or segment can order (PAL).
- Example:
  - Black Friday product:
    - SUP: Protect 1,000 units for online sales (e-commerce channel)
    - PAL: Limit each customer to max 2 units per order
- Benefit: Ensures stock is protected and fairly distributed among high-priority customers.

4. Priority During ATP Check

ATP in S/4HANA AATP considers multiple mechanisms, but when both SUP and PAL are active:

Mechanism	Role During ATP	Priority/Order
Supply Protection (SUP)	Determines reserved stock for priority segments or customers	Evaluated first; blocks stock from non-protected demand
Product Allocation (PAL)	Determines how much of available stock can be sold to a specific customer or segment	Evaluated second, after SUP; only the unblocked stock is subject to PAL limits

5. Key Insight:
   - SUP "protects" stock first, ensuring key segments are guaranteed supply.
   - PAL then applies limits on the remaining available stock to ensure fair distribution.

6. Simplified Flow During ATP Check:
   1. Check SUP → reserve stock for priority segment.
   2. Check remaining stock → apply PAL limits for customer/channel.
   3. Remaining stock → available to general demand.

7. Rule of Thumb:
   - Use SUP when you want to reserve stock for high-priority demand.
   - Use PAL when you want to control allocation per customer or channel.
   - They complement each other: SUP ensures strategic protection; PAL ensures fair distribution.

Supply protection guarantees minimum supply for priorities, while product allocation restricts maximum demand.

Supply protection enables priority customers and channels to receive guaranteed allocations during shortages without physically separating inventory.

Example: Supply Protection

Scenario#1

Customer	Supply Protection	ATP Qty
		2500 Units
1000100	1000 Units	
1000101	800 Units	
1000102	200 Units	

In this example, based on the data above, we create a sales order for customer 1000100 with an order quantity of 1200 units. The system will confirm all 1200 units because the total ATP is 2500 units and the total supply protection is 2000 units. For this customer, the current protected quantity is 1000 units, and since they requested 1200 units, the full quantity can be confirmed because sufficient additional stock is available beyond the supply protection for all 3 customers.

Scenario#2

Customer	Supply Protection	ATP Qty
		1800 Units
1000100	1000 Units	
1000101	800 Units	
1000102	200 Units	

In this example, based on the data above, we create a sales order for customer 1000101 with a requested quantity of 1000 units. The system will confirm only 600 units because the total ATP is 1800 units, while the total supply protection is 2000 units. For this customer, the current protected quantity is 800 units; however, since they requested 1000 units, the full quantity cannot be confirmed. This is because 1,200 units are already protected for customers 1000100 and 1000102, leaving only 600 units available for customer 1000101.

Scenario#3

Customer	Supply Protection	ATP Qty
		2200 Units
1000100	1000 Units	
1000101	800 Units	
1000102	200 Units	
1000103		
1000104		
1000105		

In this example, based on the data above, we create a sales order for customer 1000105 with a requested quantity of 500 units. The system will confirm only 200 units because the total ATP is 2200 units, while the total supply protection is 2000 units. For this customer, there is no current protected quantity; since they requested 500 units, the full quantity cannot be confirmed. This is because 2000 units are already protected for customers 1000100, 1000101, and 1000102, leaving only 200 units available for customer 1000105.

Also, in this example, suppose we don't receive orders from customer 1000102 for the allocated supply protection of 200 units. In that case, we can remove the supply protection (change it from 200 to 0) and reallocate it to other customers.

We can also use time-based protection in this supply protection setup. Supply protection guarantees a minimum allocation of available stock to certain customers or regions. However, demand and supply are usually spread across time periods (days, weeks, or months).

- Once the time period ends (the protection "expires"), the quantity that was reserved for that customer/time period is released.
- This means it becomes available to other customers according to ATP (Available-to-Promise) rules.
- Expired allocations free up stock for other customers or regions.
- The original customer can no longer claim their reserved quantity for that period.
- If they still want the product after expiration, they can only order what is still available in ATP, which may be less than their previous protected allocation.

Example

- Total supply = 1000 units
- Customer 1000100 has time-based protection = 200 units for January
- Customer 1000100 does not place an order in January
- January expires → 200 units are released
- Other customers can now use these 200 units in January or February

The release of expired time-based or regular supply protection is a manual activity. This means that once a protected period ends or a customer does not place an order, the system does not automatically release the reserved quantity. A user (typically a planner or superuser) must manually adjust the supply protection to free up the stock for other customers or regions.

Note-

If you have limited supply for a given month and want to ensure a minimum quantity for the respective customers or countries (e.g., Japan, India, China), you can use supply protection. This guarantees a minimum allocation, but you can allocate more if additional ATP (Available-to-Promise) quantity is available after considering the protected quantities for the customers or countries.

We can use both **product allocation** and **supply protection**, but only when there is a **strong business case**. Applying too many rules makes maintenance more difficult, and at both the **superuser and end-user levels**, it can be challenging to identify and resolve issues if problems occur.

Types of Supply Protection in S/4HANA aATP

There are two types of Supply Protection in S/4HANA aATP: Horizontal protection separates supply across groups, while vertical protection prioritizes consumption within a group. Horizontal Protection, also called Core Protection.

Horizontal (Core) Protection	Horizontal (core) protection protects complete groups against each other, for example protecting stock for Germany vs USA vs Japan so one country's demand cannot consume another's protected quota. o Protects peer groups against each other (for example, countries or regions). o Ensures one group's demand cannot consume another group's protected supply.
Vertical (Prioritized) Protection	Vertical (prioritized) protection defines priorities within a group, such as e-commerce vs retail in the same country, where higher-priority channels can consume lower-priority protection but not the other way around. o Defines priority levels within the same group (such as channels or customer types). o Allows higher-priority demand to consume lower-priority protected quantities, but not vice versa.
Combined Use of Horizontal and Vertical Supply Protection	In S/4HANA aATP, horizontal and vertical supply protection can be used together to control supply consumption across groups and within groups at the same time.

Protection Concept Overview

- Horizontal (Core) Protection is the highest-priority protection mechanism.
  It ensures that a defined quantity of supply is protected across all demands of the same (core) priority, independent of customer or sales organization.
- Vertical Protection is optional and used only if required by the business.
  It further protects supply for specific demand segments (e.g., key customers or sales orgs) within the already protected core demand.

This section defines horizontal and vertical protections clearly.

1. Core Protection

Definition:	• Core Protection is the baseline level of protection allocated at the highest level, typically for the entire customer or sales organization.
Purpose:	• It ensures that a minimum guaranteed allocation is always reserved for the customer, regardless of internal priorities among divisions or subgroups.
In the table:	• Row: Core Protection | Customer + Sales Organization • Values: 200, 100, 400 for Jan, Feb, Mar • Interpretation: The customer is guaranteed 200 units in Jan, 100 in Feb, and 400 in Mar. This acts as a safety net for overall customer supply.

2. Vertical Protection

Definition:	• Vertical Protection refers to allocations at a lower level, such as divisions, product lines, or priority segments within the customer.
Purpose:	• It distributes the core allocation further according to internal priorities. For example, some divisions may have higher priority than others.
In the table:	• Rows: Division A1 (Priority 1), Division A2 (Priority 2), Division A3 (Priority 3) • Values for Jan: 100, 100, 100 (and similarly for other months)

- Interpretation: The total allocation for divisions (vertical protection) should typically not exceed the core protection, and within that, higher priority divisions (like A1) get allocated first.

In Short-

- Core Protection = Overall guaranteed allocation for the customer (top-down view).
- Vertical Protection = Allocation breakdown by division or priority (detailed, bottom-up view).
- Priority matters in Vertical Protection: higher priority divisions get their allocation first if supply is limited.

Protection Type	Characteristic	Jan	Feb	Mar
Core Protection	Customer + Sales Organization	200	100	400
Vertical Protection	Division A1 (Priority 1)	100	40	100
Vertical Protection	Division A2 (Priority 2)	100	40	100
Vertical Protection	Division A3 (Priority 3)	100	20	200

Key Points-

- Consumption sequence: Core → Vertical (Priority 1 → 2 → 3)
- Time bucket-based: Protection applies per month; no backward/forward allocation.
- Flexibility: Core guarantees minimum supply; Vertical enables strategic prioritization for specific divisions or segments.

Summary – Horizontal vs Vertical Protection

Protection Type - ***Horizontal (Core)***

Key Purpose - Ensures a minimum quantity of supply is reserved for all demands of the same priority, independent of segment-specific characteristics.

Characteristics / Priority - Customer + Sales Organization

Example from Table (January) - 200 units protected

How It Works - Consumed first. Any demand up to 200 units in Jan is fully covered by Core Protection before Vertical Protection is considered.

Protection Type - ***Vertical (Priority)***

Key Purpose - Further protects supply for specific demand segments within the already protected core supply, based on priority.

Characteristics / Priority - Division (Priority 1 → 3)

Example from Table (January) - Division A1 = 100, A2 = 100, A3 = 100

How It Works - Consumed after Core Protection. Higher-priority divisions (A1) are fulfilled first, then lower priorities (A2, A3) until protected quantity is exhausted.

Supply Protection Configuration – SAP S/4HANA

Supply Protection is part of Advanced Available-to-Promise (aATP) and is executed during the Product Availability Check (PAC).

Prerequisites – Before Configuration, Ensure the Following Settings Are Available

1. **SAP S/4HANA with aATP Activated**
    - Advanced Available-to-Promise (aATP) must be activated in backend customizing.
    - In S/4HANA systems, aATP is usually active by default but must be verified in SPRO.

2. **Materials Use Advanced ATP Availability Check**
    - Materials must be assigned an aATP-enabled availability check group (e.g., 02 – Advanced ATP) in the Material Master (MRP 3 view) or via material type default.

3. **Sales Documents Use Product Availability Check (PAC)**

- Product Availability Check (PAC) must be configured and assigned to relevant sales document types.
- In Scope of Check, under the "Special Scenarios" tab, the indicator "With Supply Protection for Specific Document Type" must be activated.
- To enable Supply Protection during Product Availability Check, the indicator "With Supply Protection for Specific Document Types" must be activated in the Advanced Scope of Check. Without this setting, Supply Protection will not be applied during ATP confirmation.

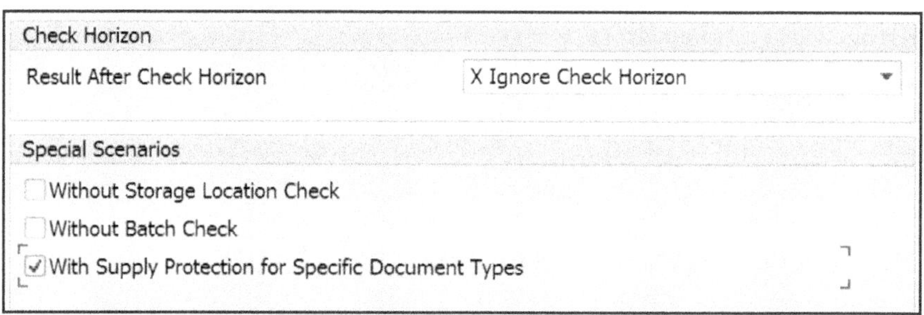

4. **Fiori Launchpad Availability**
   - SAP Fiori Launchpad must be configured and accessible.
   - The Fiori app Manage Supply Protection must be available:
     - App Name: Manage Supply Protection
     - App ID: F4569
   - This app is used to create, maintain, and activate Supply Protection objects.

5. **User Authorization / Business Role**
   - Users must be assigned the business role or an equivalent custom role.
     SAP_BR_ORDER_FULFILLMENT_MANAGER
   - The role must include authorization for aATP and Supply Protection activities.

Under Supply Protection in S/4HANA aATP (2023), Only two Fiori apps are available.

1. **Manage Supply Protection**
   - Used to create, change, and delete Supply Protection Objects

- Defines which supply elements (stock or receipts) are protected and for which demands
- Controls how protected supply is considered during ATP confirmation

2. **Monitor Periodic Supply Protection Maintenance**
    - Used to monitor the background job that periodically updates and recalculates supply protection quantities
    - Provides execution status, logs, and error details
    - Ensures supply protection remains effective and consistent

These two apps together cover the complete lifecycle of Supply Protection:

Setup & maintenance → *Manage Supply Protection*
Monitoring & logging → *Monitor Periodic Supply Protection Maintenance*

No additional Fiori apps exist for Supply Protection in S/4HANA aATP 2023.

Create Supply Protection Object.
Use Fiori App "Manage Supply Protection" → Create:

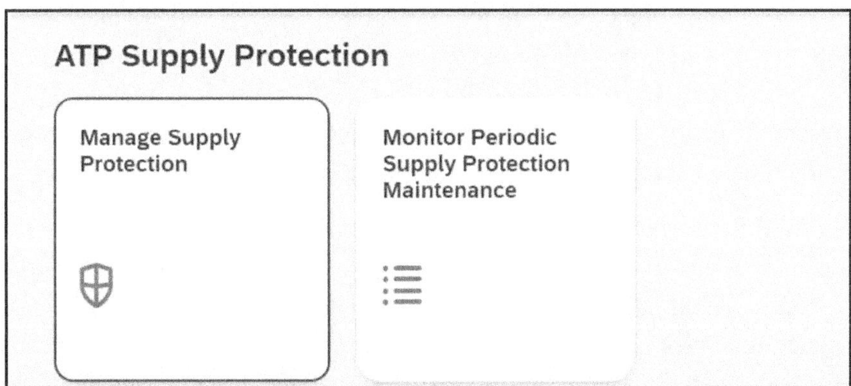

How it works in aATP-

1. Protection objects are created and maintained via the "Manage Supply Protection" / "Configure Supply Protection" Fiori apps, where you define groups, characteristics, protected quantity, time horizon, and planning buckets.
2. During the aATP check, the system evaluates the relevant protection object(s) and confirms quantities only if the protected

quota for that characteristic combination is still available, otherwise it uses unprotected stock or reduces the confirmation.

Create a Planning Object for Supply Protection-

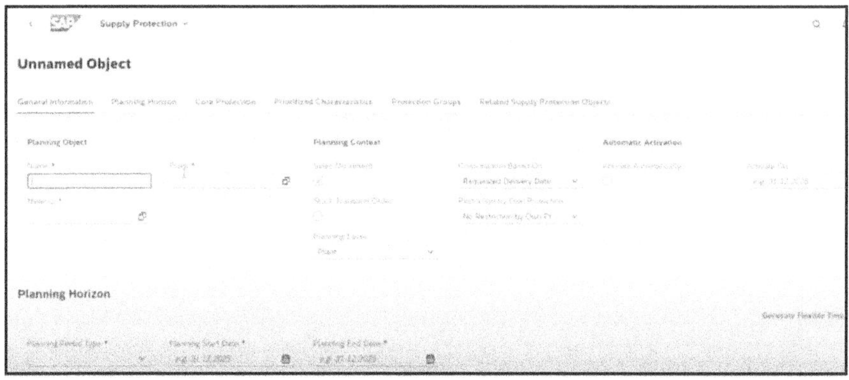

Important Fields Defined While Creating a Planning Object (Supply Protection). When creating a Supply Protection Planning Object using the Manage Supply Protection Fiori app, the following key fields must be defined carefully, as they control how Supply Protection is applied during Product Availability Check.

The planning object in Supply Protection defines which material is protected, where the protection applies (plant or storage location), when it is valid (planning horizon and time buckets), for whom the supply is reserved (based on characteristics and protection groups), and how much supply is protected for each group and time period.

In SAP S/4HANA, Supply Protection planning objects are created for a specific material and plant combination. As Supply Protection works directly within aATP and PAC, protected quantities must be defined at material level. Therefore, separate Supply Protection objects are required for each material. Multiple materials are managed by creating multiple planning objects, typically using mass upload or automation tools for efficiency.

<table>
<tr><th>#</th><th>Key Fields</th><th>Descriptions</th></tr>
<tr><td>1</td><td>Material</td><td><ul><li>Identifies the product for which Supply Protection is defined.</li><li>Only ATP-relevant materials using Advanced ATP can be selected.</li></ul></td></tr>
<tr><td>2</td><td>Plant</td><td><ul><li>Specifies the plant where supply protection applies.</li><li>Supply Protection is plant-specific and does not work across plants.</li></ul></td></tr>
<tr><td>3</td><td>Planning Context</td><td><ul><li>Defines where Supply Protection is applied:<ul><li>Sales Orders</li><li>Stock Transport Orders</li></ul></li><li>Determines which document types will consume the protected quantities.</li></ul></td></tr>
<tr><td>4</td><td>Planning Level</td><td><ul><li>Defines the level at which supply is protected:<ul><li>Plant level, or</li><li>Plant / Storage Location level (if enabled)</li></ul></li><li>Controls the granularity of protection.</li></ul></td></tr>
<tr><td>5</td><td>Consumption Based On</td><td>The “Consumption Based On” field in a Supply Protection (SuP) object determines how the system reduces the protected quantity when demand occurs.
There are two options:<ol><li>Requested Delivery Date (RDD) – The protected quantity is consumed based on the requested delivery date of the order.<ul><li>This is the standard option in SAP.</li></ul></li><li>Material Availability Date (MAD) – The protected quantity is consumed based on the material availability date, i.e., when the material is actually available to fulfill the order.</li></ol>Selecting between RDD and MAD allows you to control whether SUP consumption aligns with customer requested dates or with actual supply availability.</td></tr>
<tr><td>6</td><td>Restriction by Own Protection</td><td>Supply Protection has four restriction scenarios, which determine how protected supply interacts with demand:
Possible Scenarios<ol><li>No Restriction by Own Protection<ul><li>The demand is not restricted at all by the defined Supply Protection.</li><li>Usually occurs if characteristics do not match or the object is inactive.</li></ul></li><li>Restriction Outside Planned Protection<ul><li>Demand falls outside the defined protection group or characteristics.</li><li>Protected supply is not applied because the demand does not match the planning object.</li></ul></li></ol></td></tr>
</table>

		3. **Restriction Outside Planning Horizon** • Demand occurs **outside the start/end dates** of the planning horizon defined in the SuP object. • Supply protection is only valid within the planning horizon. 4. **Restriction Outside Time Bucket** • Demand falls **in a time bucket not covered** by the SuP object. • Protected quantities are allocated per time bucket; demand outside these buckets is not restricted.
7	*Automatic Activation*	The **"Automatic Activation"** (or **"Activate Automatically Check"**) controls whether a newly created or updated Supply Protection (SuP) object is **automatically activated** in the system after saving. Batch Program – "RATPSUP_PERIODIC_MAINTENANCE" This program is the standard SAP background job used for activation and deactivation of Supply Protection (SUP) rules based on their validity period. • Activate SUP rules that have become valid. • Deactivate SUP rules that have expired. Essentially, it automates the dynamic management of supply protection so you don't have to manually toggle rules.
8	*Activate On*	The **"Activate On"** field determines the **date from which the Supply Protection (SuP) object becomes active** and is considered in ATP/PAC checks.
9	*Planning Horizon (Start Date / End Date)*	• Determines the time period for which supply is protected. • Time buckets must fully cover this horizon with no gaps. • Cannot be changed after the planning object is activated.
10	*Time Bucket Profile*	• Defines how the planning horizon is divided: o Daily o Weekly o Monthly • Used to allocate protected quantities over time.
11	*Characteristics*	Define Characteristics (Very Important) - Supply Protection works by characteristic matching. Rules: • At least 1 characteristic is mandatory • Characteristics must exist in the Characteristic Catalog • Sales order must carry the same characteristic values • Commonly used characteristics: o Customer Group

		o Sold-to Party o Ship-to Party o Delivery Priority • Demand is protected only if characteristic values match.
12	*Protection Groups*	A Protection Group = **Characteristic combination** • Combination of characteristic values forming a protection group. • Each group represents a segment of demand eligible for protection. Example: • Customer Group = "RETAIL" • Delivery Priority = "01" For each group: • Assign **priority** (for vertical protection) • Assign **protected quantity** per time bucket Types: • **Core (Horizontal) Protection** – single group • **Prioritized (Vertical) Protection** – multiple groups with priority
13	*Priority*	• Used for prioritized (vertical) Supply Protection. • Lower number = higher priority. • Determines which demand consumes protected quantities first.
14	*Protected Quantity*	In the Protection Group, the protected quantity is maintained based on the defined planning parameters, such as weekly or monthly time buckets. The protected quantities are allocated per time bucket and are consumed according to the selected planning period. • Quantity reserved for a protection group within a time bucket. • Prevents lower-priority or non-matching demand from consuming supply.
15	*Unit of Measure*	• Unit in which protected quantities are defined. • Must be consistent with the material's base unit.
16	*Status (Draft / Active)*	• Only Active planning objects are considered during ATP. • Once activated: o Quantities can be adjusted o Characteristics and planning horizon cannot be changed

This is a detailed explanation of the field "Automatic Activation."-
The Automatic Activation field controls whether a Supply Protection rule is activated and deactivated automatically by the background job RATPSUP_PERIODIC_MAINTENANCE, based on its validity period. This program is the standard SAP background job used for activation and deactivation of Supply Protection (SUP) rules based on their validity period.

- Activate SUP rules that have become valid.
- Deactivate SUP rules that have expired.

Essentially, it automates the dynamic management of supply protection so you don't have to manually toggle rules. Steps to setup.

1. Program: RATPSUP_PERIODIC_MAINTENANCE
2. Usually scheduled as a periodic background job (via SM36) - daily, hourly, or per your business need.
3. On execution, the program:

    - Scans all SUP rules in the system.
    - Activates rules whose start date/time ≤ current date/time and are not already active.
    - Deactivates rules whose end date/time < current date/time and are currently active.

Business Scenario-
Product FG10101 has limited supply in Q1 so the business wants to: Protect stock for E-commerce in January and February and Release all stock to general customers in March. This is configured using Supply Protection (SUP) with validity dates.

SUP Configuration (Example)-

SUP Rule	Protected Segment	Validity
SUP_01	E-commerce channel	01-Jan → 28-Feb

At configuration time:

- SUP rule exists in the system
- BUT it is not automatically active/inactive by date
- That's where RATPSUP_PERIODIC_MAINTENANCE comes in

Role of RATPSUP_PERIODIC_MAINTENANCE by Month

RATPSUP_PERIODIC_MAINTENANCE - ensures that Supply Protection rules are activated and deactivated according to their validity periods so that ATP checks always use the correct, time-relevant protection status.

Month	What the program does:	System impact:
January *Date: 01-Jan* *Job runs (e.g., midnight)*	• Reads SUP rules • Sees: Start date = 01-Jan End date = 28-Feb • Activates SUP_01	• Stock is now protected for E-commerce • Retail / wholesale orders can only use non-protected stock • ATP checks respect SUP protection *Business goal achieved: Priority channel is protected*
February *Date: Any day in February* *Job runs periodically*	• Checks SUP_01 • Finds it already active and still within validity • No status change	• Supply Protection remains active • ATP continues to block protected stock from non-priority demand *Stable behavior during the protected period*
March *Date: 01-Mar* *Job runs (e.g., midnight)*	• Checks SUP_01 • Finds: End date = 28-Feb (expired) • Deactivates SUP_01	• Protected stock is released • All customers/channels can now consume the inventory • ATP checks no longer apply SUP restriction *Business goal achieved: Stock fully released after peak period*

Note –

A Supply Protection (SuP) planning object is always created for one specific Material and one specific Plant.

SAP Designed Supply Protection This Way because Supply Protection is tightly integrated with aATP and Product Availability Check (PAC), which themselves work at material and plant level. The reasons are:

1. *ATP logic is material-specific*	• Stock, receipts, and requirements are calculated per material • Protected quantities must be checked against the same ATP elements
2. *Supply is plant-specific*	• Inventory and receipts belong to a plant • Protection rules cannot span multiple plants
3. *Precise demand matching*	• Characteristics (customer group, priority, etc.) are applied to demand for a specific material • This avoids ambiguity in confirmations
4. *Performance and accuracy*	• Keeping SuP objects granular ensures faster ATP checks and correct allocation of supply

How to Manage Multiple Materials in Supply Protection. Do We Need Separate SuP Objects for Each Material? **Answer is - Yes.**

In SAP S/4HANA, Supply Protection (SuP) planning objects are always created for a specific Material–Plant combination. Consequently, each material requires its own SuP object for every plant where protection is needed. To manage multiple materials efficiently and ensure scalability, it is recommended to use mass creation tools and standardized templates, maintaining consistency in planning horizons, time buckets, characteristics, and protection rules across all objects.
If you want Supply Protection for 10 materials, you need 10 separate Supply Protection planning objects per plant. Example of Supply Protection Planning Objects as below.

Material	Plant	Number of SuP Objects
ABC	1000	1
XYZ	1000	1
ABC	2000	1

- Each row represents one Supply Protection (SuP) planning object.
- Even if the same material exists in multiple plants, a separate SuP object is required per plant.
- This ensures that protected quantities are correctly applied at the plant level during ATP.

Best Practices for Managing Many Materials in Supply Protection

Mass Creation via Upload / Automation. When managing a large number of materials, it is recommended to create Supply Protection (SuP) objects in bulk rather than manually. This ensures consistency, scalability, and efficiency. Recommended approaches include:

- Fiori Mass Upload – Available in certain S/4HANA releases for bulk creation of SuP objects.
- Custom Programs / APIs – Use BAPI or OData services to automate creation and maintenance.
- Excel-Based Templates – Prepare templates with material, plant, characteristics, time buckets, and protected quantities for bulk upload. ***This approach is highly recommended for large material volumes to maintain consistency, reduce errors, and speed up configuration.***

This is a comparison between Supply Protection & Product Allocation.

Feature	Supply Protection (SuP)	Product Allocation (PA)
Level	Material + Plant	Material + Plant
Mass Handling	Required	Required
Characteristics	Mandatory (e.g., customer, delivery priority)	Optional
Purpose	Protect specific supply for defined demand (ensures high-priority orders get supply). Represents a **minimum protection** of supply.	Limit the quantity available for certain customers, regions, or sales channels. Represents a **maximum allocation** of supply.
Integration	Works with aATP / PAC during ATP check	Works with aATP / PAC as a restriction during ATP
Time Buckets	Mandatory	Optional

Key Takeaways:

- Both SuP and PA operate at material + plant level.
- SuP requires characteristics for demand matching; PA can be simpler without characteristics.
- SuP focuses on protecting supply for prioritized customers or groups, while PA is more about capping allocations.

Also, the following parameters are defined while creating a planning object in Supply Protection:

In Supply Protection, Core Protection reserves supply for defined customers or organizational units, while Vertical Protection applies priority-based consumption to ensure higher-priority demand is fulfilled first.

1. Core Protection - In Core Protection, you can define up to two characteristics with specific values, for example:
    - Customer
    - Sales Organization / Distribution Channel / Division
2. Vertical Protection (Priority Protection)
    - Vertical Protection is maintained when prioritization is required.
    - Along with Core Protection, you can also define Vertical Protection.
    - In Vertical Protection, you can maintain one characteristic (for example, Division) and assign priorities.

Example – Core and Vertical Protection in Supply Protection.

Below are clear examples based on your Supply Protection setup to show how Core and Vertical Protection behave in different scenarios for January.

Supply Protection Setup (Monthly Buckets)

Protection Type	Characteristic	Jan	Feb	Mar
Core Protection	Customer + Sales Organization	200	100	400
Vertical Protection	Division A1 (Priority 1)	100	40	100
Vertical Protection	Division A2 (Priority 2)	100	40	100
Vertical Protection	Division A3 (Priority 3)	100	20	200

- **First row**: Core Protection based on **two characteristics** (Customer + Sales Organization) with **monthly planning buckets**.

- **Next three rows**: Vertical Protection based on **Division**, with priority applied (A1 → A2 → A3, assuming A1 is highest priority). Priority order: **A1 → A2 → A3**

Once you have defined all the parameters for a Supply Protection (SuP) object, save and activate it:

- When you save, the object is initially in Planning Status (Draft).
- Activation is required for the object to be considered in ATP/PAC checks during sales order processing.

Supply Protection first applies Core Protection, then consumes Vertical Protection based on division and priority, with any remaining demand dependent on unprotected stock availability.

Scenario 1: Order Quantity = 180 units (January)

1st Sales Order Created with assumption-

- The sales order matches the Core Protection characteristics (Customer + Sales Org).
- The sales order belongs to Division A1 (highest priority).
    - Month: January & Order Quantity: 180 units

Step 1: Core Protection	In this Core Protection check, the core protected quantity for January is 200 units and the order quantity is 180 units. Since 180 ≤ 200, the order is fully covered by Core Protection.
Step 2: Vertical Protection	Vertical Protection is not consumed because Core Protection already fully covers the order quantity. Vertical Protection is applied only when the demand exceeds the available Core Protection.
Result	• 180 units are confirmed • Remaining Core Protection for January: 200 – 180 = 20 units

Final Outcome is –

Confirmed Quantity-	180 units
Supply Protection Used-	Core Protection only
Vertical Protection-	Not triggered
Remaining Protection (Jan):	• Core Protection: 20 units • Vertical Protection: Unchanged

Key Points-

- Core Protection is always consumed first.
- Vertical Protection is used only if demand exceeds Core Protection.
- Supply Protection applies only within the January time bucket.
- No backward or forward consumption from Feb or Mar.

Scenario 2: Order Quantity = 250 units (January)

Step 1: Core Protection	• Core Protection available = 200 units • Order quantity = 250 units • 200 units are confirmed using Core Protection • Remaining demand = 250 – 200 = 50 units
Step 2: Vertical Protection	Next, system checks Vertical Protection based on priority. Assume order matches Division A1: • Vertical A1 available = 100 units • Remaining demand = 50 units → consumed from Vertical A1 • Vertical A1 remaining = 100 – 50 = 50 units Vertical A2 and A3 are not touched, since demand is already fulfilled.
Result	• Total confirmed quantity: 250 units • Remaining protected quantities: as above • All demand covered from Core + Vertical Protection

Final Outcome is -

Protection Type	Jan Qty Used	Remaining Qty
Core Protection	200	0
Vertical A1	50	50
Vertical A2	0	100
Vertical A3	0	100

Key Points-

1. Core Protection is always consumed first.
2. Vertical Protection is consumed based on priority and matching characteristics.
3. Any remaining demand after Core + Vertical would rely on unprotected stock.
4. Supply Protection applies only to the January time bucket — no backward/forward consumption.

Scenario 3: Order Quantity = 220 (Division A1)

Step 1: Core Protection	• Core Protection (Jan) = 200 units • Order Qty = 220 units • 200 units are confirmed using Core Protection • Remaining demand = 20 units
Step 2: Vertical Protection	• Remaining 20 units are covered by Vertical Protection – Division A1 • Vertical A1 remaining after consumption = 80 units
Result	• 220 units confirmed • Core Protection remaining = 0 units • Vertical Protection (A1) remaining = 80 units

Scenario 4: Order Quantity = 350 units (Division A1)

Step 1: Core Protection	• Core Protection = 200 units • Order Qty = 350 units • 200 units confirmed • Remaining demand = 150 units
Step 2: Vertical Protection	• Vertical A1: 100 units → fully consumed • Remaining demand = 50 units • Vertical A2: 50 units consumed • Vertical A2 remaining = 50 units
Result	• 350 units confirmed • Core Protection remaining = 0 units • Vertical A1 remaining = 0 units • Vertical A2 remaining = 50 units • Vertical A3 unchanged

Scenario 5: Order Quantity = 220 units (Division A2)

Step 1: Core Protection	• Core Protection = 200 units • Remaining demand = 20 units
Step 2: Vertical Protection	• Order belongs to Division A2 • 20 units consumed from Vertical Protection – A2
Result	• 220 units confirmed • Vertical A2 remaining = 80 units • Vertical A1 and A3 unchanged

Scenario 6: Order Quantity = 350 units (Division A3)

Step 1: Core Protection	• Core Protection = 200 units • Remaining demand = 150 units
Step 2: Vertical Protection	• Division A3 protection = 100 units → fully consumed • Remaining demand = 50 units • No further matching vertical protection available
Result	• 300 units confirmed (200 Core + 100 Vertical A3) • 50 units unconfirmed (unless unprotected stock is available)

Key Rules-

- Core Protection is always consumed first
- Vertical Protection is consumed based on priority and matching characteristic
- Protection is applied only within the same month bucket
- No backward/forward consumption across months

Scenario 7:

Supply Protection Setup (Monthly Buckets) How does this work if I want to set it up for 5 customers? Do I need to create separate planning objects, or can I combine them into a single planning object? What is the best practice in aATP 2023? Plant: P100, Material: FG10101, Core Protection: For 5 customers based on characteristics – Sold-to and Sales Organizations.

Recommendation for this case-

Since you have 5 customers and likely want customer-specific quantities, the best approach is:

- If quantities differ per customer: create 5 planning objects (one per customer).

- If quantities are identical and change rarely: use 1 planning object with characteristic-based protection for all 5 customers.

These are some common issues and checks – Supply Protection.

Issue	Possible Check / Cause
SuP not working	Product Availability Check (PAC) not assigned to the sales document
No restriction applied	Characteristics in the SuP object do not match the sales order / demand
Full confirmation received	Wrong availability check group assigned to the material (aATP not enabled)
Protection ignored	SuP object is not activated in the system

Important Note –
Unlike Product Allocation, Supply Protection does not support backward or forward consumption across planning periods. Protected quantities in SuP are consumed only within the defined time bucket and based on actual available supply
In **Supply Protection (SuP)**, the consumption logic **does not work like Product Allocation (PAL)** with backward or forward allocation across periods.

- In **Product Allocation (PAL)**, if a sales order is created for a specific period (for example, **February**) and the allocated quantity for that period is insufficient, the system can **consume quantities from backward or forward periods**, depending on the configuration settings.
- In **Supply Protection (SuP)**, this behavior **is not applicable**.
    - SuP **does not shift consumption across time buckets** (no backward or forward consumption).
    - The protected quantity is consumed **only within the relevant time bucket** based on the selected consumption date (RDD or MAD).
    - SuP consumption is driven by **actual available stock and receipts**, not by reallocating quantities from other periods.

Key Difference-

PAL	SuP
Allows backward and forward allocation across periods (if configured).	No backward or forward consumption; protection applies strictly within the defined time bucket and consumes from available supply.

Top of Form

Bottom of Form

ATP Protection Restriction Scenarios-

Here, additional scenario-based details are provided to clarify the concept. ATP protection is used to control the allocation, exclusivity, and timing of supply. By combining different restriction settings, organizations can guarantee minimum quantities to key customers, prevent unauthorized consumption, and strictly enforce planning horizons and time buckets during order confirmation.

In this example, material FG10101 in plant P100 has an available ATP quantity of 1000 units in January. Core Protection is defined at Customer + Sales Organization level for customer 10010000 and sales organization S100, with protected quantities of 200 units in January, 100 units in February, and 400 units in March.

1. Restriction by Own Protection

When *Restriction by Own Protection* is active, the customer can consume only up to the protected quantity within the time bucket. For January, the customer is restricted to 200 units, even though total ATP is available. Any order quantity exceeding 200 units is not confirmed. In this case, protection acts as a hard upper limit.

Orders consume their own protected quantity, and once exceeded, further confirmation is restricted.

Scenario

- Customer 10010000 + S100 has 200 units protected in Jan.

Execution-

Order	Qty	Protected Consumption	Result
SO1	50	50	Confirmed
SO2	200	Only 150 units protected left	⚠150 confirmed, 50 rejected
SO3	250	0	Rejected
SO4	500	0	Rejected

Explanation-

- Own protection limit = 200 units
- Orders beyond 200 units are restricted

2. No Restriction by Own Protection

When *Restriction by Own Protection* is inactive, the protected quantity represents a minimum guaranteed quantity, not a limit. The customer is guaranteed confirmation of 200 units in January, but can consume additional quantities from general ATP as long as stock is available. All January sales orders are fully confirmed because ATP is sufficient.

Customer can consume beyond protected qty using general ATP.

Scenario

- Own protection exists, but restriction indicator is OFF

Execution

Order	Qty in Units	Result
SO1	50	Confirmed
SO2	200	Confirmed
SO3	250	Confirmed
SO4	500	Confirmed

Explanation

- Protected qty (200) units is only a guarantee, not a limit
- Total ATP qty (1000) units is sufficient → no restriction

3. Restriction Outside Planned Protection

With *Restriction Outside Planned Protection* active, protected quantities can be consumed **only by sales orders that match the protection characteristics** (Customer + Sales Organization). Other customers or sales organizations cannot use this protected stock, even if general ATP exists. The protection is therefore **exclusive to the planned customer**.

Other customers cannot consume this customer's protected quantity.

Scenario

- Another customer **20020000** places order in Jan

Customer	Order Qty	Result
20020000	150units	Rejected

Why?

- 200 units are **reserved for Customer 10010000**
- Even though ATP exists, **protected stock is blocked**

4. Restriction Outside Planning Horizon

When *Restriction Outside Planning Horizon* is active, sales orders requested **outside the defined planning horizon** (January to March in this example) are not confirmed. The system does not allow ATP confirmation beyond the months for which protection is planned, even if unrestricted ATP is available.

Orders outside defined protection horizon are restricted.

Scenario

- Protection horizon = **Jan–Mar**
- Customer places **April** order

Month	Order Qty	Result
April	100 units	Rejected

Explanation

- No protection maintained beyond **March**
- System restricts confirmation **outside planning horizon**

5. Restriction Outside Time Bucket

With *Restriction Outside Time Bucket* active, protected quantities are valid **only within their defined time bucket** (monthly in this case). January protection cannot be used for February or any other month. Sales orders must fall within the same bucket to consume the protected quantity.

Orders must fall within defined time bucket (month).

Scenario

- January protection bucket = **Jan 1–31**
- Order requested on **Feb 2**

Requested Date	Qty	Result
Feb 02	100 units	Rejected

Explanation

- January bucket protection **cannot be used in February**
- February has only **100 units protected qty**

Summary

Scenario	Key Idea
Restriction by Own Protection	Customer limited to own protected qty
No Restriction by Own Protection	Protected qty is a guarantee only
Restriction Outside Planned Protection	Others can't use protected stock
Restriction Outside Planning Horizon	Orders beyond horizon rejected
Restriction Outside Time Bucket	Month-specific protection enforced

Perfect! Let's create a **complete table with "Meaning" and "Example"** for **all 5 scenarios**, based on your style and your example. I'll also provide clear, concise explanations for each.

Scenario	Meaning	Example
Restriction by Own Protection	The customer can only consume up to their **own protected quantity** within the time bucket. Any demand exceeding the protection is restricted.	Customer: CUST_A Sales Org: SO1 Month: January Requested Quantity: 250 Jan Core Protection = 200 **Analysis:** 250 > 200 → Only 200 confirmed, 50 restricted Result: Order partially confirmed (200), remaining restricted
No Restriction by Own Protection	The customer is **guaranteed the protected quantity**, but can consume additional ATP if available. Protection is a minimum guarantee, not a limit.	Customer: CUST_A Sales Org: SO1 Month: January Requested Quantity: 150 Jan Core Protection = 200

		Analysis: 150 ≤ 200 → Allowed Result: Order fully confirmed, no restriction
Restriction Outside Planned Protection	Only sales orders matching the **protection characteristics** (customer, sales org, etc.) can consume the protected stock. Others are restricted.	Customer: CUST_B Sales Org: SO2 Month: January Requested Quantity: 50 Jan Core Protection for CUST_A = 200 **Analysis:** Order does not match characteristics → restricted Result: Order not confirmed
Restriction Outside Planning Horizon	Protected quantities can be consumed **only within the defined planning horizon** (months planned). Orders outside this horizon cannot consume protected stock.	Customer: CUST_A Sales Org: SO1 Month: April Requested Quantity: 100 Jan–Mar protection = 200,100,400 **Analysis:** April is outside horizon → restricted Result: Order not confirmed
Restriction Outside Time Bucket	Protected quantities can be consumed **only within their assigned time bucket** (e.g., January). Orders in other months cannot use this protection.	Customer: CUST_A Sales Org: SO1 Month: February Requested Quantity: 50 Jan Core Protection = 200 **Analysis:** Feb ≠ Jan → restricted Result: Order not confirmed

Key Takeaways from Table:

1. Restriction by Own Protection → limits quantity to protection.
2. No Restriction by Own Protection → protection is minimum guarantee.
3. Restriction Outside Planned Protection → enforces customer/sales org exclusivity.
4. Restriction Outside Planning Horizon → enforces valid months only.

5. Restriction Outside Time Bucket → enforces bucket-specific allocation.

Functionality of field "Restriction by Own Protection"-
Restriction by Own Protection:

- *No Restriction by Own Protection*
- *Restriction Outside Planned Protection*
- *Restriction Outside Planned Horizon*
- *Restriction Outside Time Bucket*

In SAP S/4HANA Advanced Available-to-Promise (aATP) 2023, the field "Restriction by Own Protection" on the *Supply Protection* (SuP) planning object controls how a demand is restricted against the supply protection quantities of its own protection group — that is, whether and when the system should treat its own protected quantities as a restriction during the availability check.

"Restriction by Own Protection" is a field name on the *Supply Protection* (SuP) planning object. But it's not just a label; it has functional significance in the ATP calculation.

There are four (4) options available for this setting.

Available values for Restriction by Own Protection

01 – No Restriction by Own Protection	Own protection is not treated as a restriction for matching demand.
02 – Restriction Outside Planning Horizon	Protects demand outside the defined planning horizon of the protection object.
03 – Restriction Outside Time Bucket	Own protection in buckets other than the current one (for the requested date) restricts demand.
04 – Restriction Outside Planned Protection (introduced/enhanced in 2023)	Applies restriction logic also based on the remaining protected quantity of the bucket and other planned protection periods.

These settings influence how the ATP engine calculates *restrictions* for the demand matching a supply protection group — especially in how it considers own protection quantities across time buckets and planning horizon.

This field represents-	• It controls whether and how the demand is restricted by its own protected quantities. • In other words, when a demand tries to consume stock that is under supply protection for its own group, this field determines if that protection should block or restrict the availability.
Functionally	• It dictates how the system treats the *demand's own protected supply* during ATP.
Where it is used	• On the Supply Protection planning object (transaction / configuration for SuP). • In advanced ATP calculations, especially for multi-level protection and bucketed supply planning.
Its behavior	• The field itself doesn't store numbers, but the value you select (01, 02, 03, or 04) determines how the system considers own-protected quantities. • Each option adjusts time horizon, bucket-level, or planned protection logic during the availability check.
Values	4 options (01–04), each giving slightly different restriction behavior.
Default Value	• **01 – No Restriction by Own Protection** This means that, **by default**, the system does **not** treat the protected quantities of the *same protection group* as a restriction during ATP. Only protection from other groups or higher-level protection applies as a restriction by default. So, in a fresh system or without customization, "01 – No Restriction by Own Protection" is the standard selection.

Challenges and Shortcomings of SAP S/4HANA aATP 2023 – Supply Protection

Here's a detailed overview of the challenges and shortcomings of SAP S/4HANA aATP 2023 Supply Protection functionality.

While SAP S/4HANA aATP 2023 Supply Protection is powerful for guaranteeing allocations, prioritizing customers, and controlling order confirmation, its main challenges are:

1. Setup and maintenance complexity for multiple customers and verticals
2. Strict time bucket and planning horizon restrictions

3. Limited reporting and transparency for protection consumption
4. ATP consumption rules can be confusing (Restriction ON vs OFF)
5. Scalability/performance issues in high-volume environments

Best Practice Tip: To reduce complexity, organizations often:

1. Limit the number of vertical protection levels
2. Use characteristic-based planning objects only when necessary
3. Implement custom dashboards for protection monitoring

1	*Complexity in Setup*	• Core vs Vertical Protection: Configuring Core Protection and multiple levels of Vertical Protection (by division, region, customer priority) can be complicated. • Multiple Characteristics: When protection is maintained for several characteristics (Customer + Sales Org + Division + Material), setup and maintenance become cumbersome. • Planning Objects: For multiple customers with different protection quantities, each may require a separate planning object, increasing system complexity and effort. Example: Setting up Core Protection for 5 customers with different quantities requires 5 separate planning objects, which is difficult to maintain if quantities change frequently.
2	*Limited Flexibility*	• Time Bucket Restriction: Supply Protection applies strictly to the defined time bucket (e.g., monthly). Partial carryover or rolling allocation is not supported out-of-the-box. • Planning Horizon Restriction: Protection cannot be applied beyond the defined horizon unless manually extended. • Static Allocation: Once protection is set, it cannot dynamically adjust based on changing demand or stock shortages without manual intervention.
3	*ATP Consumption Confusion*	• Restriction ON vs OFF: Many users find it confusing when restriction is ON, because remaining ATP outside protection cannot be used. Misunderstanding leads to unexpected order confirmations.

		• Vertical Protection Priority: If multiple vertical protections exist, priority rules may not always align with business expectations. Example: If Vertical A1 is fully consumed, the system may allocate from Vertical A2, even if the business prefers certain divisions to always have priority.
4	*Reporting and Visibility Limitations*	• Limited Analytics: Out-of-the-box reporting on how protection is consumed (Core vs Vertical, remaining quantities, time bucket usage) is limited. • Complex Tracking: Tracking remaining protection for multiple customers across multiple months is cumbersome without custom reports.
5	*Maintenance Challenges*	• Frequent Updates Needed: Protected quantities often need frequent adjustment due to changing demand patterns, which increases manual effort. • Error-Prone: Incorrectly assigned characteristics or planning objects can lead to orders being restricted unexpectedly or protection being underutilized.
6	*Integration Challenges*	• Global ATP Scenarios: In multi-plant or multi-company setups, coordinating supply protection across sites can be difficult. • Interaction with Other ATP Checks: Conflicts may arise when using other ATP functions like Product Allocation, Backorder Processing, or Supply Chain Contracts.
7	*Performance Considerations*	• High Volume Demand: In scenarios with thousands of customers, materials, and verticals, ATP calculation with supply protection can be resource-intensive. • Real-Time Confirmation Delays: For high-volume sales orders, real-time checks may slow down order confirmation if many protection rules exist.

SAP Best Practices-

Area	Best Practice
Core vs Vertical	Use Core for key guarantees; minimal vertical levels
Planning Objects	Separate per customer if quantities differ; combine if identical
Time Buckets	Align with monthly planning cycles; avoid long horizons
Restriction Settings	ON for high-priority customers; OFF for flexible allocation
Monitoring	Track consumption and remaining protection regularly

Integration	Coordinate with Product Allocation, Backorder Processing, and Global ATP
Performance	Limit planning objects and verticals; schedule batch updates
Review	Regularly adjust protection to match demand patterns

14 SBC- Supply Based Confirmation

In aATP, SBC stands for Supply Creation-Based Confirmation, a function that lets aATP trigger planning (PP/DS) to create new supply when existing stock and receipts are not enough to confirm a sales order.

Supply-Based Confirmation - It is one of the newer and most important aATP capabilities introduced to improve delivery date accuracy, reduce overselling, and ensure realistic confirmations based on actual supply behavior.

Supply-Based Confirmation (SBC) is an aATP functionality that determines feasible confirmation dates based on:

- Actual supply
- Planned receipts
- Lead times
- Capacity-based supply situations
- Product availability across time buckets

It ensures that ATP does not just look at current stock but also how supply evolves over time.

Multiple planning systems are available, and in this context, we are specifically focusing on the PP/DS planning system.

In this case, If the required supply is not available, the system automatically creates supply elements with an appropriate date, allowing the requirement to be confirmed. This means the requirement receives confirmation. These supply elements - such as planned orders or purchase requisitions/purchase orders - are generated according to the material's procurement type (E for in-house production or F for external procurement).

If the procurement type is **E (in-house production)**, the system generates a **planned order**. If the procurement type is **F (external procurement)**, the system generates a **purchase requisition**.

The main function of SBC is that, instead of waiting for the full planning run, the system can give a rough estimate to the customer. Based on the actual planning run, this estimate may differ slightly - typically by 1 or 2 days.

SBC is used when you want the ATP to check itself to drive production or procurement (for MTO and MTS scenarios) instead of only consuming existing supply, similar to "capable-to-promise (CTP)" in APO.

- SBC links aATP with PP/DS so that, during the availability check, the system can not only look at current and planned supply but also plan and create additional supply elements (planned orders, purchase requisitions, stock transfer requisitions) for the missing quantity.
- The result of the check (confirmed quantity and dates) is based on both existing supply and the newly planned supply and is shown in the Review Availability Check Result and BOP monitoring apps.
- SBC uses a Supply Demand-Based Capability Check (SCC) that can run either:
    - Product Availability Check (PAC) with supply creation, or
    - Production Planning-Based Availability Check (PPAC) with supply creation.
- PAC/PPAC first check what can be confirmed from current time-series/PPDS supply; for the remaining open quantity, PP/DS plans new supply across BOM levels and capacities, returns dates to aATP, and aATP creates the final confirmation and delivery proposal.
    - *PP/DS (Production Planning and Detailed Scheduling),*
    - *PAC = Product Availability Check*
    - *PPAC = Production Planning Availability Check*

There are multiple planning systems that can be integrated with SAP S/4HANA, including PP/DS, IBP, MRP Live, and Ariba Supply Planning. APO can also integrate but is considered a legacy solution and is not part of S/4HANA.

Planning Systems-	Details
PP/DS	Embedded in S/4 HANA
MRP Live	Native in S/4 HANA
IBP	Integrated with S/4 HANA via CPI/SDI
Ariba Supply Planning	Integrated, but limited use cases

APO	*Not part of S/4HANA*; can integrate but is being phased out in favor of IBP + PP/DS

PP/DS (Production Planning and Detailed Scheduling) is used for short-term, operational planning, while IBP (Integrated Business Planning) supports strategic and tactical planning. These systems can work together with SAP S/4HANA to enable end-to-end supply chain planning and execution.

SBC is set up through configuration and master data preparation. The responsibility for this setup lies with the PP/DS consultant and the aATP consultant, who handle the respective configuration and master data activities, including the functional and technical linkage between the two.

How SBC Works - SBC creates a time-based supply profile. For each time bucket, it looks at:

- Available stock
- Incoming purchase orders
- Planned / production orders
- Supply constraints
- Demand already committed
- Lead times and delays

Using this, SBC calculates:

- Earliest feasible delivery date
- Feasible quantities
- Which supply bucket can cover the request
- Whether to split confirmation across dates

Key Features of SBC-

- Time-phased supply modeling
- Bucket-based availability check
- More accurate confirmation dates
- Respects supply constraints & variability
- Aligns with real procurement/production lead times
- Integrates with PAL, Supply Protection, BOP
- Real-time calculation using HANA engine

When Do You Use SBC? - Use SBC when:

- You face misleading / unrealistic delivery dates
- Supply is volatile (long lead-time, unreliable suppliers)

- You want promise dates tied to true supply reality
- You want to avoid backorder chaos
- You need feasible confirmations for future supply

Industries: Pharma, Electronics, Consumer Goods, MTO/CTO manufacturing.

End-to-End Flow: aATP, SBC, PP, and PP/DS Interaction Process During Order Confirmation-

1. aATP Initiates Supply & Demand-Based Capability Check (SCC):
   When a sales order is entered or saved, SBC runs either the Product Availability Check (PAC) with supply creation or the Production Planning–Based Availability Check (PPAC) with supply creation. These checks evaluate existing supply using ATP time series or PP/DS net requirements.

2. Identify Shortfall and Trigger PP/DS:
   If PAC/PPAC cannot fully confirm the requirement, aATP sends the open quantity (including multi-level BOM explosion) to embedded PP/DS for net requirements calculation, capacity checks, and scheduling simulation.

3. PP/DS Plans Temporary Supply:
   PP/DS aligns demand and supply across all BOM levels according to planning procedures. It creates temporary planning elements—such as planned orders, purchase requisitions, and stock transfers—while considering capacities and component availability. PP/DS then returns feasible dates and quantities to aATP.

4. aATP Calculates Final Confirmation:
   aATP combines the PAC/PPAC results with the supply proposed by PP/DS to determine confirmed quantities and dates. When the order is saved, these temporary planning elements are converted into permanent supply in PP/DS.

Test Scenario Example with Detailed Explanation-

As a test scenario, we created one FG (Finished Goods) material and two raw materials in the S/4HANA PP side. When these materials are transferred to PP/DS, they are referred to as Products.

Although PP and PP/DS are part of the same S/4HANA system, PP/DS runs on a separate logical box, even though it shares the same server environment.

The key objects required for SBC are the Plant, Work Center, and Production Version.

1. PLANT - Before transferring materials to PP/DS, we must ensure that the **Plant** exists in PP/DS. If a plant - for example, **1710** - is created in the S/4HANA PP system, it must also be available in PP/DS. In PP/DS, a Plant is referred to as a **Location**.

A plant is automatically created in PP/DS only if the plant is marked as PP/DS-relevant in SPRO before the plant is created.

- To transfer the Plant from S/4 PP to PP/DS, you can use the report via SE38 -**/SAPAPO/CREATE_LOCATION**
- To verify that the Location has been successfully created in PP/DS, use transaction **-/n/SAPAPO/LOC3.**

/SAPAPO/CREATE_LOCATION –

Create locations for business partners, plants and shipping points

Business Partner			
Business Partner		to	
Category		to	
Role		to	
Country		to	
Plant			
Plant		to	
Shipping/Receiving Point			
Shipping Point/Receiving Pt		to	
Country		to	
MRP Area			
MRP Area		to	

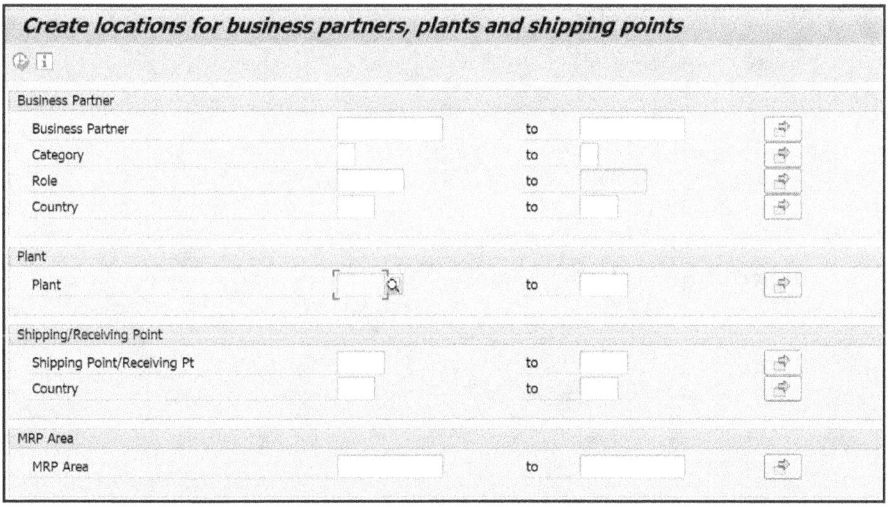

/n/SAPAPO/LOC3 –

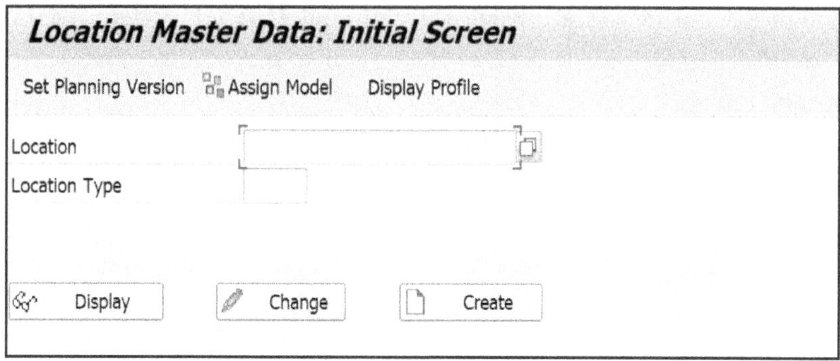

2. MATERIAL - To transfer materials automatically from S/4 PP to PP/DS, a specific material master setting must be enabled in S/4HANA. In the Material master, under the Advanced Planning view, the Advanced Planning checkbox must be activated. Once this checkbox is selected, the material master is automatically transferred to PP/DS as a Product via the RFC connection between S/4 PP and PP/DS.

To verify that the material has been successfully transferred to PP/DS, you can use transaction **/n/SAPAPO/MAT1**.

MM01/MM02/MM03-

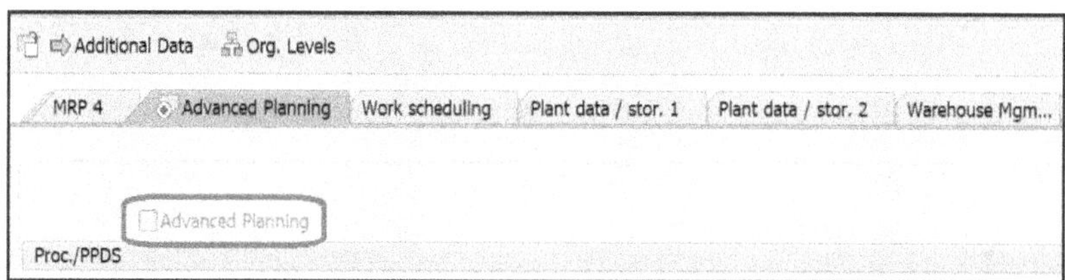

Transaction **/n/SAPAPO/MAT1**

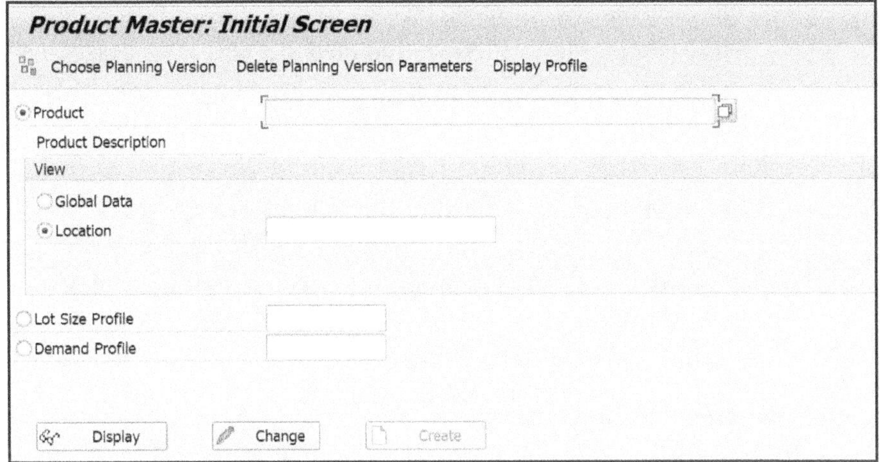

3. Work Center - Automatic Transfer of Work Centers from S/4 PP to PP/DS- Work centers are not transferred based on the Material Master. Instead, they are transferred independently through CIF and appear in PP/DS as Resources. However, for PP/DS to use these Resources during planning, the Material Master must be enabled for Advanced Planning.

In PP/DS, a Work Center is referred to as a Resource. A Work Center must exist in PP/DS as a Resource before it can be used for planning operations.

Enabling the Material for PP/DS - In the Material Master Advanced Planning view: Activate the Advanced Planning checkbox.

This enables the material for PP/DS planning and ensures that when the Production Version is transferred, the system links the material to its routing and the corresponding work center (resource).

Work Center Transfer (Independent of the Material Master)- A Work Center created in S/4 PP must be transferred to PP/DS as a Resource using one of the following methods:

- **Report:** /SAPAPO/RESOURCES_SETUP
- **CIF Model Activation:** (traditional Core Interface transfer)

After transfer, the work center is available in PP/DS as a Resource.

Verification in PP/DS- To verify that the work center has been successfully transferred and created as a Resource in PP/DS, use Transaction: /n/SAPAPO/RES01

Key Points-

- Work centers are NOT enabled or transferred from Material Master.

- They are transferred separately via CIF.
- The Material Master's Advanced Planning checkbox ensures that PP/DS planning objects (such as PDS, including the resource) can be created.
- In S/4HANA, a Work Center can be created, changed, or displayed using the following transactions: CS01/CS02/CS03.
- Work Centers define the capacity, resources, and scheduling parameters required for production operations. Once created in S/4HANA, they can be transferred to PP/DS as Resources for detailed planning and scheduling.

4. Production Version - Automatic Transfer of Production Version from S/4HANA PP to PP/DS. A Production Version in SAP S/4HANA PP defines the BOM and Routing combination used to manufacture a material. For PP/DS to plan and schedule production, this Production Version must be available in PP/DS. The transfer is handled automatically via CIF (Core Interface).

In PP/DS, a Production Version is referred to as a Production Data Structure (PDS). To create a Production Version in Sap S/4, use transaction code C223. This involves assigning a combination of Material + Plant + BOM + Routing.

Pre-requisites for Automatic Transfer of Production Version-

1. **Material must be enabled for Advanced Planning**
    - In the Material Master **Advanced Planning** view → Activate **Advanced Planning** checkbox.
    - This ensures the material is transferred to PP/DS as a Product.

2. **Work Centers must be transferred as Resources**
    - Work Centers in S/4 PP must exist in PP/DS as Resources (via CIF or report /SAPAPO/RESOURCES_SETUP).
    - Production Versions reference these work centers, so PP/DS can plan operations correctly.

3. **BOM and Routing must exist in S/4 PP**
    - The Production Version references valid BOM and Routing.

Once all the prerequisites mentioned above are fulfilled, the system can successfully perform the **automatic transfer of the Production Version** to PP/DS, enabling planning and scheduling operations. Ensure **material and plant are PP/DS-relevant.**

1. **CIF picks up the Production Version** along with:
   - Material (Product)
   - BOM
   - Routing
   - Work Centers (Resources)

2. The **Production Version** is created automatically in PP/DS and linked to:
   - The Product
   - Its BOM
   - Its Resources (from Routing)

3. PP/DS can now use the Production Version to plan and schedule production.

Verification in PP/DS is done by displaying and checking the PP/DS PDS generated from the production version, typically with /SAPAPO/CURTO_SIMU, while /SAPAPO/PRD1 is used to review PP/DS product master data; together they let you validate that the production version is correctly represented.
Reasoning:

- /SAPAPO/CURTO_SIMU is the standard transaction to display and verify the PP/DS PDS (Production Data Structure) that is generated from the production version, including BOM, routing/operations, resources, and validity.

- /SAPAPO/PRD1 is a product master display transaction in PP/DS (product-location master), not the main tool to check the PDS itself; it complements the check, but the core verification of BOM/routing consistency is via the PDS display in CURTO_SIMU.

Key Points-

- Without an active Production Version, PP/DS cannot plan the material.

- Changes in Production Version in S/4 PP are automatically reflected in PP/DS via CIF if the material is enabled for Advanced Planning.
- Multiple Production Versions allow PP/DS to select the best alternative based on lot size, priority, or availability of resources.
- In S/4HANA, a Production Version can be created using the transaction C223.
- A Production Version links material, BOM, and routing to define how a product can be manufactured.
- Once it is created, it can be transferred to PP/DS for planning and scheduling purposes.

Summary-

S/4	PP/DS	Value	Create in S/4	Verify in PP/DS-Transaction
Plant	Location	1710	Report-/SAPAPO/CREATE_LOCATION	/n/SAPAPO/LOC3
Material	Product	ABC1000	Advanced Planning Check Box	/n/SAPAPO/MAT1
Work Center	Resource	WC1001	Report-/SAPAPO/RESOURCES_SETUP (OR) CIF Model Activation	/n/SAPAPO/RES01
Production Version	PDS-Production Data Structure		Transaction to Transfer to PP/DS-CURTOADV_CREATE	/n/SAPAPO/CURTO_SIMU

If your planning system is SBP (Supply-Based Planning), then you can use this setup directly in PP/DS. However, if your planning system is SAP IBP, all relevant master data must be transferred to IBP and properly linked for planning and scheduling purposes.
In this case, PP/DS serves as the planning system, but SBP can also support integration with IBP, not just with PP/DS.

This setup can be applied to both Finished Goods and Raw Materials. Whenever there is a shortage, the system will automatically create receipt elements (such as planned orders or purchase requisitions) to cover the requirement. If the scenario involves only Finished Goods, then apply this setup specifically to the FG material.

Next Steps: Prerequisite Settings in Master Data for PP/DS-

Once all the relevant master data (Products, Resources, Production Versions, etc.) has been transferred to the PP/DS system, you need to update certain prerequisite settings in the Material Master to ensure correct planning behavior.

1. Update Product Heuristic

The heuristic defines the planning algorithm used for net requirement calculation in PP/DS. For SBP (Supply-Based Planning) scenarios, this heuristic is required to calculate supply and demand properly.

In PP/DS, a **heuristic** is a planning algorithm that determines how **net requirements** are calculated for a product. It evaluates demand and supply elements and suggests planned orders or receipts to cover shortages according to defined rules.

- Location: Material Master → *Advanced Planning* view
- Field / Value: Heuristic = SAP_PP_SBC

2. Update PP Planning Procedure

This parameter controls how the heuristic handles shortages in receipt elements. If there is a shortage, it will automatically attempt to cover it based on the selected PP planning procedure. Essentially, it defines how the heuristic allocates supply to meet demand during planning runs.

- Location: Material Master → *Advanced Planning* view
- Field / Value: PP Planning Procedure = 3 (Cover requirements immediately)

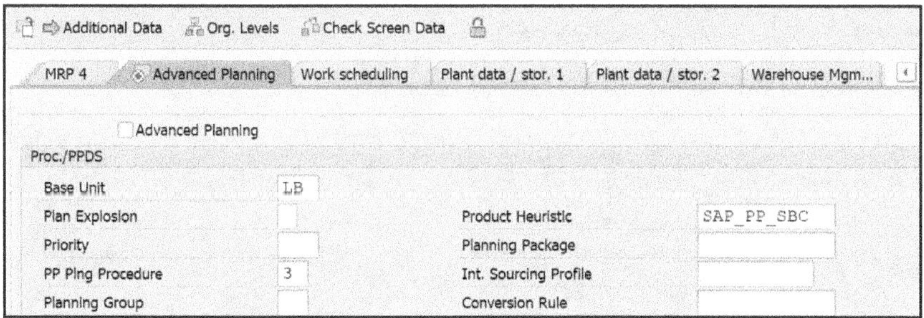

Updating the Product Heuristic and PP Planning Procedure ensures that the system correctly calculates net requirements and automatically covers shortages during PP/DS planning runs. These settings are critical for effective supply-demand planning.

3. Maintain Availability Check

Maintain Availability Check in the Material Master Record. Maintaining the availability check ensures that ATP (Available-to-Promise) or aATP (Advanced ATP) functions correctly during sales order creation and planning runs, helping to avoid overcommitment of stock.

- The Availability Check in the Material Master ensures that the system verifies whether sufficient stock or supply exists to fulfill a requirement.
- It is maintained in the Material Master → MRP 3 view.

4. Enable the Supply Creation for Availability Check

Enable this setting so the system can automatically create supply elements—such as planned orders or purchase requisitions—when a requirement cannot be fulfilled from available stock. This ensures the Availability Check can confirm requirements even in shortage situations. These two settings are required to be activated:

a. Supply Creation is active for the Availability Checking Group

b. Availability Check Type is maintained and enabled.

Supply Creation is Active-

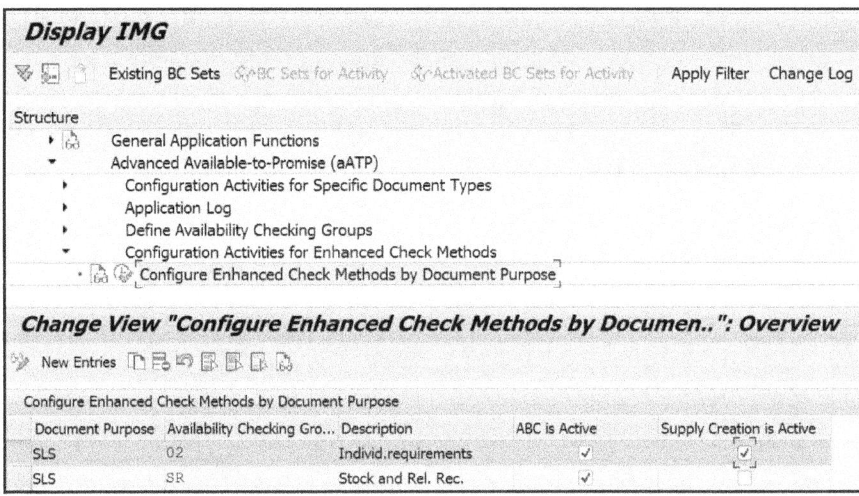

Availability Check Type-

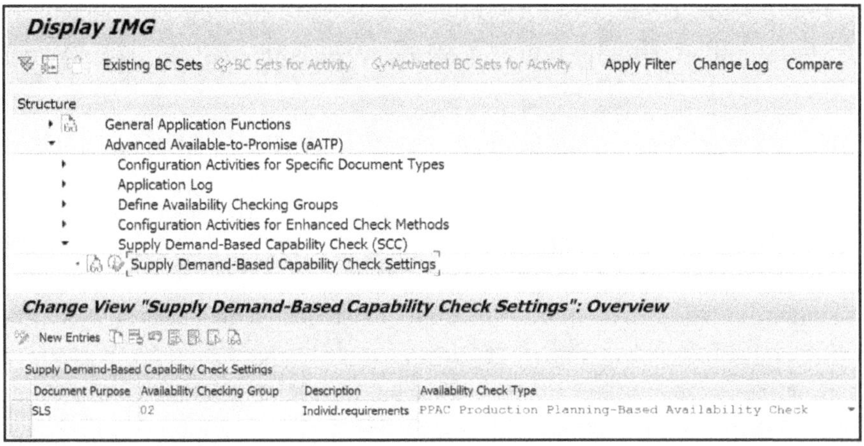

Display IMG

Existing BC Sets BC Sets for Activity Activated BC Sets for Activity Apply Filter Change Log Compare

Structure
- General Application Functions
- Advanced Available-to-Promise (aATP)
  - Configuration Activities for Specific Document Types
  - Application Log
  - Define Availability Checking Groups
  - Configuration Activities for Enhanced Check Methods
  - Supply Demand-Based Capability Check (SCC)
    - Supply Demand-Based Capability Check Settings

Change View "Supply Demand-Based Capability Check Settings": Overview

New Entries

Supply Demand-Based Capability Check Settings

Document Purpose	Availability Checking Group	Description	Availability Check Type
SLS	02	Individ.requirements	PPAC Production Planning-Based Availability Check

5. Transaction Data

We have three materials with the stock situation shown below. This Finished Goods has two Raw Materials associated with it as part of the BOM.

When a stock receipt (such as a goods receipt for a planned order or purchase order) is posted in PP, the updated stock levels are automatically reflected in PP/DS, ensuring that planning data remains synchronized.

- **Transaction Code: /n/SAPAPO/RRP** is used to display the **Product View** in PP/DS. There are three transaction codes available, each serving a different purpose as described below.

Transaction	View Type	Scope	Best Use Case
/SAPAPO/RRP	Single-Level Product View	One material, one location	Detailed planning of one product
/SAPAPO/RRP2	Multi-Product View	Many materials, single level	Plan multiple materials together
/SAPAPO/RRP3	Multi-Level Product View	Full BOM structure	Multi-level pegging and shortage analysis

Transaction "/SAPAPO/RRP3" is similar to **MD04** in SAP ERP.

- **MD04 (SAP ERP)** shows the **stock/requirements list** for material, including existing stock, receipts, and demands.
- **/SAPAPO/RRP3 (PP/DS)** provides a **multi-level product view**, displaying supply and demand across the full BOM structure, including pegging relationships.

Transaction Code: /SAPAPO/CCR is used to perform a **Consistency Check and Reconciliation (CCR)** between SAP ERP (S/4HANA) and SAP APO/PP-DS.

It is used to compare stock levels between PP (S/4HANA) and PP/DS to ensure that both systems are synchronized.

Both PP and PP/DS are part of the same system landscape, but they run in **separate technical environments (different boxes)**, which is why data synchronization is required. Since inconsistencies can occur between the PP and PP/DS systems, SAP provides this transaction to perform a **consistency check** and ensure both systems remain aligned.

If there are any synchronization or technical issues, they will be handled by the PP/DS consultant.

- Transaction data between **PP** and **PP/DS** is transferred via **queues**, which can be monitored using the transactions **SMQ1** (outbound queue) and **SMQ2** (inbound queue). Transactional data—such as planned orders, purchase requisitions, and stock changes - is transferred between PP and PP/DS through **qRFC queues**. These transactions help identify and troubleshoot any CIF-related queue issues.
    - **SMQ1** - Monitor and manage **outbound** queues (from PP/S4 to PP/DS)
    - **SMQ2** - Monitor and manage **inbound** queues (from PP/DS to PP/S4)

Material	Current Stock in PC
FG1000	50
RAW100	100
RAW101	100

So now, if you create a sales order for FG100 with a quantity of 100 PC, the system performs an availability check and detects a shortage of 50 PC. To cover this shortfall, the planning run (PP/DS, depending on your setup) generates a planned order to fulfill the demand.

In this case, the sales order is created in S/4HANA and then transferred to PP/DS, where the net requirements are calculated, and the necessary receipt elements (such as planned orders) are created.

Automatic synchronization and linkage are the key elements that enable seamless integration between SBC and the planning system (PP/DS).

Test Scenario - How SBC Works Between S/4HANA (PP) and PP/DS

1. Sales Order Created in S/4HANA
    - You create a Sales Order for FG100 = 100 PC.
    - aATP + SBC runs and checks:
        - Available stock
        - Existing planned receipts
        - Whether supply can be automatically created

Since only 50 PC are available, the system detects a 50 PC shortage.

2. SBC Decides Whether to Create Supply

If Supply Creation is active in the availability check type & checking group, SBC can initiate creation of supply elements.

Depending on your setup, SBC can trigger supply creation in PP/DS, not in classical MRP. SBC asks:

"To cover the shortage, should I create new supply (planned order or purchase requisition)?"

If yes → It signals PP/DS to generate supply.

3. Sales Order Requirement Is Transferred to PP/DS

Through CIF integration:

- The Sales Order requirement is sent from S/4HANA → PP/DS
- PP/DS receives the demand and performs net requirements calculation in the detailed planning layer. Tools involved:
- SMQ1/SMQ2 (queues)
- /SAPAPO/RRP (single-level view)
- /SAPAPO/RRP3 (multi-level pegging)

4. PP/DS Creates the Supply Elements

PP/DS generates:

- Planned Orders (for in-house production)
- Purchase Requisitions (for external procurement)

This decision depends on:

- Production Version
- Procurement Type (E/F)
- Heuristic used (e.g., SAP_PP_SBC)
- PP Planning Procedure

This is the actual "automatic supply creation" that SBC triggers.

5. Created PP/DS Receipts Flow Back to S/4HANA

The new supply elements are transferred back to S/4HANA:

- Planned Orders → S/4HANA Planned Orders
- Purchase Requisitions → S/4HANA PRs

Now S/4HANA sees that the shortage is covered.

6. aATP Updates the Sales Order Confirmation

Once supply exists (even if future-dated), aATP:

- Updates the confirmation
- Assigns new dates
- Provides reliable order promise

This is an example and summary of how SBC, PP, and PP/DS interact.

Example:

- Shortage: 50 PC → Planned order created → Confirmation updated to the planned order's availability date.

Step	System	What Happens
1	*S/4HANA*	Sales Order created → shortage identified
2	*aATP/SBC*	Decide if supply must be created
3	*CIF*	Demand transferred to PP/DS
4	*PP/DS*	Planned orders / PRs created to cover shortage
5	*CIF*	Supply elements transferred back to S/4
6	*aATP*	Sales Order receives updated confirmation

Roles and responsibilities for each consultant in an aATP + SBC + PP/DS + IBP integrated planning landscape.

1. aATP Consultant – The aATP consultant is responsible for all availability-checking and allocation functionalities in S/4 HANA.

2. PP/DS Consultant - The PP/DS consultant focuses on detailed planning, heuristics, and supply generation logic in PP/DS.

3. IBP Consultant – The IBP consultant is responsible for global, mid- and long-term planning done in SAP IBP.

Simplified Role Summary

Area	aATP Consultant	PP/DS Consultant	IBP Consultant
Availability Check	✓	(supplies data)	(provides plan)

Supply Creation	✓ (SBC logic)	✓ (planning orders)	(provides supply plan)
Master Data Ownership	aATP fields	PP/DS fields	IBP model master data

SBC is applicable for Both MTO and MTS-

SBC (Supply-Based Confirmation) in aATP is designed to improve confirmation logic by integrating with PP/DS for supply creation and recalculation. It works with any process where supply may need to be created or reassessed, including:

1. MTS (Make-to-Stock)
    - SBC checks existing supply from ATP time series or stock.
    - If stock is insufficient, it can trigger PP/DS to propose planned orders to cover shortages.
    - Useful when MTS items still need dynamic supply creation due to demand spikes.
2. MTO (Make-to-Order)
    - SBC is highly relevant for MTO, where production/supply is created only after receiving a sales order.
    - PP/DS creates supply elements such as planned orders specifically for that sales order.
    - SBC drives detailed availability and scheduling for customer-specific items.

Process	Is SBC Applicable?	Why
MTS	Yes	To confirm stock or create replenishment supply through PP/DS
MTO	Yes (Primary Use Case)	To create order-specific supply and perform precise scheduling

Why SBC Exists (the problem it solves) - Classic ATP sometimes confirms orders too optimistically because it:

- Assumes planned receipts will arrive exactly on time
- Doesn't consider supply constraints over future time buckets
- Overcommits early dates

SBC fixes this by performing a time-phased supply-based simulation before confirming.

Example of SBC Behavior

Situation:	Classic ATP:	SBC:
• Today stock = 0 • PO coming in 10 days = 500 units • Customer order = 400 units	It might confirm 400 units today if settings allow min/max dates.	Will confirm 400 units with delivery date based on PO in 10 days, because supply is only available then. This prevents unrealistic or misleading confirmations.

Difference Between PAC and SBC in SAP S/4HANA aATP-

Aspect	PAC (Product Availability Check)	SBC (Supply Creation-Based Confirmation)
Scope	Existing stock + planned receipts	Stock + creates new supply (planned orders/PRs)
Engine	Standard aATP time series	PP/DS integration (finite capacity)
Result	Confirms from current supply	Confirms + schedules production
Use Case	MTS with existing inventory	MTO (make-to-order) scenarios
Purpose	Check availability	Confirm based on supply feasibility
Focus	Current stock + receipts	Time-phased supply constraints
Logic	Snapshot check	Dynamic bucket-based future check
Output	Confirmation qty/date	Feasible promise date

Key Difference is -

- PAC = "What's available now?"
- SBC = "What's available + what can we make?"

SBC is more accurate and forward-looking. Supply-Based Confirmation, a time-phased, supply-sensitive ATP check ensuring realistic confirmations and delivery dates.

Real Business Scenarios Where SBC is Essential-

Make-to-Order (MTO) Finished Goods	Customer orders custom bike (FG1000) - 0 stock. SBC creates planned order + component planned orders, confirms delivery 01/15 after capacity check.

Engineer-to-Order (ETO) Projects	Large machinery order triggers multi-level BOM planning. SBC simulates production feasibility across 5 plants before promising.
High-Value Spare Parts	Aircraft engine part (no stock) - SBC confirms based on manufacturing lead time + capacity, creates dependent requirements instantly.
Fashion/Seasonal Products	New collection launch - SBC allocates capacity during peak demand, rejects overbooked slots automatically.
Contract Manufacturer	Customer requests 5000 widgets. SBC checks component availability + line capacity before confirmation.
Medical Devices	Custom implants - SBC integrates sterile processing capacity + regulatory lead times for realistic promises.

Key Pattern: SBC shines when "no stock" ≠ "cannot deliver" - it proactively creates supply while respecting capacity constraints.

- Without SBC: Manual MRP runs + capacity firefighting
- With SBC: One-click confirmation + automatic planned orders

15 BPS – Business Process Scheduling

Business Process Scheduling (BPS) is a configurable scheduling framework introduced with SAP S/4HANA that works as part of the advanced Available-to-Promise (aATP) solution to improve how logistical dates and times across business processes are calculated and managed.
BPS schedules complex business processes (not just single operations) during order confirmation, respecting multi-level dependencies and finite capacities.

Business Process Scheduling (BPS) is an advanced confirmation mechanism in aATP that schedules entire end-to-end business processes—not just single production operations—at the time of order confirmation.

Unlike classic ATP or simple scheduling, BPS considers multi-level dependencies, finite capacities, and cross-functional process steps to deliver realistic and reliable confirmation dates.
BPS ensures *everything required to fulfil the order is planned and feasible before a promise is made to the customer*.

- Schedules **complete process chains**, not isolated activities
- Respects **dependencies across procurement, production, and logistics**
- Uses **PP/DS heuristics** to calculate feasible dates
- Works together with **SBC (Supply Creation-Based Confirmation)**

Why BPS Matters in aATP-
In the broader aATP process — which includes **Product Availability Check (PAC), Backorder Processing (BOP), Alternative-Based Confirmation (ABC), Product Allocation (PAL), and Supply Protection — BPS stands out by focusing on *when* the activities in a fulfilment chain actually occur.
BPS provides:

- **Greater scheduling precision** versus standard date determination.
- **Flexible activity modelling** so companies can adapt to their logistics processes.
- **Better utilization of working time and calendars** rather than simple date arithmetic.

- **Alignment of confirmed delivery dates with realistic logistics flows**, improving customer satisfaction.

BPS goes beyond standard scheduling by:

- Decoupling business documents (like sales orders) from the scheduling logic.
- Providing flexible configuration of **activities** (pick, load, transport, etc.) and their **attributes** (calendar, duration, time zone).
- Enabling scheduling precision down to **days or seconds**.
- Integrating into aATP's order promising logic, so confirmed dates account for realistic logistical timelines.

Big picture (one sentence)-

When you create a sales order, aATP determines when the product is available, and BPS determines when it can actually be delivered by adding realistic logistics activities like pick, pack, load, and transport.

- aATP answers **"When is the product available?"**
- BPS answers **"Given our real logistics process, when can we actually deliver it?"**

BPS scheduling logic (forward scheduling)
Since the material is available on Jan 05, BPS does forward scheduling:

Material Available Date (MAD): Jan 05
+ Picking / Packing: 2 days
→ Ready for Shipment: Jan 07
+ Transportation: 3 days
→ Delivery Date: Jan 10

Final result in the sales order:
Schedule line confirmed delivery date = Jan 10

What dates you actually see in the sales order
In S/4HANA with aATP + BPS active, you typically get:

- Material Availability Date (MAD) → Jan 05
- Confirmed Delivery Date (schedule line date) → Jan 10

So, the system is basically saying: ***"I can get the product on Jan 05, but the customer will realistically receive it on Jan 10."***

How this differs from classic ATP (important!)-

Classic ATP	**aATP + BPS**
Fixed lead times	Activity-based scheduling
Date math	Calendar & time-zone aware
Limited flexibility	Fully configurable logistics flow
One-size-fits-all	Process-specific schemas

In classic ATP, pick/pack/transport was often:	• Hard-coded • Shipping-point based • Less transparent
With BPS, it's:	• Explicit • Configurable • Visible in scheduling results

Where This Is Configured (High Level)

aATP (Advanced Available-to-Promise)	• Determines the Material Availability Date (MAD) based on stock, receipts, and supply elements.
Business Process Scheduling (BPS) Schema	• Defines the logistics activities and timing, including: o Pick / pack duration o Transportation duration o Working calendars and time zones
Schema Assignment	• The BPS schema is assigned based on business design to: o Sales document type o Shipping point o Item category
Fallback Behaviour	• If no BPS schema is assigned, the system uses classic delivery and transportation scheduling.

In short: *BPS replaces static shipping point durations with dynamic, characteristic-based, and capacity-aware durations, giving more accurate and realistic delivery dates.*

Here's why:

Static shipping point durations:	• In standard SAP, Pick, Pack, Load, and transport times are fixed per shipping point. They cannot vary by order type, product, or sales characteristics, so ATP dates are less flexible.
BPS makes durations dynamic:	• Activity durations are defined in schemas and can be overridden by characteristics combinations (Sales Org, Distribution Channel, Division, Shipping Point, etc.). • This allows ATP calculations to consider order-specific factors.
Realistic ATP dates:	• By integrating with PP/DS finite scheduling, BPS calculates activity durations that reflect capacity constraints and actual timelines, not just static defaults.

BPS Key Features and Characteristics-

Key Features	• End-to-end process scheduling Schedules complete business process chains from procurement → production → delivery • PP/DS integration Uses PP/DS heuristics to calculate feasible and realistic confirmation dates • SBC integration Works with Supply Creation-Based Confirmation (SBC) to create and schedule required supply
Key Capabilities	• End-to-end scheduling Links procurement, production, and delivery into a single confirmation process • Finite capacity planning Considers capacity constraints to avoid unrealistic confirmations based on infinite resources • Multi-level planning Supports complex product structures, including assemblies, sub-assemblies, and external procurement • Reliable confirmations Confirms orders based on actual planning constraints, lead times, and available capacities

Core Concepts & Components of BPS-

1. Scheduling Schema

You build a *BPS scheduling schema* to define:

- A sequence of activities (e.g., goods issue → transport → delivery).
- Date types (start/end dates) associated with those activities.
- How scheduling should calculate these based on calendars, durations, and time zones.

SAP provides standard schemas, but you can customize them.

2. Backward & Forward Scheduling

BPS supports:

- Backward scheduling: calculates required start dates from a requested delivery date by subtracting activity durations and considering calendars.
- Forward scheduling: adjusts dates if backward scheduling produces dates in the past (e.g., due to holidays or non-working times).

This helps ensure confirmed dates are realistic given actual logistics flow, not just theoretical ones.

3. Attribute Determination

Each activity in a schema can use attribute determinations such as:

- Calendar source (e.g., shipping point or customer calendar).
- Duration (e.g., how long loading or transport takes).
- Time zone determination (critical for global operations).

This increases accuracy by adapting scheduling to operational realities.

How BPS Actually Gets Activated in S/4HANA aATP-

To use BPS:

1. Activate scheduling for relevant documents (e.g., sales orders).
2. Assign BPS schemas to document types, shipping points, item categories.
3. Optionally configure advanced attributes via the "Configure Activity Attributes" Fiori app when licensed for advanced aATP.

Documents not assigned to BPS fall back to standard shipment & transportation scheduling.

BPS is activated via sales document / item category / shipping point schema assignment and aATP configuration.

To enable and use BPS in SAP S/4HANA (including 2023):

Activate advanced ATP / BPS functions in aATP configuration	Use the Customizing to assign scheduling schema for specific combinations of sales document type, item category, and shipping point. This controls whether BPS is used for a given sales order.
Define BPS scheduling schema	Create the sequence of activities and time logic for backward and forward scheduling. This is the core of the BPS configuration.
Configure characteristics combination and activity durations	Detailed activity attributes are maintained via Fiori apps (e.g., *Configure Activity Attributes*). Characteristic-based overrides are used for dynamic durations.
Integrate with PP/DS for finite scheduling	The PP/DS engine drives the actual timetable calculations when BPS is in scope.

Business Scenarios-

Business Process Scheduling (BPS) is ideal for:

- Complex Make-to-Order (MTO) and Engineer-to-Order (ETO) scenarios
- Long fulfilment cycles spanning weeks or months
- Orders involving multiple departments, systems, and handoffs
- Situations where reliable and realistic confirmation dates are critical

Multi-Step Assembly	Order → Raw Material PR → Sub-assembly → Final assembly → Delivery *Business Process Scheduling (BPS) confirms the sales order only after all dependent steps in the process chain can be successfully scheduled.* *This ensures that procurement, sub-assembly, final assembly, and delivery are all feasible before a customer commitment is made.*
Engineer-to-Order	Design approval → Procurement → Testing → Customer acceptance *Business Process Scheduling (BPS) links all technical and logistical milestones into a single end-to-end confirmation flow, confirming the order only when each dependent step can be scheduled.*
Repair/Overhaul	Disassembly → Parts procurement → Reassembly → Quality gate → Shipping

> *With Business Process Scheduling (BPS), each step in the repair process is scheduled in sequence, with all dependencies and finite capacity constraints respected, ensuring realistic and reliable order confirmation.*

The Material Master allows ATP, but whether BPS runs are determined by system configuration. "BP" is a conceptual term, not a standard field value—you won't find it in the F4 help for the Material Master *Availability Check* field. BPS is triggered by the combination of aATP, Supply-Based Confirmation (SBC), PP/DS integration, and the assigned BPS schema - not by any single material setting.

In the context of SAP S/4HANA BPS (Business Process Scheduling), "BP" is not a real field or value - it's just a conceptual shorthand that was sometimes used in older documentation or discussions to refer to "Business Process" scheduling or logic.

Transportation Activities-

On top of material availability date, we can add these transportation activities days to give final confirmation to the customer.

Example-

If sufficient stock is available, the system may determine the material availability date as today. However, if an additional 4 days are required for picking, packing, and transportation, the system automatically adds these durations and proposes today + 4 days as the confirmed delivery date to the customer.

To support this kind of end-to-end, process-based date determination, Business Process Scheduling (BPS) is the appropriate tool to implement.

Advantage of BPS over ECC / TM-

In ECC or basic TM scenarios, delivery dates are typically calculated using standard pick/pack and route times.

If additional or customized business steps are required, these are often handled outside ATP or through manual adjustments.

With Business Process Scheduling (BPS) in S/4HANA aATP, custom business objects and process steps can be configured and scheduled as part of the ATP confirmation itself. This allows all relevant activities—standard and customized—to be planned, scheduled, and date-calculated together within aATP.

As a result, BPS provides:

- Greater flexibility for complex business requirements
- More accurate and reliable customer delivery confirmations
- A single, integrated scheduling logic instead of fragmented calculations

Configuration & Setup Steps for BPS in SAP S/4HANA aATP-

Below is a practical, end-to-end configuration checklist for Business Process Scheduling (BPS) in SAP S/4HANA aATP. It is split into GUI (SPRO) and Fiori steps, with clear guidance on what is mandatory versus what is supporting configuration.

An important truth upfront: there is no single "BPS ON" switch.

Business Process Scheduling works only when aATP is active, Supply-Based Confirmation is configured, PP/DS is integrated and active, and the required sales order configuration supports BPS.

Business Process Scheduling requires PP/DS for finite, time-based scheduling. Without PP/DS, BPS cannot work. Why this is correct:

- BPS does not have its own scheduling engine
- PP/DS provides the finite, capacity-aware, time-phased scheduling logic
- aATP and SBC only orchestrate confirmations; they do not schedule supply on their own
- Without PP/DS, aATP falls back to non-BPS confirmations only

BPS configuration and setup is primarily done via the SAP GUI, and it's all about defining the activities, their durations, and assigning them to the relevant business objects. Here's a detailed, high-level summary:

Configuration Steps-

1. Prerequisites (Must Be in Place)

- S/4HANA with embedded PP/DS
- Advanced ATP (aATP) license activated
- Relevant plants and materials enabled for PP/DS

2. BPS Configuration Overview

For BPS to work, define Scheduling Schema along with Activities Date Types.

Then Activate BPS for Sales Documents (also applies to Outbound Deliveries and Stock Transport Orders).

1. Define Activities for BPS

Define required activities: Arrive, Load, Pick, Plan, etc.

2. Define Date Types for BPS

Assign date types to activities (e.g., delivery date, material staging date).

3. Define Scheduling Schema for BPS

Create schema structure: activities + sequence + date types.
Standard Sales Schema: SALES_DAYS_ADV_ATTRIB

4. Activate BPS for Sales Documents

Sales Doc Type (OR) + Item Category (TAN) + Shipping Point (1710) → Assign Schema

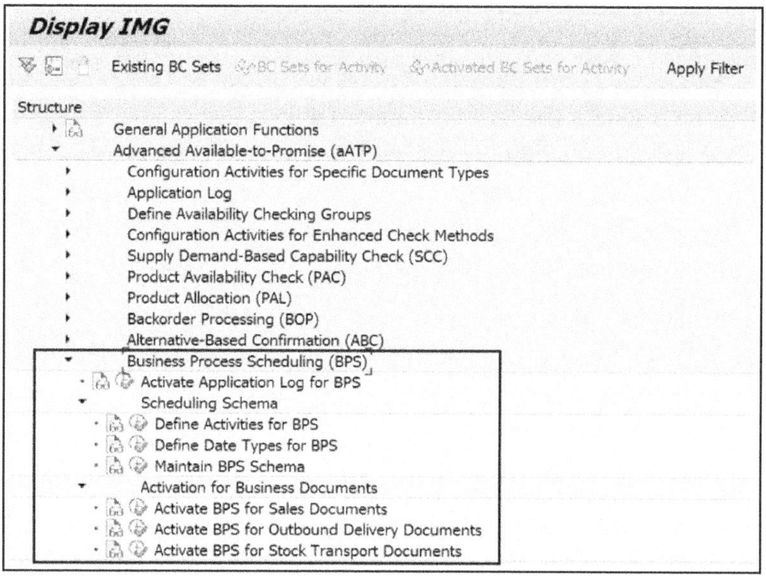

Display IMG

Existing BC Sets   BC Sets for Activity   Activated BC Sets for Activity   Apply Filter

Structure
- General Application Functions
- Advanced Available-to-Promise (aATP)
  - Configuration Activities for Specific Document Types
  - Application Log
  - Define Availability Checking Groups
  - Configuration Activities for Enhanced Check Methods
  - Supply Demand-Based Capability Check (SCC)
  - Product Availability Check (PAC)
  - Product Allocation (PAL)
  - Backorder Processing (BOP)
  - Alternative-Based Confirmation (ABC)
  - Business Process Scheduling (BPS)
    - Activate Application Log for BPS
    - Scheduling Schema
      - Define Activities for BPS
      - Define Date Types for BPS
      - Maintain BPS Schema
    - Activation for Business Documents
      - Activate BPS for Sales Documents
      - Activate BPS for Outbound Delivery Documents
      - Activate BPS for Stock Transport Documents

Activate Application Log for BPS	Activating the BPS Application Log ensures you can trace each scheduling calculation, including pick/pack/transportation steps, and quickly identify why the confirmed schedule line date was determined the way it was.
Scheduling Schema & Activation for Business Documents	**Activity:** Single logistics step (Picking, Packing, Loading, Transportation) used to calculate schedule line dates. **Parameters:** Activity ID, Duration, Calendar, Time Zone, Forward/Backward scheduling. **Schema Assignment:** Linked to Sales Document Type, Shipping Point, Item Category **Purpose:** Converts MAD from aATP into realistic confirmed delivery dates by applying planned logistics steps.

1. Define Activities for BPS-

An **Activity** represents a single logistical step in a supply chain or fulfilment process. It is the basic building block that SAP BPS uses to calculate schedule line dates.

Examples of Activities: Picking, Packing, Loading, Transportation

Each activity has **parameters** that define:

- How long it takes (duration)
- How it affects schedule dates (start/end times, dependencies)

The **actual duration** can be defined or adjusted using the **SAP Fiori App**. Activities are **combined in a schedule** to calculate overall timelines and workflow sequences.

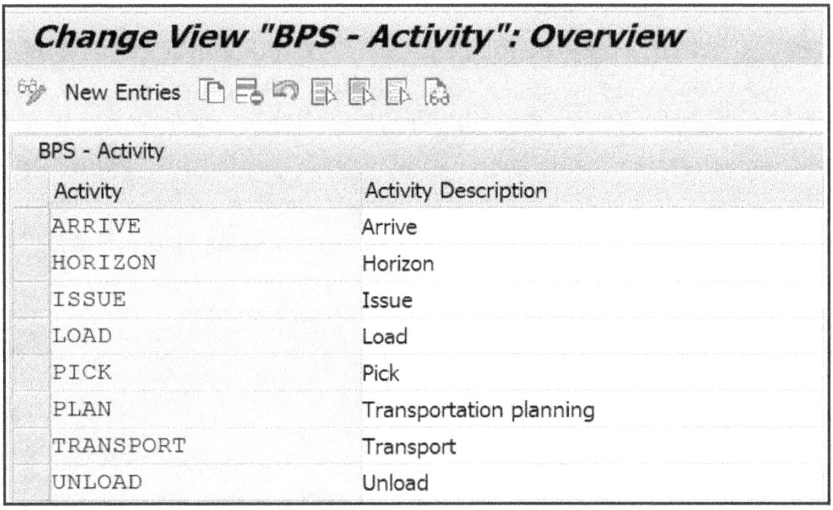

Change View "BPS - Activity": Overview

New Entries

BPS - Activity

Activity	Activity Description
ARRIVE	Arrive
HORIZON	Horizon
ISSUE	Issue
LOAD	Load
PICK	Pick
PLAN	Transportation planning
TRANSPORT	Transport
UNLOAD	Unload

2. Define Date Types for BPS-

Here you can configure date types that represent key points in time within your business process and are relevant for scheduling.

These date types can later be assigned to the start or end of specific activities when defining the schema.

Example:

- MBDAT – Material availability date
- LFDAT – Delivery date

Change View "BPS - Activity date type definition": Overview

New Entries

BPS - Activity date type definition

Date Type	Date Type Description
ELDAT	Unloading Date/Time
HRZN_END	End of Check Horizon
HRZN_START	Start of Check Horizon
LDDAT	Loading Date/Time
LFDAT	Delivery Date/Time
MBDAT	Material Availability Date/Time
TDDAT	Transportation Planning Date/Time
WADAT	Goods Issue Date/Time

3. Maintain BPS Schema-

Here you define a BPS scheduling schema to control how activities in your business process are scheduled. The schema includes the activities, their attributes, sequence, possible delegation, and assigned date types. Before creating a custom schema, activities and date types must be defined. You also choose the scheduling time granularity (days or seconds), select the relevant activities, and maintain attribute determination and priorities for each activity.

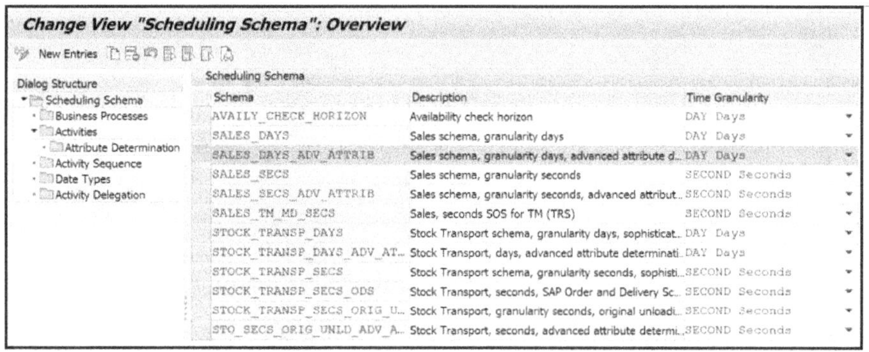

Change View "Scheduling Schema": Overview

New Entries

Dialog Structure
- Scheduling Schema
  - Business Processes
  - Activities
    - Attribute Determination
  - Activity Sequence
  - Date Types
  - Activity Delegation

Scheduling Schema

Schema	Description	Time Granularity
AVAILY_CHECK_HORIZON	Availability check horizon	DAY Days
SALES_DAYS	Sales schema, granularity days	DAY Days
SALES_DAYS_ADV_ATTRIB	Sales schema, granularity days, advanced attribute d...	DAY Days
SALES_SECS	Sales schema, granularity seconds	SECOND Seconds
SALES_SECS_ADV_ATTRIB	Sales schema, granularity seconds, advanced attribut...	SECOND Seconds
SALES_TM_MD_SECS	Sales, seconds SOS for TM (TRS)	SECOND Seconds
STOCK_TRANSP_DAYS	Stock Transport schema, granularity days, sophisticat...	DAY Days
STOCK_TRANSP_DAYS_ADV_AT...	Stock Transport, days, advanced attribute determinati...	DAY Days
STOCK_TRANSP_SECS	Stock Transport schema, granularity seconds, sophisti...	SECOND Seconds
STOCK_TRANSP_SECS_ODS	Stock Transport, seconds, SAP Order and Delivery Sc...	SECOND Seconds
STOCK_TRANSP_SECS_ORIG_U...	Stock Transport, granularity seconds, original unloadi...	SECOND Seconds
STO_SECS_ORIG_UNLD_ADV_A...	Stock Transport, seconds, advanced attribute determi...	SECOND Seconds

"**SALES_DAYS_ADV_ATTRIB**" is a standard SAP BPS schema intended for Sales documents.

4. Activate BPS for Sales Documents-

Here you activate Business Process Scheduling (BPS) for sales documents by assigning a BPS schema to combinations of sales document type, item category, and shipping point. This activation is required to use BPS for scheduling.

Wildcard logic can simplify maintenance by leaving fields empty, but only from right to left (shipping point → item category → document type). The schema field is mandatory.

The schema's time granularity (days or seconds) determines how BPS interprets shipping point durations. For accurate results, shipping point duration granularity should match the schema granularity. Avoid wildcard shipping points if different shipping points use different granularities.

Before activation, any custom schema must be configured and assigned to the Sales business process.

Create the schema → Attach it to Sales → Then activate it for specific sales documents.

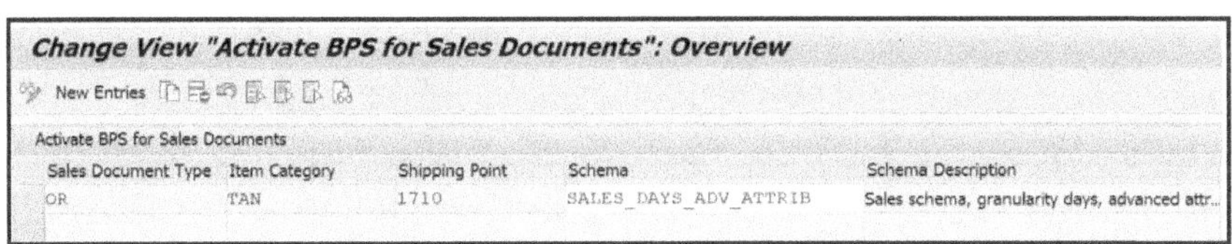

Change View "Activate BPS for Sales Documents": Overview

New Entries

Activate BPS for Sales Documents

Sales Document Type	Item Category	Shipping Point	Schema	Schema Description
OR	TAN	1710	SALES_DAYS_ADV_ATTRIB	Sales schema, granularity days, advanced attr...

4. Fiori Apps to Create Characteristics Combination and Assign It in "Configure Activity Attributes".

After activating the BPS schema, you define **scheduling durations and working times** with **Configure Activity Attributes**.

The **Fiori app "Configure Activity Attributes" (App ID F6439)** is used to define **durations and working times** for scheduling activities in Business Process Scheduling (BPS). This app lets you set different duration values based on characteristic combinations for advanced scheduling logic.

To define different combinations of characteristics (like sales organization, distribution channel, priority, route, etc.), you use the **Fiori app "Manage Characteristic Catalogs"** and then create characteristic combinations (in apps like *Manage Characteristic Combinations*). Those combinations are then used in *Configure Activity Attributes* to assign values.

Manage Characteristic Catalogs

Standard

Business Process Scheduling | Editing Status: All | Catalog Use Type: | Catalog Type:

Characteristic Catalogs (2)

Catalog Use Type	Catalog Type	Enabled for Sales	Enabled for Stock Transport
Business Process Scheduling	Base (Sales Document)	Yes	No
Business Process Scheduling	Base (Stock Transport Order)	No	Yes

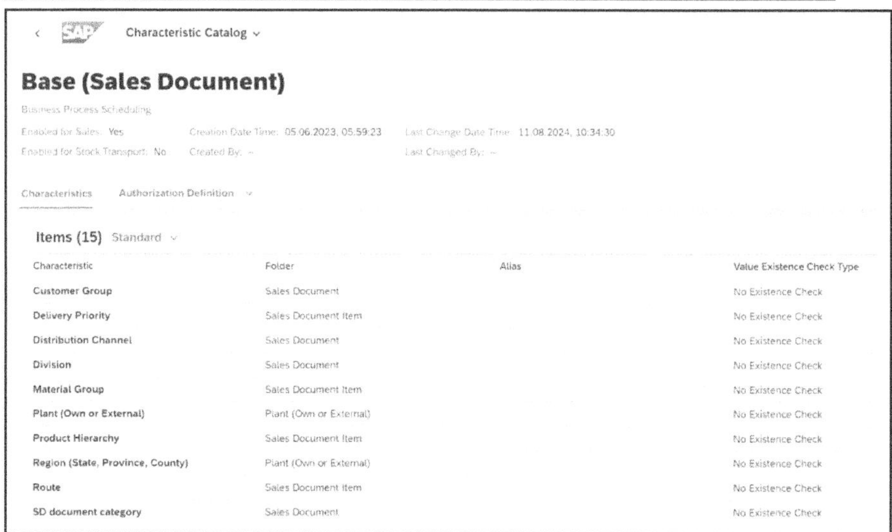

Use this app to create a Characteristics Combination: Go to the Fiori app "Manage Characteristics Catalog."
Here you maintain the characteristics you want to use (for example, Sales Org, Distribution Channel, Shipping Point, or any custom characteristic).

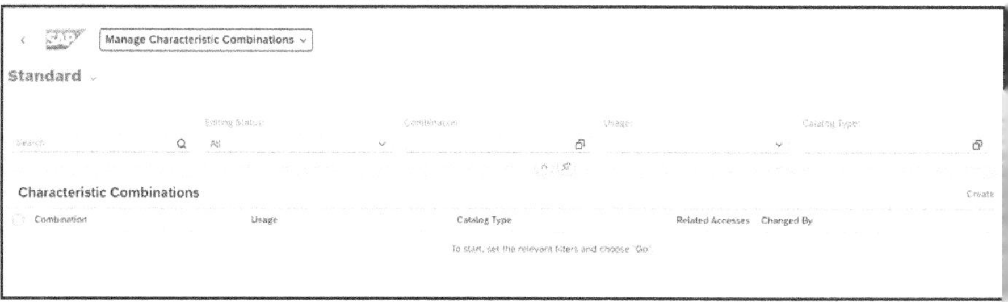

Then click on 'Create' and create a characteristics combination.

- Combination Name – ZSSP_BPS01
- Usage – Duration for Scheduling
- Catalog Type – Base (Sales Document)

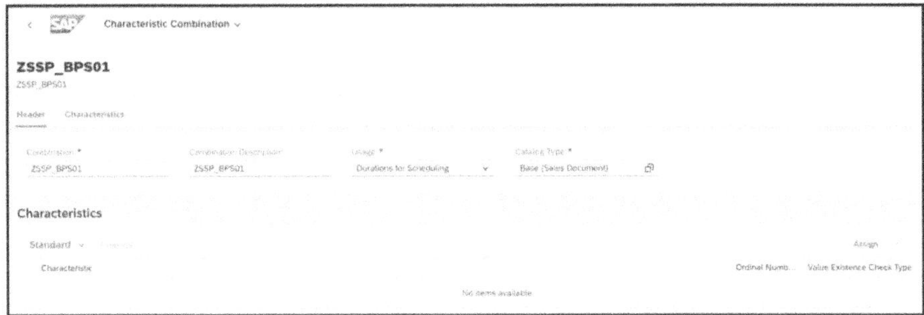

Then click on 'Assign' to add characteristics. Added 4 characteristics:

- Sales Organization
- Distribution Channel
- Division
- Shipping Points

Then use the Fiori app **"Configure Activity Attributes"** to assign a **characteristics combination** (e.g., ZSSP_BPS01) to a specific activity and maintain **durations, priorities, or other scheduling attributes**.

The system applies these values only when the activity/document matches the assigned characteristics combination.

Important Note-

1. When you create a sales order, the system uses the BPS activity durations (such as *Pick* and *Pack*) that are defined in the system.
If an activity is active in BPS, the standard durations maintained at the shipping point are ignored. Only the activities that are defined as part of BPS will have their durations considered during the ATP date calculation; all other standard durations are ignored.
For example, if *Pick/Pack* times are maintained at the shipping point level, they will not be considered when BPS is active. Instead, the system uses the BPS-defined activity durations.

2. Fiori – RACR Screen & BPS
"In RACR, BPS-managed activities appear as hyperlinks to show activity-level scheduling details."

- When BPS is active for a sales order, the RACR Fiori screen shows hyperlinks for activities maintained in BPS (Pick, Pack, Load, etc.).
- Clicking the hyperlink allows users to see detailed activity-level information: durations, start/finish times, and resources.
- Activities not maintained in BPS or using standard shipping point durations do not appear as hyperlinks.
- Key point: The hyperlink visually indicates that BPS is controlling the activity timing for ATP calculation.

Why Use BPS Activities (Pick, Pack) Instead of Standard Shipping Point Durations

1. Flexible, Characteristic-Based Durations
    - Standard shipping point durations are static: the same Pick/Pack time applies to all orders from a shipping point.
    - BPS allows durations to vary based on characteristics like sales organization, distribution channel, division, product type, or order type.
    - Example: High-volume products may take longer to pick than low-volume products from the same shipping point.
2. Integration with aATP and PP/DS
    - Pick/Pack activities become time-phased operations in ATP calculation.

- Delivery dates consider finite capacity, resource constraints, and other scheduled activities, not just a flat duration.

3. Dynamic, Order-Level Scheduling
   - Classic shipping point durations apply the same time to all orders.
   - BPS dynamically calculates durations per order, reducing over- or under-estimation of delivery dates.
4. Improved Visibility and Planning
   - Each activity (Pick, Pack, Load, Inspection, etc.) is visible in Fiori apps.
   - Planners can adjust or optimize timelines per activity instead of using a single static duration.

Summary Table-

Aspect	Standard Shipping Point	BPS Activities
Granularity	Static	Flexible, characteristic-based
Scheduling	No capacity consideration	Integrated with PP/DS, finite scheduling
Dynamic ATP	Limited	Realistic, order-specific
Visibility	Total duration only	Each activity visible and adjustable
Flexibility	Low	High (per product, order type, etc.)

In short: BPS turns flat, static shipping durations into dynamic, order-aware, capacity-conscious, and characteristic-specific timelines, which gives much more accurate delivery dates and better planning.

Test Scenario#1

- Sales Order: 1000 units of Product FG10101
- Shipping Point: SP01
- Standard shipping durations at SP01:
  - Pick = 1 day
  - Pack = 0.5 day
- BPS activities defined:
  - Pick = 2 days
  - Pack = 1 day

Case 1: BPS (inactive)

- ATP calculation uses shipping point durations:
  - Total = Pick (1) + Pack (0.5) = 1.5 days

- Expected delivery date = Order date + 1.5 days

Case 2: BPS (active)

- ATP calculation uses BPS-defined durations:
    - Total = Pick (2) + Pack (1) = 3 days
- Expected delivery date = Order date + 3 days
- Shipping point durations are ignored

Key takeaway:

When BPS is active, the system overrides standard shipping point durations and relies only on BPS-defined activity durations for ATP calculation.

Test Scenario#2

- Sales Order: 1000 units of Product **FG10101**
- Shipping Point: SP01
- Standard shipping durations at SP01:
    - Pick = 1 day
    - Pack = 0.5 day
- BPS activities defined:
    - Pick = 2 days
    - Pack = not maintained

Behaviour when BPS is active:

1. Pick activity (active in BPS)
    - System uses BPS-defined duration (2 days) instead of shipping point duration.
2. Pack activity (not maintained in BPS)
    - System falls back to the standard shipping point duration (0.5 day).

ATP calculation:

- Total = Pick (BPS, 2 days) + Pack (shipping point, 0.5 day) = 2.5 days
- Expected delivery date = Order date + 2.5 days

Key takeaway:

- Only activities maintained in BPS override standard durations.
- Activities not maintained in BPS will continue using standard shipping point durations.
- You don't need to define all activities in BPS—partial coverage is allowed, and the system merges BPS durations with standard durations.

Test Scenario#3 – Integration with TM Module

In sales order processing, aATP determines MAD, BPS adds planned logistics time to promise a delivery date, and TM later optimizes and confirms transportation dates based on real routes and carriers.

aATP + BPS + TM

- aATP → determines *when the product is available* (MAD)
- BPS → calculates a *planned delivery date* by adding pick/pack/transport time
- TM → performs *real transportation planning* and can refine or override transportation dates based on actual routes, carriers, and schedules
    - *BPS gives a realistic promise*
    - *TM executes and optimizes that promise*

Sales order example - Business scenario

- Plant: 1000
- Shipping point: SP01
- Customer location: New York
- **TM is active and integrated with S/4HANA**

Step 1: Sales order creation (VA01 / Fiori)

Customer orders:

- Material: M-100
- Quantity: 10
- Requested Delivery Date: Jan 08

Step 2: aATP determines MAD

aATP checks:

- Stock
- Planned receipts
- Allocations, etc.

Result

- Material Availability Date (MAD) = Jan 05

At this stage: "Material is available in the plant on Jan 05."

Step 3: BPS adds logistics activities (planned dates)

BPS schema assigned to the sales order defines:

- **Activity "Picking & Packing" -> Duration 2 days**
- **Activity "Transportation (planned)" -> Duration 3 days**

BPS forward scheduling-

MAD: Jan 05
+ Pick / Pack (2 days): Jan 07
+ Transport (3 days): Jan 10
Confirmed delivery date in sales order

- Schedule line delivery date = Jan 10

This date is only a planned transportation duration, not yet optimized by TM.

Step 4: Delivery creation triggers TM integration

When you create the outbound delivery (or it's created automatically):

- Delivery is relevant for TM
- System creates a Transportation Requirement (TOR) in TM

Transferred to TM:

- Source location (plant / shipping point)
- Destination (customer)
- Requested delivery date (Jan 10)
- Weight, volume, Incoterms, etc.

Step 5: TM planning & optimization

In TM, the planner:

- Selects a real route
- Assigns carrier
- Considers:
    - Transit time
    - Schedules
    - Cut-off times
    - Capacities

TM result:

- Actual transportation time = 4 days (instead of planned 3)

Example:

Goods issue: Jan 07

Transport time: 4 days

Delivery date: Jan 11

Step 6: TM sends dates back to S/4HANA

TM sends confirmed transportation dates back to S/4:

- Planned goods issue date
- Planned delivery date

Sales & delivery documents are updated

- Delivery date changes from Jan 10 → Jan 11

This ensures: ***The sales order promise is now aligned with real transportation execution***

Key integration principle (very important)-

- BPS does NOT replace TM
- TM does NOT determine MAD

They complement each other.

Component	Role
BPS	Strategic / promise-level scheduling
TM	Operational / execution-level scheduling
aATP	Availability & supply confirmation

What Happens if TM is NOT Used-

- BPS transportation duration is final
- No recalculation based on routes or carriers
- Delivery date = BPS result

In Short

- aATP still determines when the material is available (MAD)
- BPS calculates planned pick/pack/transport time to provide a promised delivery date
- Delivery creation does not involve TM
- Transportation is simplified, static, and not route/carrier optimized
- Confirmed delivery date is less precise for complex logistics scenarios

BPS alone can give a realistic promise, but TM is required for dynamic, optimized, and execution-based transport dates.

BPS Overview – How Schema and Characteristics Work Together-
BPS allows dynamic, order-specific activity durations, overriding static shipping point times, while merging schema defaults and characteristic-based overrides for realistic ATP dates.

1. BPS Activation
    - Controlled by Sales Document Type, Item Category, and Shipping Point.
    - Determines whether BPS is active for a particular sales order.

2. Activity Schema
    - Defines activities (Pick, Pack, etc.) and default durations.
    - Example: Pick = 1 day, Pack = 0.5 day.

3. Characteristics Combination
    - Provides more granular overrides based on business fields like Sales Organization, Distribution Channel, Division, etc.
    - Example: Pick = 2 days for a specific Sales Org / Division.

4. How System Calculates Durations
   - If BPS is active:
     1. Checks sales order against characteristics combination.
     2. Applies override durations from the combination if matched.
     3. Falls back to schema default durations for activities not defined in the combination.
   - Shipping point durations are ignored for activities maintained in BPS.

5. Key Principle
   - BPS = Schema defaults + Characteristics combination overrides
   - Only activities maintained in BPS are used for ATP calculations; everything else falls back or is ignored.

16 aATP BOP (Backorder Processing) & Strategies

In **SAP S/4HANA aATP**, **BOP (Backorder Processing)** is the mass, rule-based rescheduling of both confirmed and unconfirmed requirements—such as sales orders and stock transport orders—to reallocate limited supply based on business priorities rather than a simple *first-come, first-served* approach.

In S/4HANA aATP, sales orders that cannot be confirmed at the time of saving (due to insufficient stock or allocation constraints) are marked as backorders, but they do not automatically enter BOP. Instead, they remain unconfirmed until a scheduled BOP run explicitly selects and processes them.

In short, a backorder occurs when a sales order cannot be fully confirmed at the time of creation because sufficient stock or supply is not available on the requested date, and the order is later confirmed for a future date by running the BOP procedure.

In SAP S/4HANA aATP, there are specific processes, steps, and configuration settings designed to handle the Backorder Processing (BOP) procedure. These deliver the best and most efficient way to manage backorders within aATP.

BOP is available in both Classic ATP and aATP; the main difference is in how availability is recalculated and stock is reassigned.

- Classic ATP BOP (V_V2) is static and limited in functionality – requires enhancements for custom logic, with very restricted scope and criteria.
- aATP BOP is dynamic and flexible – configure rules, strategies, and logic via Fiori apps without enhancements, offering full control over prioritization (PAL, BOL, customer hierarchies).

Expanded explanation:

Classic BOP:	aATP BOP (Advanced BOP):
• Based on classic ATP logic • Limited rule flexibility • Processes backorders mainly by rescheduling dates and quantities • Less optimization and prioritization options	• Uses advanced ATP logic • Rule-based and priority-driven (customer, product, confirmation strategy, etc.) • Supports simulation, segmentation, and prioritization • More flexible and powerful compared to classic BOP

BOP is a key aATP function and strongly recommended, but it is not technically mandatory in the sense that aATP cannot work without it, and you are not forced to run at least one strategy like Redistribute.

BOP is a central and highly recommended aATP component for mass reprioritization and reallocation of confirmations, and when you use BOP, you must define at least one confirmation strategy, but aATP itself can technically run without scheduling BOP.

- aATP can perform online confirmations using PAC, PAL, ABC, SUP, etc. without any BOP run; confirmations are then based only on the original ATP result at order entry.
- BOP is provided as a mass re-prioritization tool: it rechecks many requirements together and redistributes limited supply according to confirmation strategies (WIN, GAIN, REDISTRIBUTE, FILL, LOSE, IMPROVE), but the system does not enforce that you schedule or execute BOP.
- In practice, many customers treat BOP as "backbone" for shortage situations because it is the only standard way to systematically realign confirmations when supply or priorities change, but this is a design best practice, not a hard technical requirement.
- To run BOP, you must indeed use at least one confirmation strategy in your BOP variant; without strategies BOP would have no rules for how to adjust confirmations.
- However, you can decide not to run BOP at all; in that case, no strategies (including REDISTRIBUTE) are used, and your confirmations stay as initially created by ATP/aATP.

Common Business Scenarios for BOP in S/4HANA

Backorder Processing in SAP S/4HANA is used to ensure that limited supply is allocated optimally across open sales orders based on business priorities and predefined confirmation strategies.

aATP BOP is used to decide "who gets the available stock first" when not everyone can be fulfilled immediately.

Common business scenarios for Backorder Processing (BOP) in SAP S/4HANA aATP include:

Scenario	Details
Short Supply / Allocation Shortage	• Demand exceeds available stock. • When demand exceeds available stock, BOP reallocates supply to the most important customers or orders. • BOP ensures limited stock is assigned to high-priority customers or orders first.
New Stock Arrival (Goods Receipt / Production Completion)	• When new supply becomes available, BOP redistributes stock to open backorders based on priority rules. • After goods receipt or production completion, BOP redistributes newly available stock to open backorders.
High-Priority Customer Fulfillment	• Ensures key or strategic customers are fulfilled first based on customer priority or hierarchy. • BOP reallocates stock from lower-priority orders if required.
Seasonal or Peak Demand Management	• During peak seasons (e.g., holidays), many orders go into backorder. • BOP helps confirm orders systematically using predefined strategies. • Manages large volumes of backorders during peak periods by confirming orders according to predefined strategies.
Frequent Supply Changes	• Supply changes multiple times a day (partial receipts, cancellations). • BOP can be run multiple times daily to keep confirmations up to date.
Reallocation Due to Order Changes or Cancellations	• When orders are reduced or canceled, freed-up stock is reassigned to other backorders. • Takes stock from lower-priority orders and reallocates it to higher-priority backorders.
Fair Distribution of Limited Stock	• Ensures no single customer consumes all available supply. • Supports business rules such as customer hierarchy and prioritization.
Integration with PAL / BOL	• BOP works together with Product Allocation (PAL) and Backorder Levels (BOL) to enforce allocation and prioritization logic.

Situations-

1. New high-priority order arrives but stock is already committed
Example: A VIP customer places an urgent order; BOP is used to take quantity away from lower-priority customers and reallocate it to the VIP order based on WIN/LOSE or WIN/GAIN strategies.

2. New supply or production becomes available after initial ATP checks
Example: Extra production is planned or an inbound delivery is posted; BOP is run to confirm or improve previously unconfirmed or partially confirmed orders using GAIN or FILL strategies.

3. Supply reduction or production delay after orders were confirmed
Example: A production line fails and planned receipts are delayed; BOP reprioritizes and may deconfirm low-priority orders while protecting key customers and near-due orders via WIN and REDISTRIBUTE.

4. Profit or margin-based allocation of scarce stock
Example: Company prefers domestic/high-margin customers over export/low-margin customers; BOP segments and strategies are set so domestic segment runs in GAIN or WIN and export in REDISTRIBUTE or LOSE.

5. Periodic re-prioritization for peak seasons or campaigns
Example: Weekly BOP runs during a promotion to continuously shift stock to time-critical, high-priority, or campaign-relevant orders while pushing out less urgent ones.

In such environments, BOP is run periodically (daily or even hourly) to re-optimize all open order confirmations using strategies such as:

- WIN: Secure confirmations first.
- GAIN: Improve confirmations if possible.
- IMPROVE: Shifts confirmations to earlier dates when possible.
- REDISTRIBUTE: Shift quantities between orders.
- FILL: Confirm as much as available.
- LOSE: Remove confirmations from lowest-priority orders.

BOP strategies define how the system redistributes available supply to backordered sales orders when availability changes. In aATP, they help prioritize important customers or orders and reassign stock so that the most important orders are fulfilled first.

Example:

There may be many sales orders in backorder on a daily basis. When stock availability changes (for example, due to goods receipt or production completion), the system needs to confirm quantities for these backorders. Based on business requirements, the BOP process can be run multiple times a day (for example, 2–3 times daily). During each run, the system assigns the available stock to backordered sales orders based on order priority, since the available stock may not be sufficient to fulfill all backorders.

As a result, the system confirms backorders only up to the limited available stock, prioritizing the most important orders first. This cycle continues and is repeated every day as stock availability changes.

In aATP, BOP provides multiple BOP strategies for confirmation. These strategies can be used to run BOP and confirm backorders based on business-defined rules and logic, such as customer priority, product importance, or allocation rules.

Use of BOP Strategies in aATP

BOP strategies control how backorders are processed and ensure that limited supply is assigned to the most important orders first, based on business rules.

They determine:

- Which sales orders are selected
- The priority sequence in which orders are processed
- How available stock is reallocated
- Whether confirmations are increased, reduced, or rescheduled
- Reassignment of stock from lower-priority orders to higher-priority orders
- Confirmation or re-confirmation of sales orders based on:
    - Customer priority
    - Product priority
    - PAL / BOL rules
    - Requested or confirmed delivery dates

Key Capabilities of aATP BOP Strategies-

- Rule-based prioritization (no enhancements required)
- Simulation before execution
- Flexible confirmation strategies (increase, reduce, reschedule)

Business Example-

- Customer 1001 (high priority) and Customer 1002 (low priority) both have backorders.
- New stock of 1000 units is becoming available.
- BOP strategy ensures:
    - Customer 1001 is confirmed first
    - Remaining stock is allocated to Customer 1002

How it is configured and used-

In S/4HANA 2023, BOP is configured and used through aATP Fiori apps such as:

- ***Create BOP Segment***
- ***Create BOP Variant***
- ***Schedule BOP Run***
- ***Monitor BOP Run***

Business users define selection criteria and strategies in segments, bundle them into variants, schedule runs (either simulation or update), and then review the results to ensure that confirmations align with business rules and service-level priorities.

Confirmation Strategies in SAP S/4HANA aATP BOP

A total of six strategies is available to control how sales orders gain, lose, or retain ATP confirmations during Backorder Processing (BOP) runs. It is not mandatory to use all six strategies—businesses can apply any single strategy or a combination of multiple strategies, depending on prioritization and fulfillment requirements.

These strategies control how orders gain/lose confirmations during Backorder Processing runs:

During BOP processing, strategies are executed in a predefined priority sequence: WIN, GAIN, IMPROVE, REDISTRIBUTE, FILL, and LOOSE.

Sales orders are assigned to BOP segments first based on selection criteria (customer group, order type, delivery priority, ATP category, etc.), then each segment gets a confirmation strategy from the BOP variant.

There are six available BOP strategies, and it is not mandatory to use all of them. Based on business requirements, you can use a combination of two, three or four strategies or even just one strategy if that meets the business need.

1. Strategy WIN-

Key Points	• This is the highest priority strategy • Full confirmation only • On-time confirmation • Only increases confirmations (+) • No decrease (–) is allowed • Raises an exception if full confirmation is not possible
Example	• Sales Order Quantity: 1,000 units (Confirmed – 0) After BOP Run- • Sales Order Quantity: 1,000 units (Confirmed – 1,000) If the system is not able to fully confirm the order quantity on time, it will raise an exception and stop BOP processing for that specific material and plant combination. On-time means that if the Sales Order RDD (Requested Delivery Date) is, for example, Jan 15, the system will try to confirm the order on for Jan 15 and not for a future date. If it cannot do so, an exception will be raised. In such cases, the exception must be reviewed manually, and a business decision is required on how to proceed. • Sales Order Quantity: 1,000 units (Confirmed – 800) • Sales Order Quantity: 1,000 units (Confirmed – 1,000) In this case, the system will try to increase the confirmed quantity from 8,00 units up to 1,000 units using any additional available stock. If the additional stock is still not sufficient, it will reallocate stock from lower-priority sales order strategies. However, it will not reduce the confirmed quantity below 800 units, such as to 5,00 or 0 units.
Details	For the WIN strategy, the system always tries to increase confirmations to achieve full and on-time confirmation. • Additional stock can come from sources such as purchase order goods receipts, customer returns, or other supply elements. • If the additional stock is still not sufficient, the system will reallocate stock from sales orders of lower-priority strategies (such as Redistribute or Fill and Loose) by reducing the confirmed quantities of those sales order line items and using that stock to increase full and on-time confirmations for the WIN strategy, for the same material and plant combination.

- The system first uses the available stock for the current strategy. If full and on-time confirmation is not achieved, it then looks to lower-priority sales order strategies (such as Redistribute, Fill, and Loose) to reallocate stock and fulfill the confirmation.
- This process reduces confirmations from lower-priority sales order line items and increases confirmations for the WIN strategy sales order line items.

The system evaluates all strategies for the relevant material/plant combination, reduces confirmations from sales order line items of lower-priority strategies as needed, and uses that stock to fully and on-time confirm sales orders assigned to the WIN strategy.

If the system cannot achieve full and on-time confirmation for the WIN strategy, it will raise an exception. The material/plant combination will be put on hold for that WIN processing. The stock will not be reallocated to other strategies until the issue is resolved. This ensures that VIP or high-priority orders under WIN are treated with the highest priority, and other strategies cannot consume their stock. If the system cannot satisfy the WIN criteria, the order will not be processed by any other strategies.

WIN is the highest-priority strategy, and stock is not allocated to lower-priority strategies unless the WIN strategy is fully satisfied.

Full and on-time confirmation is based on the customer's Requested Delivery Date (RDD). If the RDD is in the past, the system will try to confirm the order with today's date; if this is not possible, an exception will be raised. The WIN strategy does not allow confirmation on a future date.

A report is available in Fiori to view the exception list. This report shows the affected material/plant combinations and the number of customers impacted by each exception.

If a sales order line item with a specific material/plant combination falls under the WIN strategy and cannot be fully and on-time confirmed, the system will raise an exception, put the combination on hold, and prevent it from being used by other strategies.

This means the order cannot be shipped until the issue is resolved. Users must check the exception list and take action to achieve confirmation for the earliest possible delivery, as these are critical orders for VIP customers.

The WIN strategy will never reduce the confirmed quantity, and an exception is always generated if the quantity cannot be fully confirmed.

2. Strategy GAIN-

Key Points	• This is the 2nd priority strategy • Full confirmation only • In Future confirmation • Only increases confirmations (+) • No decrease (–) is allowed • Raises an exception if full confirmation is not possible
Example	• Sales Order Quantity: 1,000 units (Confirmed – 800) • Sales Order Quantity: 1,000 units (Confirmed – 1,000) If the system is not able to fully confirm the order quantity on the Requested Delivery Date (RDD) or any future date, it will raise an exception and stop BOP processing for that specific material and plant combination. "In Future" means that if the Sales Order RDD is, for example, Jan 15, the system will try to confirm the order on Jan 15 or on a later date. If it cannot, an exception will be raised. In such cases, the exception must be reviewed manually, and a business decision is required on how to proceed. • Sales Order Quantity: 1,000 units (Confirmed – 8,00) • Sales Order Quantity: 1,000 units (Confirmed – 0) In this case, the system will try to increase the confirmed quantity from 800 units up to 1,000 units using any additional available stock. If the additional stock is still not sufficient, it will reallocate stock from lower-priority sales order strategies. However, it will not reduce the confirmed quantity below 8,00 units, such as to 5,00 or 0 units.
Details	The GAIN strategy aims to increase confirmations to achieve full and "in future" confirmation. If the system cannot confirm the full quantity, even considering future availability, it will raise an exception and stop further processing for that material/plant combination. This ensures that orders under GAIN are prioritized for full fulfillment, similar to WIN, but with the flexibility of "future" confirmation dates. • Additional stock can come from sources such as purchase order goods receipts, customer returns, or other supply elements. • If the additional stock is still not sufficient, the system will reallocate stock from sales orders of lower-priority strategies (such as Redistribute or Fill and Loose) by reducing their

confirmed quantities and using that stock to increase full and on-time confirmations in future for the GAIN, for the same material and plant combination.

- The system first uses the available stock for the current strategy. If full confirmation, even considering future availability, is not achieved, it then looks to lower-priority sales orders strategies (such as Redistribute, Fill, and Loose) to reallocate stock and fulfill the confirmation.
- This process reduces confirmations from lower-priority sales order line items and increases confirmations for the GAIN strategy sales order line items.

The system evaluates all strategies for the relevant material/plant combination, reduces confirmations from sales order line items of lower-priority strategies as needed, and uses that stock to fully and in future confirm sales orders assigned to the GAIN strategy.

If the system cannot achieve full and in future confirmation for the GAIN strategy, it will raise an exception. The material/plant combination will be put on hold for that GAIN processing. The stock will not be reallocated to other strategies until the issue is resolved.

This ensures that VIP or high-priority orders under WIN are treated with the highest priority, and other strategies cannot consume their stock. If the system cannot satisfy the GAIN criteria, the order will not be processed by any other strategies.

GAIN will be the highest-priority strategy if not using WIN, and stock is not allocated to lower-priority strategies unless the GAIN strategy is fully satisfied.

Full and In Future confirmation is based on the customer's Requested Delivery Date (RDD). If the RDD is in the past, the system will try to confirm the order with today's date or future date; if this is not possible, an exception will be raised. The GAIN strategy does allow confirmation on a future date.

A report is available in Fiori to view the exception list. This report shows the affected material/plant combinations and the number of customers impacted by each exception.

If a sales order line item with a specific material/plant combination falls under the GAIN strategy and cannot be fully and in future confirmed, the system will raise an exception, put the combination on hold, and prevent it from being used by other strategies.

This means the order cannot be shipped until the issue is resolved. Users must check the exception list and take action to achieve

	confirmation for the earliest possible delivery, as these are critical orders for VIP customers. The system evaluates all strategies for the relevant material/plant combination, reduces confirmations from lower-priority strategies as needed, and uses that stock to fully and in future confirm orders assigned to the GAIN strategy. If full and in future confirmation still cannot be achieved, the system will raise an exception and stop further processing for that material/plant combination. The stock will not be used for other strategies. GAIN is the 2nd highest-priority strategy, and stock is not allocated to lower-priority strategies unless the GAIN strategy is fully satisfied.

Additional Key Points-

1. Main Difference between WIN and GAIN: The key difference is "On Time" vs. "In Future".

- WIN supports "In Full" and "On Time" confirmations.
- GAIN supports "In Full" and "In Future" confirmations.

2. For WIN, if the sales order line item is not confirmed In Full and On Time, an exception will be raised.

3. For **GAIN**, if the sales order line item is **not confirmed In Full**, an **exception will be raised**, but
confirmation may still be scheduled for a **future date**.

4. If both WIN and GAIN strategies are used and there are two sales orders with the same material/plant combination - one under WIN and one under GAIN - the system always gives priority to the WIN strategy to fulfill the sales order before considering GAIN.

5. It is not mandatory to use both WIN and GAIN strategies - you can decide based on business needs. You may choose to use GAIN without WIN, and in this case, GAIN will be treated as the highest-priority strategy, since WIN is not being used.

6. In basic ATP checks, the future confirmation date for a sales order is determined by cumulative ATP quantities, which are calculated from future receipts such as purchase orders, production orders (released or unreleased), or planned orders during forward scheduling. The future receipts drive the confirmation date—the system confirms the order when enough cumulative supply becomes available.

3. Strategy IMPROVE-

Key Points	• This is the 3rd priority strategy • Only increases confirmations (+) • No decrease (–) is allowed, except in some specific situations. • In Future confirmation • Full (100%) confirmation is not mandatory. • No exception is generated.
Example	• Sales Order Quantity: 1,000 units (Confirmed – 8,00) After BOP Run- • Sales Order Quantity: 1,000 units (Confirmed – 1,000) • Sales Order Quantity: 1,000 units (Confirmed – 9,00) • Sales Order Quantity: 1,000 units (Confirmed – 5,00) reduction only in exceptional ATP-decrease scenarios. The IMPROVE strategy has no exceptions; it primarily increases confirmed quantities and generally does not decrease them. It supports "In Future" confirmations, allowing orders to be confirmed on a future date if stock becomes available. For example, if a sales order was originally created with 1,000 units but initially confirmed for 800 units, during BOP the system may adjust the confirmation: • Increase from 800 → 900 or 800 → 1,000 units • In some situations, a decrease is possible from 800 → 500 units The IMPROVE strategy can reallocate stock from lower-priority strategies (Redistribute, Fill, and Loose), but it will not provide stock to WIN or GAIN from the IMPROVE strategy.
Details	The IMPROVE strategy is used to optimize order confirmations, increasing partially confirmed orders using available or reallocated stock while respecting higher-priority orders like WIN and GAIN. The IMPROVE strategy does not guarantee 100% (full) confirmation. It can result in confirmations at any percentage level, depending on available stock. If you are using all three strategies—WIN, GAIN, and IMPROVE—the system will prioritize confirmations in the following order: 1. WIN – highest priority, full and on-time confirmation for VIP customers. 2. GAIN – next priority, full confirmation "in future" if needed.

	3. IMPROVE – last, increases confirmed quantities using available or reallocated stock without affecting WIN or GAIN. If you are using IMPROVE without WIN or GAIN, then IMPROVE becomes the highest-priority strategy and will be used first for confirming sales orders. In some situations, a decrease is possible (for example, from 800 → 500 units). This can occur when stock was available earlier but becomes blocked or unavailable at the time the BOP is executed. No exception is triggered for the IMPROVE strategy. However, you can still review the Fiori dashboard to see whether quantities were increased, reduced, or remained unchanged. The Improve strategy continuously attempts to optimize confirmed quantities and delivery timing, including rescheduling to future dates when required. It does not reduce quantities to reallocate to WIN or GAIN; however, confirmed quantities may decrease due to reduced ATP availability, and no exception is generated.

Key Points-

The main difference between WIN/GAIN and IMPROVE is that WIN/GAIN cannot reduce the confirmed quantity and always raise an exception, whereas in IMPROVE the confirmed quantity may be reduced and no exception is raised.

4. Strategy REDISTRIBUTE-

Key Points	• This is the 4th priority strategy • Increases Confirmations (+) • Decrease Confirmation (–) • In Future confirmation • No exception is generated.
Example	• Sales Order Quantity: 1,000 units (Confirmed – 8,00) • Sales Order Quantity: 1,000 units (Confirmed – 1,000) • Sales Order Quantity: 1,000 units (Confirmed – 9,00) • Sales Order Quantity: 1,000 units (Confirmed – 0) The REDISTRIBUTE strategy has no exceptions; it can both increase and decrease confirmed quantities. It supports "In Future" confirmations, allowing orders to be confirmed on a future date if stock becomes available. If a sales order was originally created for 1,000 units but initially confirmed for 800 units, during BOP the system may adjust the confirmation: • Decrease: 800 → 0 units to give stock to WIN, GAIN, or IMPROVE • Increase: 800 → 900 or 800 → 1,000 units The REDISTRIBUTE strategy can reallocate stock to WIN, GAIN, or IMPROVE; if not needed by these strategies, it can use the stock to improve its own confirmations.
Details	• REDISTRIBUTE is similar to IMPROVE in that it can adjust confirmations to optimize stock. • It can decrease confirmed quantities to give stock to WIN, GAIN, or IMPROVE; if no other strategy needs it, it can use the stock to improve its own confirmations. • Within REDISTRIBUTE, reallocation among sales orders is possible for example, stock from one sales order can be given to another. • The main importance of REDISTRIBUTE is that decreasing confirmed quantities is allowed, which is not possible in WIN, GAIN, or IMPROVE. • IMPROVE cannot reduce confirmed quantities to give stock to WIN or GAIN, whereas REDISTRIBUTE can reduce confirmed quantities and reallocate stock to WIN, GAIN, or IMPROVE. • REDISTRIBUTE is the most widely used strategy in AATP Backorder Processing (BOP).

5. Strategy FILL-

Key Points	• This is the 5th priority strategy • Retain Confirmation • Decrease Confirmation (–) • NO Increases Allowed (+) • In Future confirmation • No exception is generated
Example	• Sales Order Quantity: 1,000 units (Confirmed – 8,00) • Sales Order Quantity: 1,000 units (Confirmed – 8,00) • Sales Order Quantity: 1,000 units (Confirmed – 5,00) • Sales Order Quantity: 1,000 units (Confirmed – 0) The FILL strategy has no exceptions. It can retain confirmed quantities and reduce confirmations if needed, but it cannot increase or improve confirmations. If a sales order was originally created for 1,000 units but initially confirmed for 800 units, during BOP the system may adjust the confirmation: • Retain: 800 → 800 • Decrease: 800 → 0/500 units to give stock to WIN, GAIN, IMPROVE or REDISTRIBUTE The FILL strategy can reallocate stock to WIN, GAIN, IMPROVE, or REDISTRIBUTE. If no other strategy requires the stock, the confirmed quantity remains unchanged, and FILL cannot use the stock to improve its own confirmations.
Details	• FILL retains the confirmed quantity; it can decrease confirmations but cannot increase them. • Any decrease happens only to reallocate stock to WIN, GAIN, IMPROVE, or REDISTRIBUTE. • FILL never increases confirmations on its own. • In the FILL strategy, fulfillment does not happen immediately from the FILL bucket itself. Orders will move to earlier strategies (such as REDISTRIBUTE, IMPROVE, WIN, or GAIN) based on dynamic conditions, like the delivery date, for actual confirmation. Example – • A sales order with a delivery date 2 months from now will initially remain in the FILL bucket. • After 1 month, it may move to the REDISTRIBUTE bucket based on dynamic conditions, and later to

IMPROVE, GAIN, or WIN as the schedule line delivery date approaches and priority increases.
    - This behavior occurs because of dynamic conditions.
    - If static conditions are used (e.g., Sales Organization = 1710), the order will remain in the FILL bucket and will never improve its confirmation.
    - Therefore, FILL requires dynamic conditions based on the schedule line delivery date to move orders to other buckets over time.
- FILL can decrease confirmed quantities to reallocate stock to WIN, GAIN, IMPROVE, or REDISTRIBUTE.

If no other strategy requires the stock, the confirmed quantity remains unchanged; FILL does not increase confirmations on its own.

6. Strategy LOOSE-

Key Points	• This is the 6th priority strategy • No Retain • No Increase • Only decrease: confirmations are reduced, and always down to 0 • No exception is generated
Example	• Sales Order Quantity: 1,000 units (Confirmed – 8,00) • Sales Order Quantity: 1,000 units (Confirmed – 0) The LOOSE strategy has no exceptions. It can only reduce confirmations to 0, but it cannot retain or increase confirmations. If a sales order was originally created for 1,000 units but initially confirmed for 800 units, during BOP the system will reduce to 0 only: • Decrease: 8,00 → 0 unit to give stock to WIN, GAIN, IMPROVE or REDISTRIBUTE LOOSE strategy always clears confirmed quantities (to 0) so that stock can be reallocated to higher-priority strategies such as WIN, GAIN, IMPROVE, or REDISTRIBUTE.
Details	• LOOSE always reduces confirmations to 0; it cannot retain or increase quantities. • Any decrease happens only to reallocate stock to WIN, GAIN, IMPROVE, or REDISTRIBUTE. • LOOSE never increases confirmations on its own. • Fulfillment does not happen immediately from the LOOSE bucket itself. Orders move to earlier strategies (REDISTRIBUTE, IMPROVE, WIN, or GAIN) based on dynamic conditions, such as the delivery date, for actual confirmation. Example- • A sales order with a delivery date 6 months from now will initially remain in the LOOSE bucket. • The order will not be confirmed in this bucket because LOOSE always reduces confirmations to 0. • Therefore, dynamic conditions are required to move the sales order from LOOSE to higher-priority strategies; otherwise, the order would never be confirmed.

- LOOSE can decrease confirmed quantities to reallocate stock to WIN, GAIN, IMPROVE, or REDISTRIBUTE.
- If no other strategy requires the stock, the confirmed quantity remains at 0; LOOSE does not increase confirmations on its own.

Key Points-

1. **For the FILL and LOOSE strategies, dynamic conditions are always required.** Without dynamic conditions, sales orders in these buckets **will never be confirmed** and will remain in the bucket without any confirmation.

2. If no other strategy requires the stock, the confirmed quantity remains at 0, and LOOSE cannot use the stock to improve its own confirmations.
    - LOOSE always reduces confirmation to 0
    - It never retains or increases confirmation

So "remains unchanged" means it stays at 0, not that it keeps any previous quantity

7. Strategy SKIP-

Key Points	• SKIP does not participate in BOP. • Any order assigned to SKIP is excluded from the BOP run, and the system skips these orders. • Sales orders with the Fixed Indicator are assigned to SKIP and are not considered in BOP processing.
Example	Example – SKIP in AATP BOP • A sales order is created for 1000 units. • The order is manually confirmed, and the Fixed Indicator is set on the schedule line. • Because the confirmation is fixed, the order is assigned to SKIP. Result: • During the BOP run, the system skips this order completely. • The confirmed quantity (1000 units) and delivery date remain unchanged. The order is not re-evaluated, reduced, increased, or rescheduled by any BOP strategy (WIN, GAIN, IMPROVE, REDISTRIBUTE, FILL, or LOOSE).
Details	Sales order items marked with the "Fixed Date & Quantity" indicator are excluded from AATP Backorder Processing (BOP) runs. Typical Use Cases • A customer-critical order is manually confirmed by the business and must not be changed by BOP. • Orders already agreed with the customer • Orders that must not be rescheduled or reallocated by BOP Setting the Fixed Indicator ensures the order is placed in SKIP and excluded from BOP.

Recommended Usage of aATP BOP Strategies

1. **REDISTRIBUTE alone can be sufficient** and is used in most implementations.
2. **FILL and LOOSE are typically used together with REDISTRIBUTE** to manage low-priority or future-dated demand.

3. From **WIN, GAIN, and IMPROVE**, **only one strategy should be used**, as these strategies serve similar purposes.
4. **WIN and GAIN are used very selectively** because they are specific to VIP or critical customers and **always raise exceptions**. Managing a high volume of exceptions requires **manual review and resolution**, which can be challenging.
5. **WIN/GAIN are suitable for special customers or critical business scenarios** where full quantity and on-time fulfillment are mandatory.
6. **Recommendation:** Use **IMPROVE** (instead of WIN/GAIN) in combination with **REDISTRIBUTE, FILL, and LOOSE** for most implementations, as this provides optimization without excessive exception handling.

This is a **comprehensive comparison table** of all standard SAP S/4HANA aATP BOP strategies:

Strategy	Priority	Increase Confirmation	Decrease Confirmation	Retain Confirmation	Exception Generated	Fulfillment Timing
WIN	1	YES – In Full/On Time	NO	YES	YES	Immediate
GAIN	2	YES – In Full/In Future	NO	YES	YES	Immediate
IMPROVE	3	YES	YES – If needed	YES	NO	DYNAMIC
REDISTRIBUTE	4	YES	YES	CONDITIONAL	NO	DYNAMIC
FILL	5	NO	YES	YES	NO	DYNAMIC
LOOSE	6	NO	YES – Always to 0	NO	NO	DYNAMIC

Key Points:

- **FILL and LOOSE** require **dynamic conditions** (e.g., schedule line delivery date) to move orders to other strategies; otherwise, orders remain unconfirmed.
- **WIN/GAIN** never decrease confirmations and raise exceptions if stock is insufficient.
- **REDISTRIBUTE** is the most flexible strategy and can increase, decrease, or reallocate stock.

- **IMPROVE** tries to optimize without generating exceptions but can reduce confirmations if ATP decreases.

The priority and execution (evaluation) sequence in a Backorder Processing (BOP) run in SAP S/4HANA Advanced ATP follows a defined order of confirmation strategies. The system begins by **deconfirming** orders assigned to the lowest-priority strategy (**LOSE**) to free up available stock. It then proceeds to **confirm or reconfirm** orders associated with higher-priority strategies, following this evaluation sequence:

Strategy	Priority / Execution Sequence in BOP Run	What Happens / Meaning
Lose	1 (lowest)	First, the system revokes confirmations from low-priority orders to free stock. Requirements assigned "Lose" lose their existing confirmation and are not confirmed — freed-up supply becomes available for other orders.
Win	2	Next, it confirms or reconfirms the highest priority orders to secure supply. Requirements are (re)confirmed fully on the requested date (or immediately if date has already passed); i.e. highest priority.
Gain	3	Then, orders under this strategy are confirmed or improved after WIN orders. Requirements retain their existing confirmation (or — if possible — get a better confirmation). They should not be worsened.
Improve	4	Orders try to maintain or improve confirmation but may lose quantity if limited supply. The system tries to retain or improve the confirmation for these requirements, but they might also lose confirmation if supply is constrained.
Redistribute	5	Available supply is distributed fairly among similar priority orders. Requirements may get a better, equal, or worse confirmation depending on availability — supply may be redistributed across requirements.
Fill	6	Finally, orders maintain existing confirmation levels without gaining more.

		Requirements will not get earlier confirmations or more quantity; they may get equal or worse confirmation. Fill is often used for less-critical orders.

How a Sales Order is picked up by BOP in SAP S/4HANA aATP (2023), how it goes into the right bucket strategy, and what happens after that.

1. How a Sales Order becomes BOP-relevant

A sales order does not automatically go to BOP. It becomes BOP-relevant based on aATP configuration.

- A sales order does not automatically go to BOP.
- Advanced ATP (aATP) must be active (not classic ATP).
- BOP relevance must be enabled in aATP configuration (e.g. via Supply Protection / aATP check).
- During ATP check at order creation/change, the system evaluates the result.
- If the order is partially confirmed, unconfirmed, date-shifted, or redistributable, it is flagged as BOP-relevant.
- BOP relevance is set at item level, not header level.
- The flag is stored, and the order is picked up later when BOP is executed.

In short: During sales order creation or change, the following process occurs:

Sales Order → aATP check → BOP relevance flag set → included in BOP run

At this point, the sales order item is flagged as BOP-relevant and stored for reprocessing. "Stored for reprocessing" means the sales order item is internally marked as eligible so that future BOP runs can select it and reassign ATP confirmations.

BOP relevance is technically persisted within the aATP framework; however, there is no user-facing or documented field in standard SD tables (VBAK, VBAP, VBEP) that explicitly indicates "BOP-relevant. This information is stored internally and consumed by BOP processing logic.
"Stored for reprocessing" does not correspond to a visible SD table entry. It means the sales order item is internally marked as eligible so that future BOP runs can select it and reassign ATP confirmations.

Key Points-

- When a sales order item passes the aATP check, it may be flagged as BOP-relevant.
- This relevance is persisted internally in aATP tables.
- There is no documented or visible BOP flag in standard SD tables.
- The item is stored so that future BOP runs can reprocess ATP confirmations.
- Verification is done via BOP simulation or execution, not by checking database tables.

2. How the Sales Order goes into a BOP Segment (Bucket Strategy)

In S/4HANA aATP, a **bucket strategy** is essentially **BOP Segmentation**, which is used to prioritize and allocate available stock across different orders.

Step-by-step flow:

BOP Run is Triggered	• BOP can run manually or via background job.
	• Transaction / Fiori apps used:
	    o *Schedule Backorder Processing*
	    o *BOP Simulation*
	At this stage, the system will select only those sales order items flagged as BOP-relevant.

System Evaluates BOP Selection Criteria	The system determines which sales order items should enter BOP based on: • Plant • Material / Material group • Sales organization / Distribution channel • Sales document type • Requested delivery date • ATP confirmation status (partial/unconfirmed) Only items matching the criteria proceed to segmentation.
Sales Order Item is Assigned to a BOP Segment	Segmentation (Bucket Strategy) groups orders based on business rules. Typical criteria for segments include: • Customer or Customer Group • Material or Material Group • Sales Organization / Distribution Channel • Priority (VIP vs Regular Customers) Important: Each sales order item belongs to only one segment. The system evaluates segments top-down and assigns the item to the first matching segment.

Segment Strategy Determines Allocation	Once in a segment, the allocation strategy applies: • Sort rules (e.g., earliest requested delivery date, order priority) • Confirmation strategy (Win/Lose, Gain/Redistribute) • Confirmation mode (fully, partially, or zero confirmation) Example: VIP customers may get stock first; lower-priority customers get the remainder.

Stock & Supply Are Reassigned Within the Segment	• Existing ATP confirmations may be deconfirmed.
	• Stock is reassigned according to segment priorities.
	• Supply sources used include:
	  ○ On-hand stock
	  ○ Planned receipts (POs, Production Orders)
	  ○ Subcontracting or transfer stock

Remaining Segments Are Processed	• The next lower-priority segment only receives stock left after higher-priority segments are fulfilled. • This ensures business priorities are enforced.
Sales Order Is Updated	After BOP processing: • Confirmed quantity may increase/decrease. • Confirmed delivery dates may shift. • ATP category/status is updated. • Results can be monitored in VA03 or Fiori BOP apps.

Simple Visual Example

Stock = 100 units

Segment	Customer Type	Demand
10	VIP	70 units
20	Regular	50 units

BOP Result:

- VIP Customer -> 70 units confirmed
- Regular Customer -> 30 units confirmed

BOP Segment Strategy –

- SAP S/4HANA aATP allows multiple BOP strategies (e.g., 6 strategies available).
- It is not mandatory to use all strategies — you can use 1 or a combination of 2–3, based on business requirements.
- Only the assigned strategies are applied, in the defined sequence.

- This flexibility ensures priority-based stock allocation without overcomplicating BOP processing.

Confirmation strategies in SAP S/4HANA aATP Backorder Processing (BOP) are predefined business rules that determine how sales order and stock transport order requirements are handled and prioritized during a BOP run. These strategies allocate available supply by sorting orders into categories such as WIN, GAIN, IMPROVE, REDISTRIBUTE, FILL, and LOSE, each specifying whether orders retain, gain, redistribute, or lose confirmation quantities based on priority and supply changes.

In SAP S/4HANA AATP Backorder Processing (BOP), the available supply during a BOP run is determined from ATP (Available-to-Promise) checks, which consider multiple sources of stock and supply.

Unrestricted Stock	Stock in the plant or storage location that is freely available.
Stock in Transit	Stock that is being transferred between plants or storage locations.
Planned Receipts	Incoming purchase orders, production orders, or planned stock from MRP.
Sales Orders	Existing confirmed sales orders are considered to avoid overcommitment.
Reservations	Stock reserved for other orders or production.
Quality or Blocked Stock	Usually excluded unless configured.

- During the BOP run, the system performs ATP checks for each sales order line to determine if the requested quantity can be confirmed.
- The BOP strategies (WIN, GAIN, IMPROVE, REDISTRIBUTE, FILL, LOOSE) then allocate or reallocate available supply based on priorities and dynamic conditions.
- The system uses the current available supply in real-time, including incoming stock and unconfirmed orders, to optimize confirmations.

Confirmation strategies are assigned to BOP segments in the "Configure BOP Variant" app and dictate how orders are confirmed or de-confirmed as stock availability changes.

1	*WIN*	Highest priority orders that are fully confirmed first.
2	*GAIN*	Orders that retain or gain confirmations after WIN orders.
3	*IMPROVE*	Orders that try to retain or improve confirmations may lose quantity if supply is insufficient.
4	*REDISTRIBUTE*	Orders redistribute confirmations fairly among orders of similar priority.
5	*FILL*	Orders that do not gain additional confirmations but may receive equal or worse confirmation.
6	*LOOSE*	Lowest priority orders that lose confirmation to free supply for higher-priority orders.

These strategies help prioritize high-value or critical orders and improve supply chain efficiency by dynamically reallocating inventory during availability fluctuations. The system can revoke confirmations from low-priority orders (LOSE) and reallocate to WIN or GAIN orders, ensuring optimal fulfillment per business priorities. Confirmation strategies allow businesses to implement complex prioritization logic beyond simple first-come-first-served approaches.

Thus, confirmation strategies serve as essential configuration elements in BOP runs to maintain or improve order confirmations aligned with priority rules, maximizing customer service levels under constrained supply scenarios.

These strategies enable dynamic reallocation of supply by de-confirming low-priority orders and reallocating to higher-priority ones, with each strategy defining specific rules on whether an order can gain, retain, redistribute, or lose confirmation quantities during a BOP run. This sophisticated approach allows better fulfillment of valuable orders while adapting to fluctuating supply and demand conditions.

These strategies can be configured in the "Configure BOP Variant" app, where segments are linked with confirmation strategies and executed in sequence during a BOP run. This setup provides businesses with flexibility to implement complex prioritization and adjustment of order confirmations dynamically during availability fluctuations.

At the time of a Backorder Processing (BOP) run in SAP S/4HANA Advanced ATP (aATP), the first BOP confirmation strategy that starts is the LOSE strategy. The system begins by revoking or deconfirming all confirmations assigned to low-priority orders classified under the LOSE strategy. This action frees up stock and supply that can then be reallocated to higher-priority orders based on the subsequent strategies like WIN, GAIN, and REDISTRIBUTE.

This approach ensures stock is first taken away from the lowest priority demands to enable optimal fulfillment of more critical or high-priority orders early in the BOP process.

BOP Setup Configuration Steps - (Fiori-Only Approach)

In SAP S/4HANA, Backorder Processing (BOP) can be fully configured and executed through Fiori Launchpad—no SAP GUI or classic transactions are required. Using Fiori apps such as Configure BOP Segment (F2158) and Configure BOP Variant (F2160), users with roles like Order Fulfillment Manager can perform the entire setup and monitoring workflow.

The BOP Fiori configuration setup starts with the **Characteristics Catalog** - In the Characteristics Catalog, we define the required fields (characteristics) that will be used to build segments and conditions in Backorder Processing.

These characteristics are derived from sales order, customer, material, and schedule line data (for example: Sales Organization, Requested Delivery Date, Customer Group, Material, etc.).

Fiori App – Manage Characteristics Catalog

- The Manage Characteristics Catalog Fiori app is used to activate and manage fields (characteristics) for multiple AATP applications.
- Currently, there are 13 Characteristics Catalogs available.
- These catalogs are shared across several applications, including:
    - Product Allocation
    - Backorder Processing (BOP)
    - Alternative-Based Confirmation (ABC)
    - Supply Protection
    - Business Process Scheduling (BPS)

Characteristics Catalogs for BOP

- For Backorder Processing, three catalogs are available:
    - Sales Order
    - Stock Transport Order (STO)
    - Mixed
- The characteristics that can be used in segments and conditions must be selected from the relevant catalog, depending on the document type being processed.
- In this case, the catalog "Backorder Processing – Base (Sales Document)" is selected.

Manage Characteristic Catalogs

Standard

Characteristic Catalogs (13)

Catalog Use Type	Catalog Type	Enabled for Sales	Enabled for Stock Transport
Product Allocation	Base (Sales Document)	Yes	No
Product Allocation	Base (Stock Transport Order)	No	Yes
Product Allocation	Mixed	Yes	Yes
Backorder Processing	Base (Sales Document)	Yes	No
Backorder Processing	Base (Stock Transport Order)	No	Yes
Backorder Processing	Mixed	Yes	Yes
Alternative-Based Confirmation	Base (Sales Document)	Yes	No
Order Fulfillment Responsibility	Base (Material Master Data)	No	No
Supply Protection	Base (Sales Document)	Yes	No
Supply Protection	Base (Stock Transport Order)	No	Yes
Supply Protection	Mixed	Yes	Yes
Business Process Scheduling	Base (Sales Document)	Yes	No
Business Process Scheduling	Base (Stock Transport Order)	No	Yes

1. Characteristics-

- In this section, there are 38 characteristics available.
- Additional characteristics can be added by selecting the "**Add**" icon.
- With SAP S/4HANA AATP 2023, the number of available characteristics has significantly increased, with over 200+ fields available for configuration and selection.

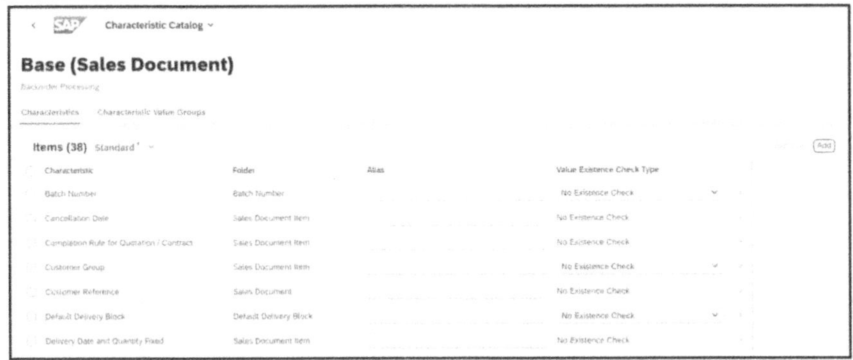

From this pop-up below, you can select the required fields. These fields are maintained at the client level, so care must be taken when selecting them. Do not select fields indiscriminately; only choose those that are truly required.

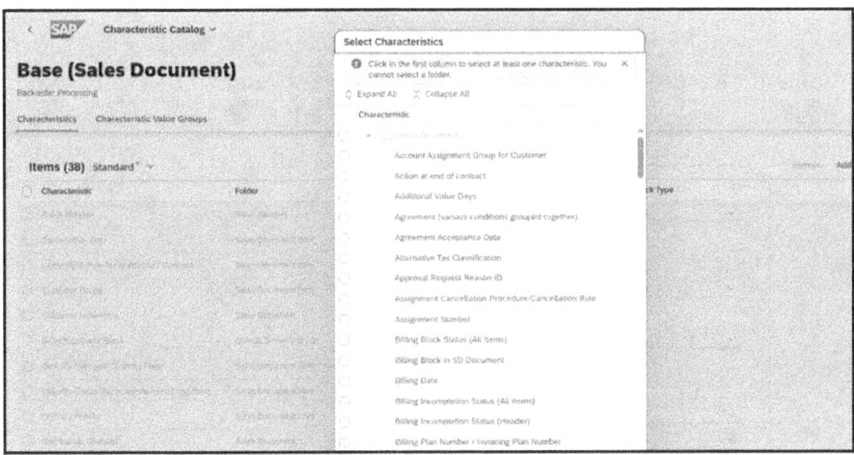

Key Points-

1. Adding a New Custom Field for BOP / aATP

- To add a new custom field, it must first be created by an ABAP consultant using the Custom Fields and Logic (User Extensibility) application.
- The custom field is not specific to aATP; once created, it becomes available across the SAP system.
- During creation, the field must be assigned to the relevant application area (for example, Sales Order – Header or Item level).

- After the field is added at the sales order header or item level, it will automatically appear in the Characteristics Catalog.
- The field can then be activated and used for aATP BOP segments and conditions.

2. Characteristics Value Group-

- Characteristics Value Groups are a separate master data object used specifically for aATP.
- They do not have a direct association with standard master data such as the Material Master or Customer Master.
- These value groups are used to group characteristic values for use in aATP features like Backorder Processing, Product Allocation, and Supply Protection.

Scenario:

1. You want to prioritize a group of customers in BOP, but you don't want to maintain multiple customer values in every segment.

Steps / Concept:

- Create a Characteristics Value Group called VIP_CUSTOMERS.
- Add the following values to this group:
    - Customer Group = A1
    - Customer Group = A2
    - Customer Group = A3
- In the segment selection condition, instead of selecting Customer Group A1, A2, A3 individually, you select the Value Group: VIP_CUSTOMERS.

2. Characteristics Value Group

- Value Group: FAST_MOVING_MATERIALS
- Characteristic: Material Group (MG1, MG2, MG3, MG4, …)

Value Mapping:

- MG1 → Materials M1, M2
- MG2 → Materials M3, M4, M5
- MG3 → Materials M6, M7

Usage: Instead of maintaining separate records for materials M1 and M2, you can use the Value Group "FAST_MOVING_MATERIALS" and select Material Group = MG1.

Key Points-

The value group is independent of the Customer Master.
It simply maps characteristic values for use in AATP logic.
If business rules change, you only update the Value Group, not every BOP segment.

aATP BOP Fiori Configuration Flow-

1. "Characteristics Catalog" -> (Maintain required or additional fields)
2. "Segment" -> (Define conditions based on business criteria)
3. "Variant" -> (Assign BOP strategies such as WIN, GAIN, IMPROVE, REDISTRIBUTE,
   FILL, LOOSE)
4. BOP Run -> (Execute Simulation or Batch Job)
5. Monitor BOP Run -> (Analyze results, confirmations, and exceptions)

Flow Diagram

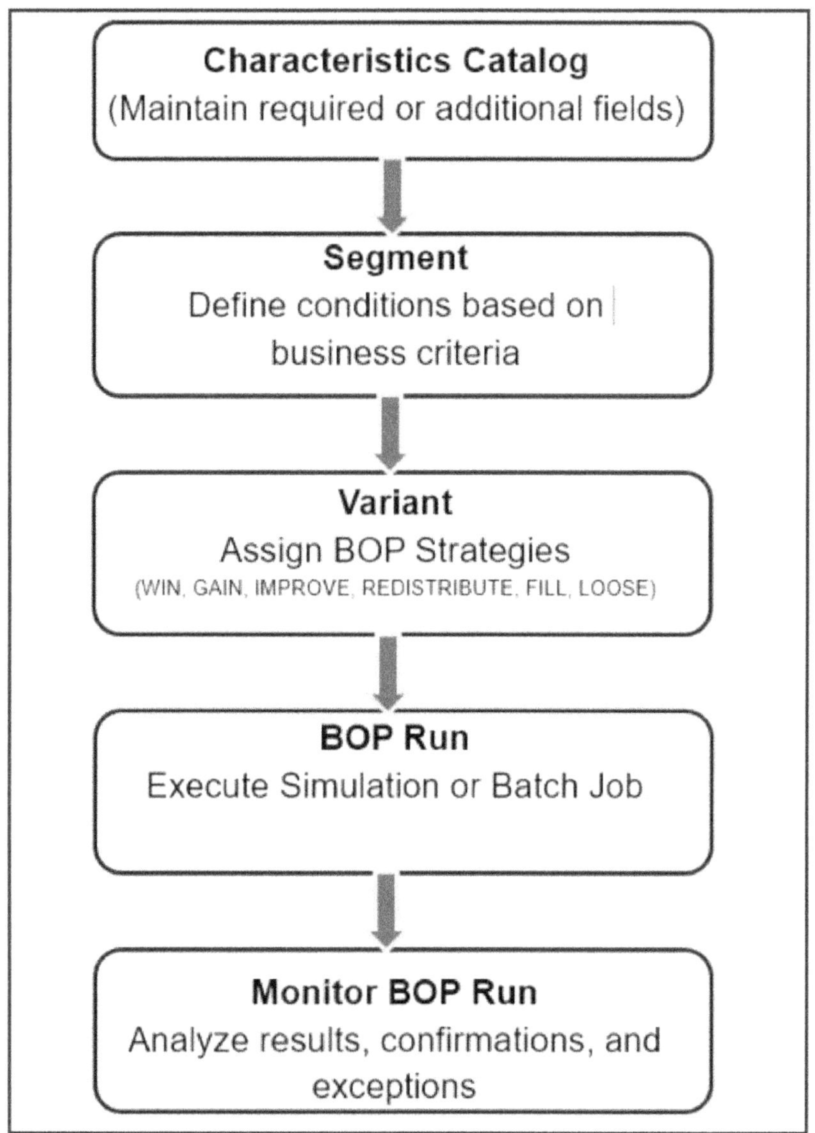

In SAP S/4HANA, BOP requires the material to be assigned to an aATP check group. Only materials with an active aATP check group are considered during the BOP run.

- *This is maintained in the material master (MRP 3 view).*
- *If the material is not aATP-relevant, it will be excluded from BOP.*

#	Step	App	Details

1	*Configure BOP Segment*	*App-F2158*	A segment is defined by a set of conditions used to group sales order items during a BOP run.
			
			Conditions can include fields such as Sales Organization, Requested Delivery Date (RDD), schedule line delivery date, customer group, material, or other business fields.
			
			Example:
			• A segment can be defined with conditions like:
			  • Sales Organization = 1710
			  • Requested Delivery Date within the next 10 days
			  • Customer Group = A1 to A5
			• Based on these conditions, sales orders that meet the criteria are grouped into the same segment.
			• Each segment can then be assigned one or more BOP strategies (WIN, GAIN, IMPROVE, REDISTRIBUTE, FILL, LOOSE) and processed accordingly.

2	*Configure BOP Variant*	*App-F2160*	BOP variants determine *how* segments are processed during a BOP run.
			
			• Create a new variant.
			• Add one or more segments created in F2158.
			• Assign confirmation strategies, such as:
			  • WIN – secure or prioritize confirmations
			  • GAIN – improve confirmations when possible

			◦ IMPROVE - retain or improve confirmations, may lose quantity if supply is insufficient. ◦ REDISTRIBUTE – reallocate from lower to higher priority ◦ FILL – confirm available quantities ◦ LOSE – remove confirmations from low-priority items • Define processing options (e.g., simulation run vs. active run). • Save and activate the variant.
3	*Schedule BOP Runs*	*App-F2665*	Once variants are defined, the execution phase ensures orders are processed and ATP confirmations are updated. Schedule BOP Run (App F2665) • Choose a BOP variant. • Schedule a manual run, background job, or recurring job. • Optionally run simulation mode to preview changes without updating orders.
4	*Monitor BOP Run*	*App-F2159*	• Review processing status and logs. • Drill into affected sales orders, confirmation changes, or exceptions. • Validate the final results of reallocation.

Fiori Apps-

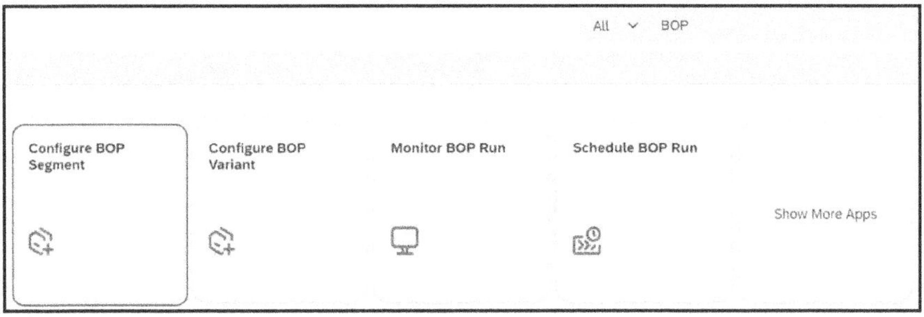

All BOP

Configure BOP Segment

Configure BOP Variant

Monitor BOP Run

Schedule BOP Run

Show More Apps

Step 1: Configure BOP Segment (Fiori App: Configure BOP Segment - F2158)

Two BOP segments were created: one for sales orders with a delivery date within the next 10 days, and another for sales orders with a delivery date beyond 10 days, both with the selection condition Sales Organization.

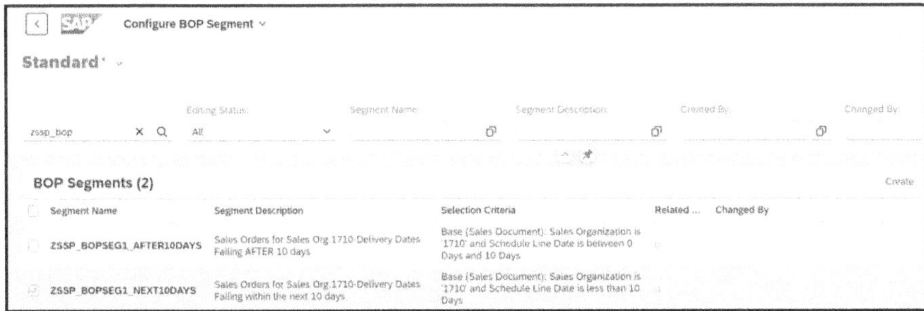

1. BOP Segment with delivery date within next 10 days-

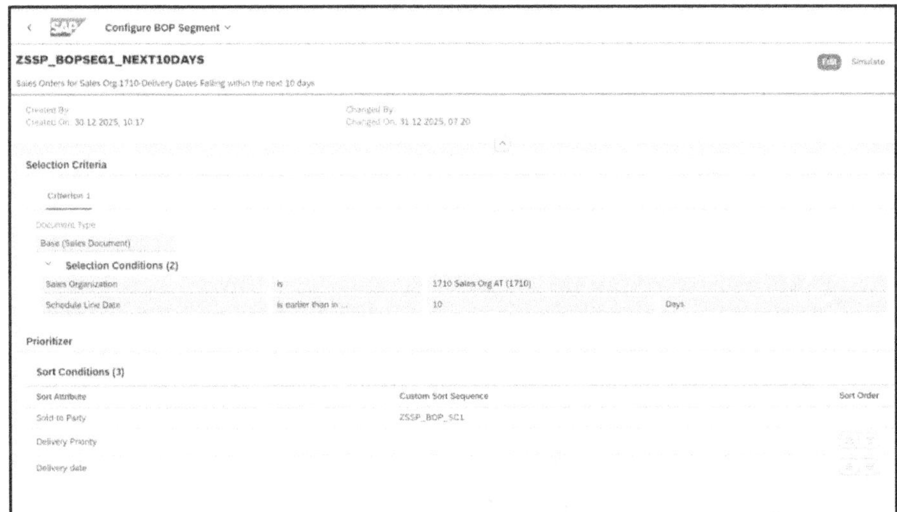

2. BOP Segment with delivery date after 10 days-

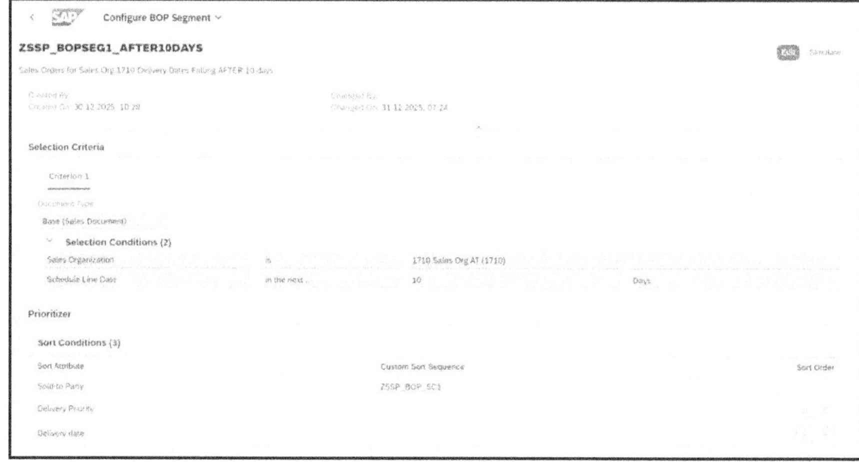

Multiple Selection Criteria – Examples

1. AND Logic

In SAP S/4HANA aATP BOP, when the selection criteria contain multiple entries (for example, Criterion 1 and Criterion 2), the system evaluates them using AND logic.

- All criteria must be fulfilled
- Only sales order items that meet every defined criterion are selected for the BOP run

Example – Selection Criteria:

1. Criterion 1 – Sales Organization = ZSSP
2. Criterion 2 – Document Type = OR

Result: Only sales order items that belong to Sales Organization ZSSP AND Document Type OR are included in the BOP run.

2. OR Logic

OR logic means that at least one condition must be true for an item to be selected or processed.

- AND logic → All conditions must be true
- OR logic → Any one condition can be true

OR Logic in SAP BOP – Simple Example

If a BOP variant or configuration uses OR logic, a sales order item is selected if it meets any one of the criteria.

Example – Criteria:

- Sales Organization = ZSSP
- Sales Organization = ZSSQ

OR Logic Result: Sales orders from ZSSP OR ZSSQ are selected.

Where OR Logic Is Used in BOP

Area	Logic Used
Selection criteria (multiple fields)	AND
Criteria within one segment	AND
Multiple segments in a variant	OR

Steps Followed to Create a BOP Segment-

1. Use Fiori App - "Configure BOP Segment"
2. Enter Segment Name (no spaces, use underscores) and Description.
3. Define Selection Conditions (Criteria): Add conditions (e.g., Sales Organization, Document Type, Customer Group, etc…) using attributes, operators (=, >), values. Use + Add Condition for multiple.
4. Add Exclusion Conditions if needed (e.g., exclude specific customers) via Add Exclusion Condition.
5. Set Prioritize: Choose sort attributes (e.g., Delivery Priority, Requested Date) with Ascending/Descending order; optionally use custom sorting from Configure Custom BOP Sorting app.
6. Simulate selection via Simulate button to preview requirements.
7. Save and activate the segment (Draft status changes to Active).

While creating a new segment, we considered the following important points:

1. **Segment Name/Description** – Define the segment name and its description.
2. **Document Type** – Specify whether this applies to Sales Order documents, STO documents, or both (Mixed).
3. **Selection Conditions** – Selection Conditions define the criteria used to select sales order documents for a specific BOP segment. By combining these fields, you can filter and group sales orders into segments to apply different BOP strategies. Multiple fields are available for defining selection conditions, such as:
    a. Sales Organization
    b. Requested Delivery Date (RDD) or Schedule Line Delivery Date

   c. Customer Group
   d. Material / Material Group
   e. Plant / Storage Location
   f. Other relevant business attributes

4. **Sort Conditions** – Determine the fields or parameters to sort the documents.
5. **Sort Order** – Specify ascending or descending order for sorting.
6. **Custom Sort Sequence** – A custom sort sequence can be defined to control the order in which sales orders are processed during BOP. Example as below.
   - Sort Conditions might include Sold-to Party, Delivery Priority, Delivery Date, etc.
   - During BOP, the system may select 100 sales orders for a group of customers.
   - To prioritize specific customers (e.g., 3 key customers), a Custom Sort Sequence can be created. The system will assign highest priority to these customers for confirmation.
   - The Custom Sort Sequence is assigned under Sort Conditions → Sort Attributes, ensuring that the selected sales orders are processed first.
   - Custom Sort Sequences allow **fine-grained control** over BOP order processing, especially when default sorting does not meet business requirements.
   - A Custom Sort Sequence is not limited to customers; it can be created for any characteristic, such as Customer, Material, Customer Group, Material Group, or other relevant fields. This provides flexibility to prioritize sales orders in BOP based on different business requirements, not just customer priority.
   - Custom sort sequences are applicable only to sort attributes with discrete values, not to date fields. For example, if the sort attribute is *Delivery Date*, you cannot create a custom sort sequence based on dates because dates are dynamic. Custom sort sequences are suitable for fields such as Customer, Material, Customer Group, Material Group, or other characteristics with predefined value lists.

Use this Fiori app to create a custom BOP sort sequence: *Configure Custom BOP Sorting*

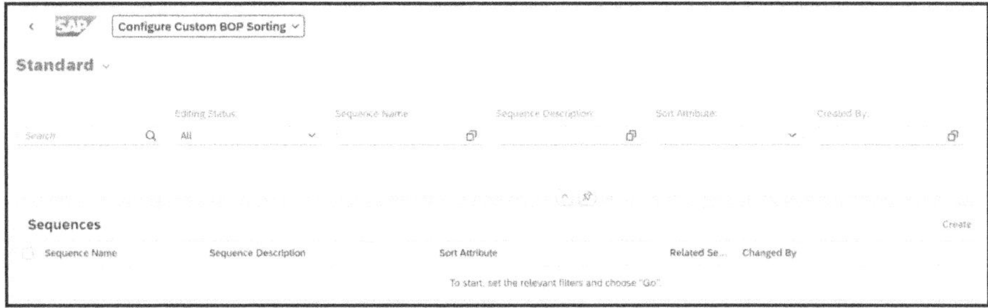

A Custom BOP Sort Sequence named "ZSSP_BOP_SC1" was created. It includes 2 specific customers: 17100010 and 17100021, plus all other customers. This allows fine-grained control over which sales orders are processed first during BOP, ensuring critical customers are prioritized.

- The system gives highest priority to the two specified customers.
- Remaining customers are processed next.
- Within these groups, other sort conditions such as Delivery Priority and Delivery Date are applied.

Assigning Custom BOP Sorting to Segment - The Custom BOP Sort Sequence (e.g., ZSSP_BOP_SC1) is assigned to a BOP Segment. By doing this, sales orders in that segment are processed according to the defined sort sequence during the BOP run. This ensures that priority customers or other key criteria are handled first, while the remaining orders follow the standard sort conditions.

Key Points-

1. The main function of BOP segmentation is to filter and select sales order documents based on defined selection conditions, enabling them to be processed in the BOP run according to the assigned strategies.

BOP Segment Test Scenario –

1. Using Two BOP Strategies (Improve & Redistribute)

One sales order was assigned to the Improve strategy without confirmation, and another sales order was assigned to the Redistribute strategy with confirmation. After executing the BOP run, the confirmed quantity in the Redistribute strategy was reduced, while the quantity in the Improve strategy increased (changing from unconfirmed to confirmed).

1. Material Setup: Create a new material FG10005. Current stock: 500 units in Plant 1710.

2. Sales Order 1: SO#:5963
    - Requested Delivery Date (RDD): Jan 15, 2026 (15 days in the future)
    - Confirmed Qty: 500 units (ATP available)
    - Segment Assignment:
        - Segment: ZSSP_BOPSEG1_AFTER10DAYS ☑
        - Segment: ZSSP_BOPSEG1_NEXT10DAYS ✖ (not included)
    - Simulation: Use the Simulate icon to verify order placement in the segment.

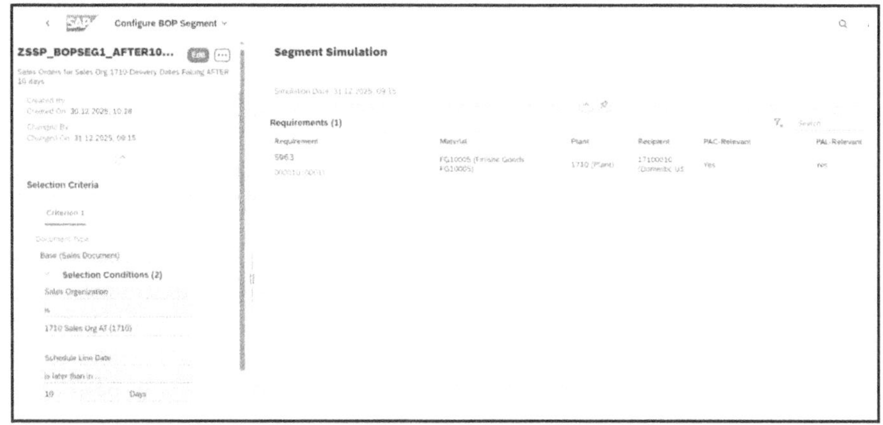

3. Sales Order 2: SO#:5964
    - RDD: Jan 06, 2026 (within the next 10 days)
    - Confirmed Qty: 0 units (ATP not available)
    - Segment Assignment:
        - Segment: ZSSP_BOPSEG1_NEXT10DAYS ☑
        - Segment: ZSSP_BOPSEG1_AFTER10DAYS ✖ (not included)
    - Simulation: Use the Simulate icon to verify order placement in the segment.

4. Results
    - Sales orders are assigned to the correct BOP segments based on their RDD:
        - SO#5963 → After 10 days → confirmed 500 units
        - SO#5964 → Next 10 days → unconfirmed (0 qty)

- The scenario works as expected: orders are segmented correctly, and confirmations align with available ATP stock. Sales orders are assigned to the correct segments based on the configuration.

Step 2: Configure BOP Variant (Fiori App: Configure BOP Variant)

Create Variant refers to the configuration step where you define how a Backorder Processing run behaves.

Steps Followed to Create a BOP Variant-

1. Open the Fiori app *Configure BOP Variant* and create a new variant.
2. Choose "Create" to start a new variant and enter Variant Name and Description on the Variant Settings screen.
3. Maintain variant settings: document category, validity, exception behavior, (optional) Global Segment Name, requirement categories, and other control options.
4. Add one or more existing BOP segments to the variant so their selection criteria define which requirements fall into each segment.
5. For each segment, assign the required confirmation strategy (for example Win, Gain, Improve, Redistribute) and maintain the sort order if relevant.
6. Save the variant so it is no longer in draft status and can be used in processing runs.
7. From the BOP Variant app or the scheduling app, trigger a BOP run (simulate or execute) using this variant, and then review the results in the monitoring app.

Note – A BOP run is executed using a **BOP Variant**, which controls the overall processing. And not segments. Each variant contains one or more **Strategies** (e.g., WIN, GAIN), and each strategy is assigned **Segments**. Segments within a strategy are used to group sales orders into buckets, enabling the system to apply confirmation strategies and determine prioritization.

BOP is executed based on the combination of **Material** and **Plant**. Since Material is always defined at the item level, the system checks each **sales order line item** individually.

- A single line item can be processed in a specific **BOP Variant**, while another item may be processed in a different variant.
- Within the same variant, different line items of the **same sales order** can be assigned to **different strategies**. This allows the system to apply the appropriate confirmation logic and prioritization at a granular level.

Use the Fiori app *Configure BOP Variant*. Click the + icon to create a new variant.

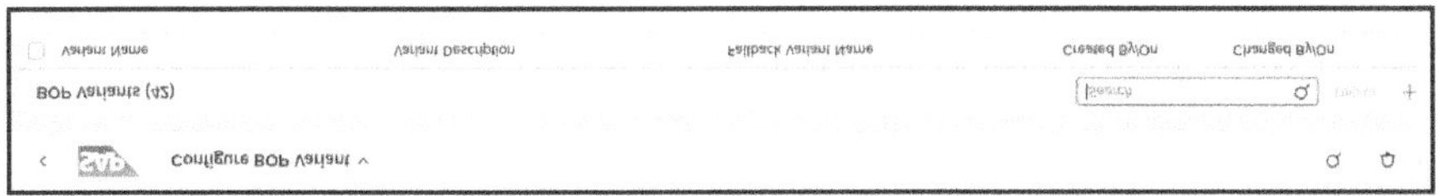

Here, you can define the variant name and variant description, select the document type (for example, Sales Documents or Stock Transport Orders), and choose the checking mode (two options are available: A and B).

Exception Behavior - Defines how the system reacts when a BOP run encounters errors (for example, document locks or missing data). The recommended best practice is *Continue Processing*, which ensures that the BOP run is not stopped due to individual issues. The available options are:

- Continue Processing – BOP continues processing the remaining documents and skips those that encounter errors.
- Terminate Processing – BOP stops entirely when an error occurs.

Fallback Variant Name - Specifies an alternative BOP variant that the system uses if the current variant cannot be executed. The fallback variant ensures business continuity by allowing Backorder Processing to complete using a backup configuration. Typical use cases include:

- The primary variant fails due to configuration issues.
- Temporary execution problems prevent the primary variant from running.

Global Segment Name – The global segment acts as a *pre-segmentation* or initial filter that is applied before the individual BOP segments in the variant are evaluated. Only requirements that pass this global segment filter are then processed further by the segment logic and sorting rules of the variant.

In scenarios with multiple sales organizations, the global segment can be defined (for example) on sales organization, so only orders from those sales orgs are selected into the BOP run. After this initial restriction by the global segment, the system continues with the standard variant behavior: applying the defined segments (Win/Gain/Redistribute/etc.) and their sort sequences to distribute confirmations among the remaining requirements.

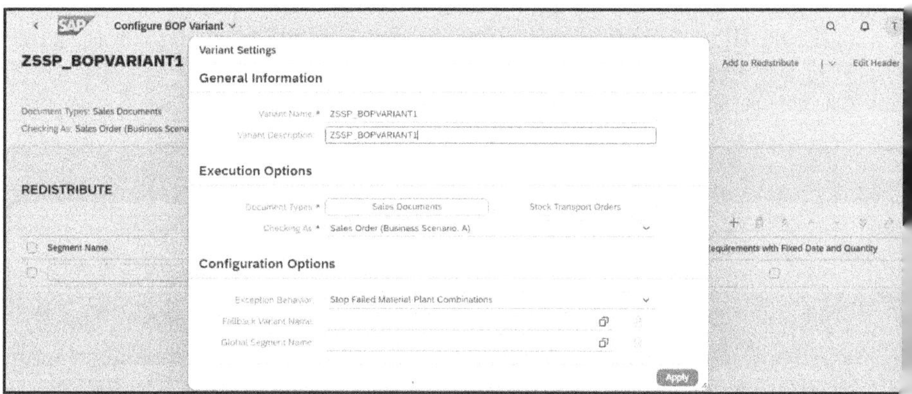

Use the **down-arrow** icon to select the strategies. To change or modify any variant settings, click **Edit Header**.

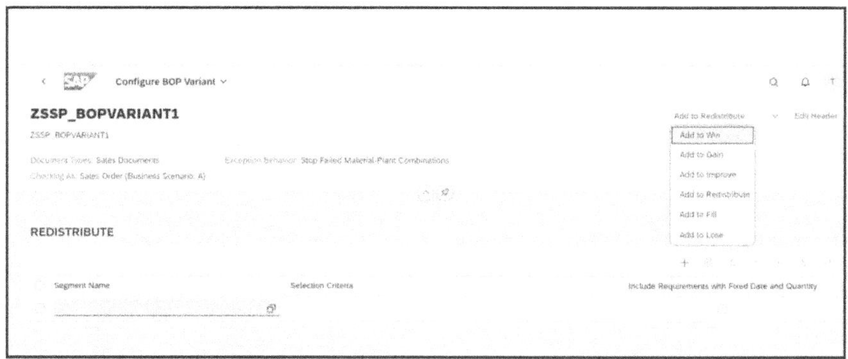

Here, we have added two strategies – ***Improve*** **and** ***Redistribute.***

When you create a new BOP Variant, the *Redistribute* strategy is added by default. You can then include additional strategies based on business requirements.

In this example, we have added one segment to each strategy. However, it is also possible to assign **one or multiple segments to a single strategy** as needed.

In the *Improve* strategy, the segment **ZSSP_BOPSEG1_NEXT10DAYS** has been added. In the *Redistribute* strategy, the segment **ZSSP_BOPSEG1_AFTER10DAYS** has been added.

"Include Requirements with Fixed Date and Quantity" - This setting in a **BOP Variant** controls whether the system includes **sales order schedule lines that have a fixed confirmation date and/or fixed quantity** during a BOP run.

- If **checked**, these requirements are **included** in the BOP run and considered for processing, but their **date and/or quantity will not be changed**.
- If **unchecked**, these requirements may be **ignored** or skipped by the BOP run, depending on other rules.
- Important for high-priority customers or critical orders where delivery date and quantity must remain unchanged.
- Allows BOP to focus on flexible orders while respecting constraints on fixed orders.

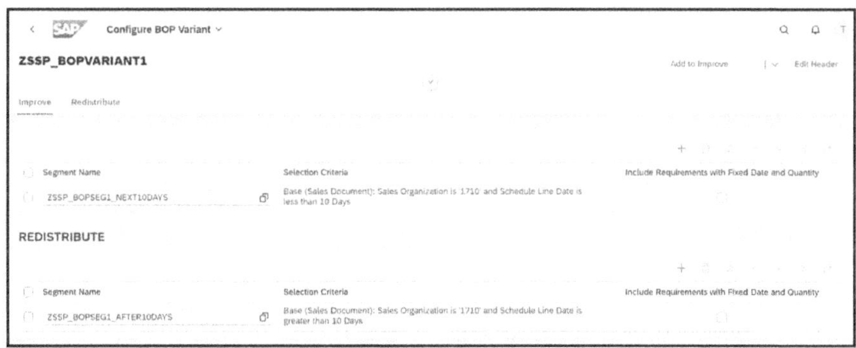

To perform a simulation, use the Simulation icon. For an actual BOP run, use the Run icon; to execute the run, click the down-arrow button next to it. Use the Simulation icon to test the BOP variant without affecting actual data.

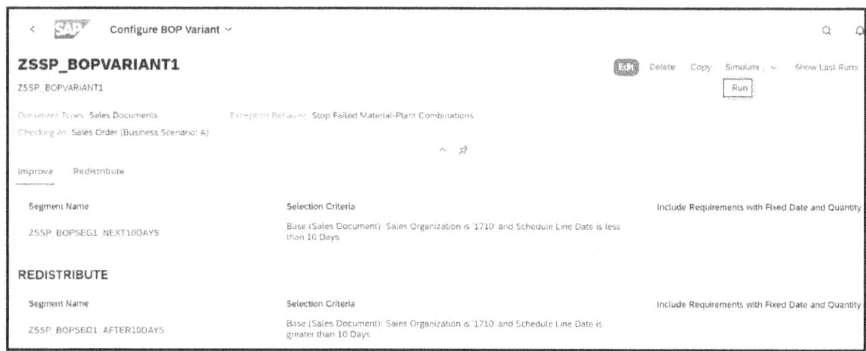

Simulation-

Processing Status -> Completed.

The BOP run is currently in **Simulation** mode, meaning no actual changes are made to the system data.

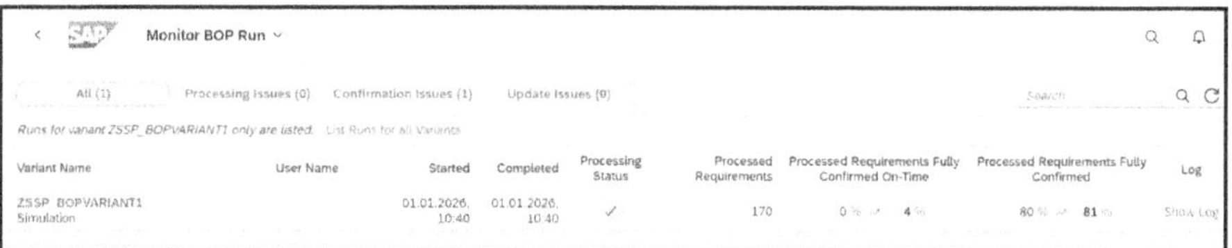

Material – FG10005

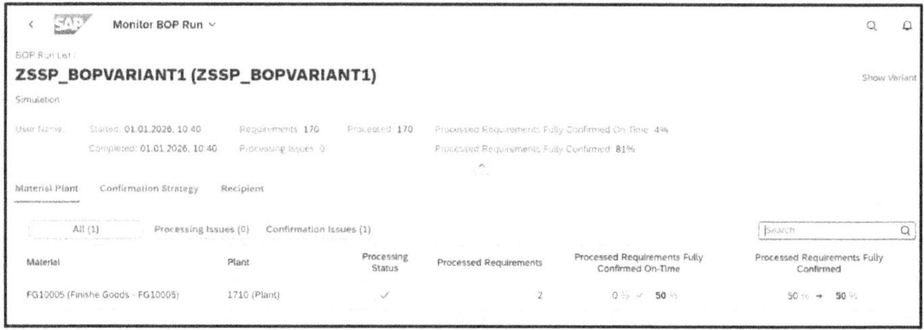

The confirmation status for two orders is as follows: Order 5963 – Unconfirmed; Order 5964 – Confirmed.

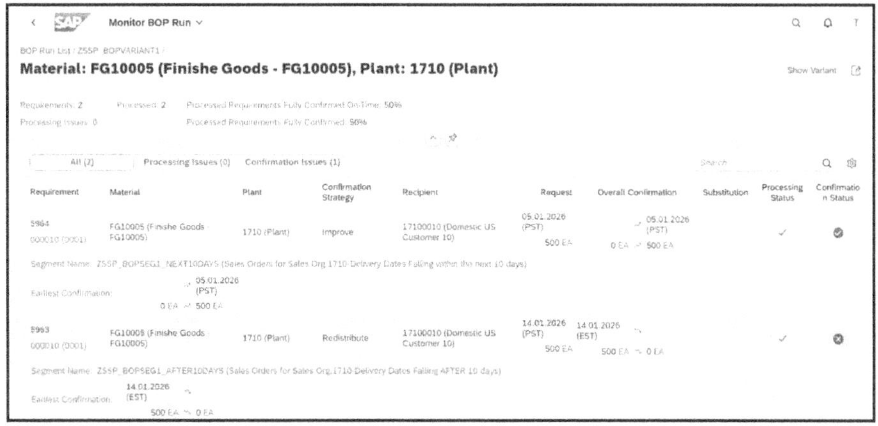

Actual RUN –

The actual BOP run has been executed and after executing the actual BOP run, the order confirmation status is: 5963 – Unconfirmed, 5964 – Confirmed.

Note: The system indicates simulation runs with the label **Simulation**. Actual BOP runs are executed without this label.

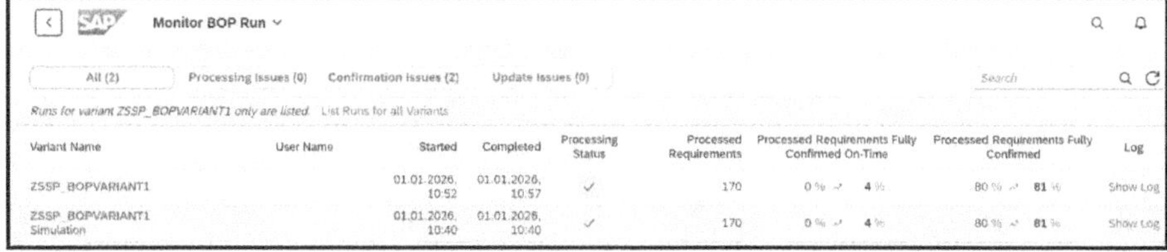

Material -> FG10005

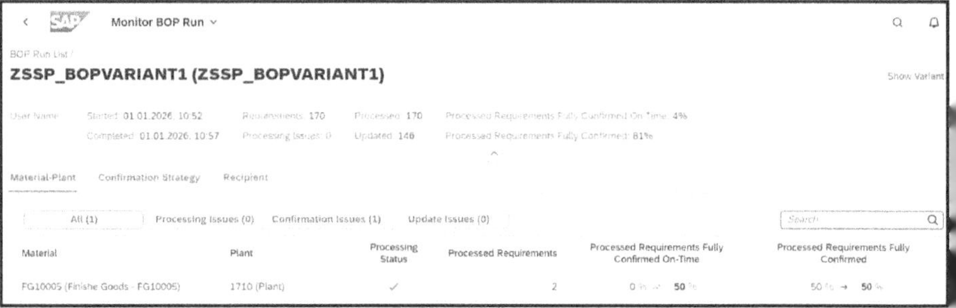

The confirmation status for two orders is as follows: Order 5963 – Unconfirmed; Order 5964 – Confirmed.

Overall Confirmation - results show the following changes after the BOP run.

- **Order 5964** – Confirmed quantity changed from 0 units to 500 units.
- **Order 5963** – Confirmed quantity changed from 500 units to 0 units.

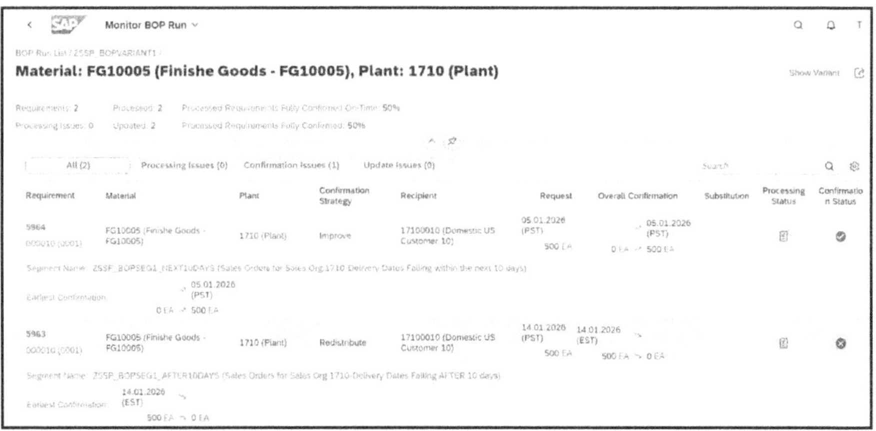

After the BOP run, the confirmation status of the sales orders was verified in the backend using **VA03**.

- Sales Order 5963: Confirmed quantity changed to 0 units (previously 500 units)
- Sales Order 5964: Confirmed quantity changed to 500 units (previously 0 units)

SO#5963

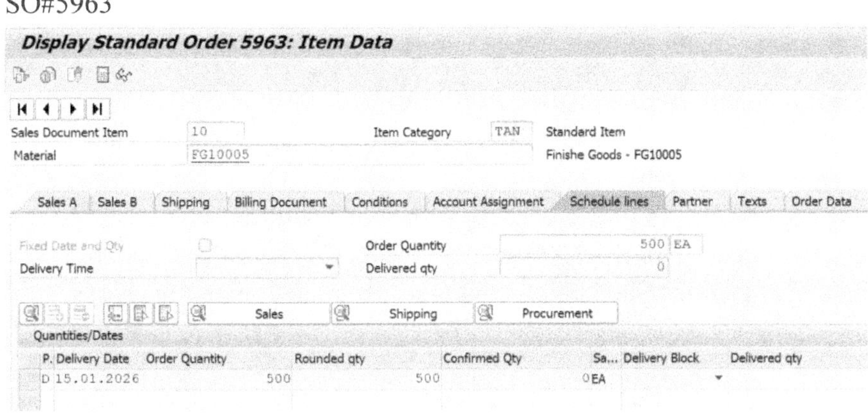

SO#5964

Step 3: Schedule and Run BOP (Fiori App: Schedule BOP Run)

Fiori App-

The Schedule BOP Run is the process where the system executes Backorder Processing (BOP) according to a defined BOP variant. It can be run in simulation mode to test the variant or as an actual run to update confirmations in the system.

In short – Schedule BOP Run: schedule a batch job with the required frequency (daily, weekly, etc.) to execute the BOP variant automatically.

Steps Followed to Schedule BOP Run-

1. Open the Schedule BOP Run Fiori app and choose Create to start a new job.
2. Select the activated BOP Variant that defines the processing logic for this run.
3. Define selection criteria (for example, date range, plants, materials, sales organizations) to filter the requirements included in the run.
4. Choose the execution mode: Simulate (test without updating documents) or Execute (live confirmation updates).
5. In Scheduling Options, either select Start Immediately for testing, or set up as a background job with start date/time and recurrence pattern –
   (one-time, minutes/hours/daily/weekly/monthly, end conditions).
6. Optionally configure log level, validity of lock objects, and time zone/calendar settings; then Save and activate the job.

Create an application job to run the BOP variant by clicking **Create**.

In **Template Selection -** define the job template name and description.

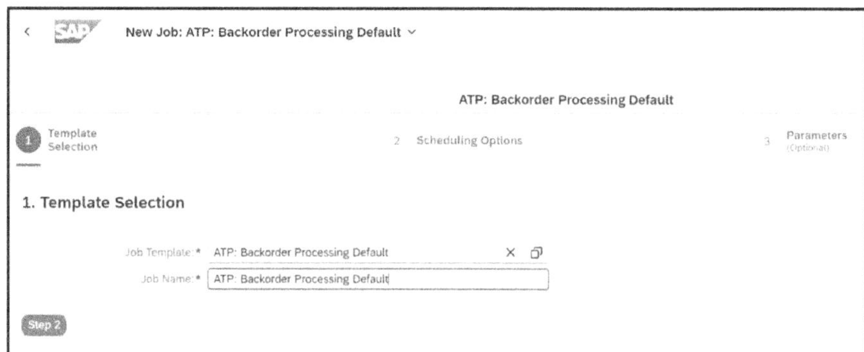

In **Scheduling Options**, select the **recurrence pattern** such as Single Run, Daily, Weekly..

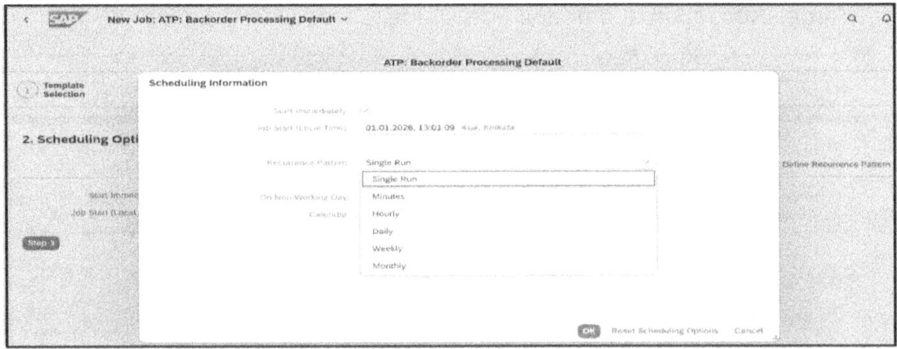

For this test, the recurrence pattern was set to **Daily**.

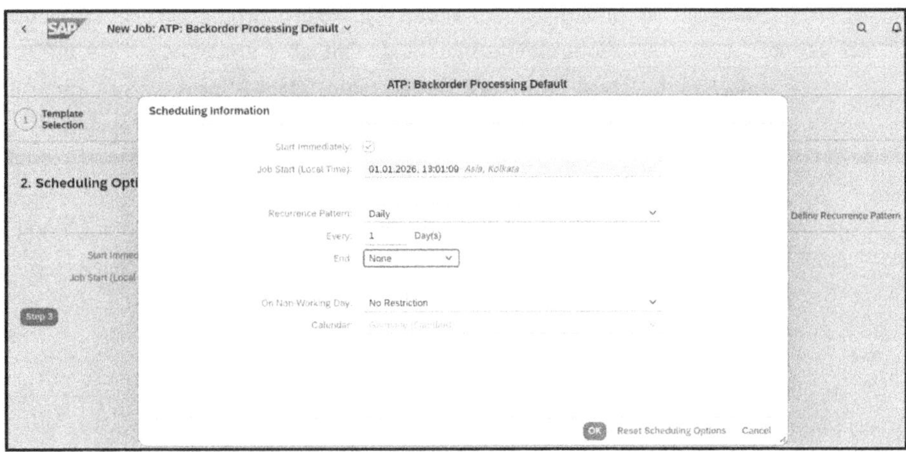

Actual jobs update confirmed quantities based on ATP availability and BOP rules.

Step 4: Monitor BOP (Fiori App: Monitor Backorder Processing)
Fiori App-

The Monitor BOP Run Fiori app is used to track, analyze, and validate the results of a Backorder Processing (BOP) run—whether executed as a simulation, manual run, or scheduled batch job.

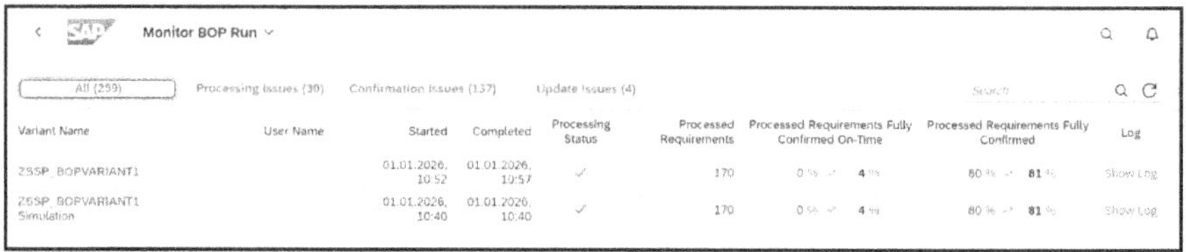

Monitor BOP Run – Meaning of the Tabs

Tab	Purpose
All	Full visibility of the BOP run - Shows all processed requirements, regardless of outcome
Processing Issues	Identify technical or data problems - Displays requirements that had technical or processing
Confirmation Issues	Analyze confirmation shortfalls - Shows requirements where confirmation could not be updated as expected.
Update Issues	Find orders not updated due to locks - Displays requirements where the system calculated a result but could not update the document. These items often require re-running BOP.

When you open a BOP run in the Monitor BOP Run app, it displays the header details of the variant along with run-specific information:

- Variant Name – The BOP variant that was used for this run.
- Started / Completed – Start and end date/time of the BOP run.
- Processing Status – Overall run result:
    - *Finished* – Completed successfully
    - *Finished with Warnings* – Completed but some issues occurred
    - *Failed* – Did not complete successfully
- Processed Requirements – Number of sales order schedule lines processed during the run.
- On Time – Number of requirements confirmed on or before the requested date.
- Confirmed – Number of requirements that received an ATP confirmation.

- Log – Detailed processing messages generated during the run, including errors, warnings, and informational messages.

Example of an Exception – Using Two BOP Strategies (WIN & Redistribute)

In this test, we will create two sales orders:

- SO#1 – Assigned to the WIN strategy, with an order quantity of 700 units and a confirmed quantity of 0.
- SO#2 – Assigned to the REDISTRIBUTE strategy, with an order quantity of 500 units and a confirmed quantity of 500.
- **ATP quantity is 0.**
- In this case, the system would attempt to allocate quantity from Redistribute to WIN, and WIN could receive 500 units. However, the WIN strategy only works with full confirmation, not partial confirmation. As a result, an exception or process issue is raised. The BOP run stops for this order due to the exception.
- Manual intervention is required to resolve the issue, typically using **Release for Delivery** to reallocate the quantity manually.

Sales Order 5966 (Redistribute strategy) - Order quantity: 500 units, Requested Delivery Date: 01/15/2026 - The order was **confirmed** based on available ATP quantity of 500 units.

Sales Order 5967 (WIN strategy) - Order quantity: 700 units, Requested Delivery Date: 01/09/2026 - The order remained **unconfirmed** due to **ATP quantity of 0 units**.

BOP Segment-

During the BOP simulation run for segment **ZSSP_BOPSEG1_AFTER10DAYS**, Sales Order **5966** was processed and included in the output results.

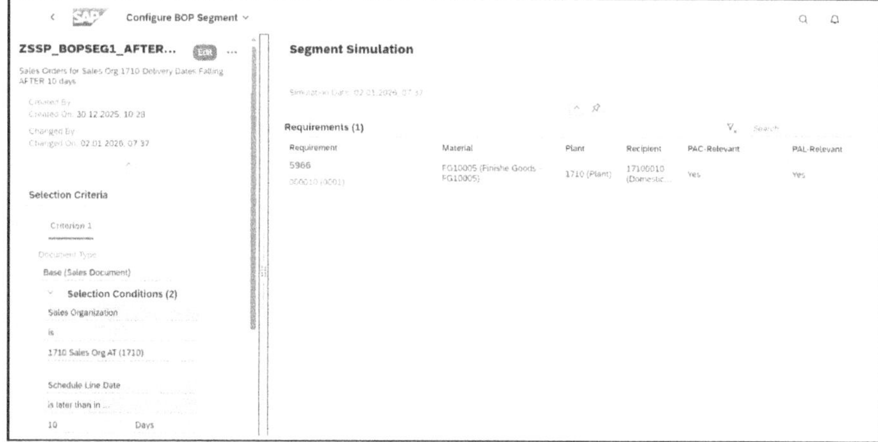

During the BOP simulation run for segment **ZSSP_BOPSEG1_NEXT10DAYS**, Sales Order **5967** was processed and included in the output results.

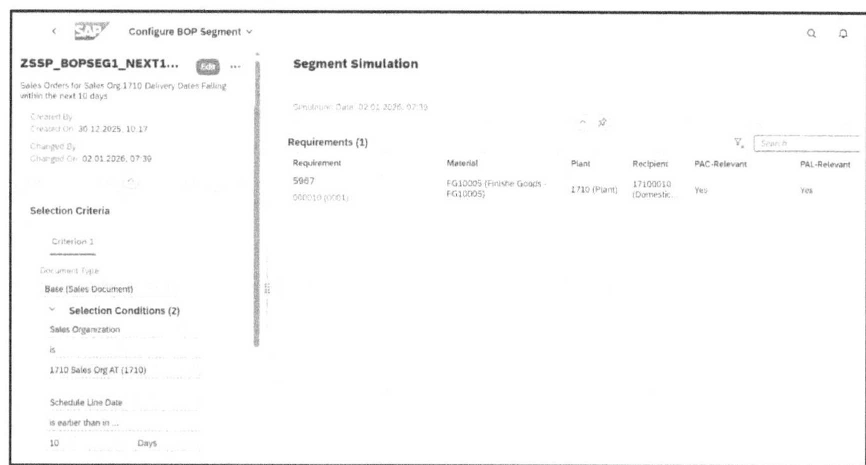

BOP Variant-

Fiori App-

The BOP Variant **ZSSP_BOPVARIANT1** includes two strategies, **WIN** and **REDISTRIBUTE**, which control how sales orders are processed during the BOP run.

Simulation Log-

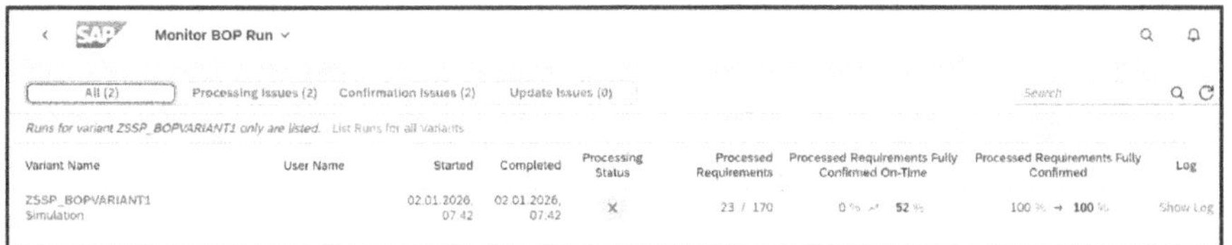

Material – FG10005

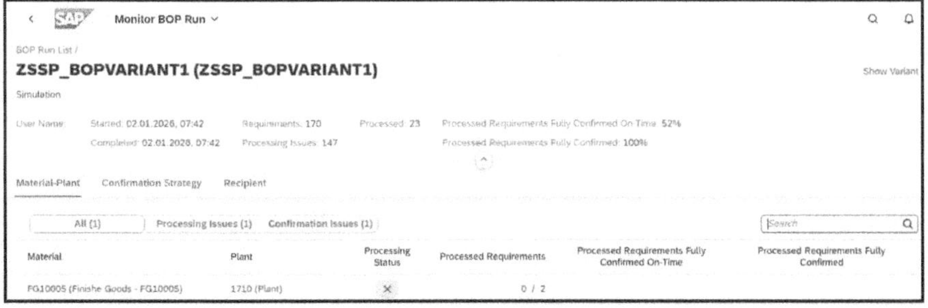

Processing Status – Failed

The BOP run did not complete successfully. Processing was stopped due to one or more critical errors or exceptions.

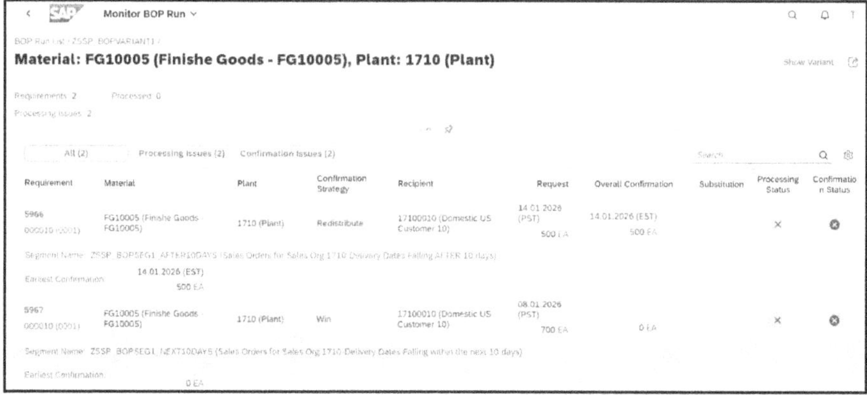

Actual Run-

Check "Show Last Runs" - Actual Run Failed

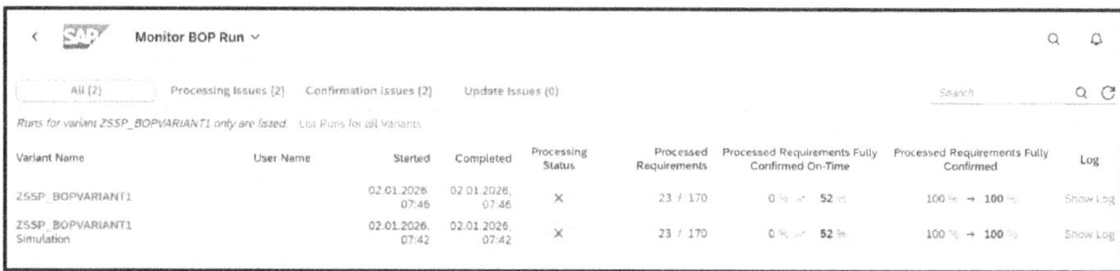

Material – FG10005

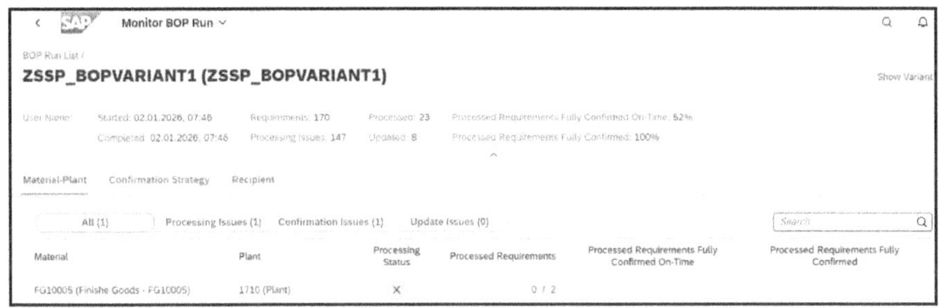

Processing Status – Failed

The BOP run did not complete successfully. Processing was stopped due to one or more critical errors or exceptions.

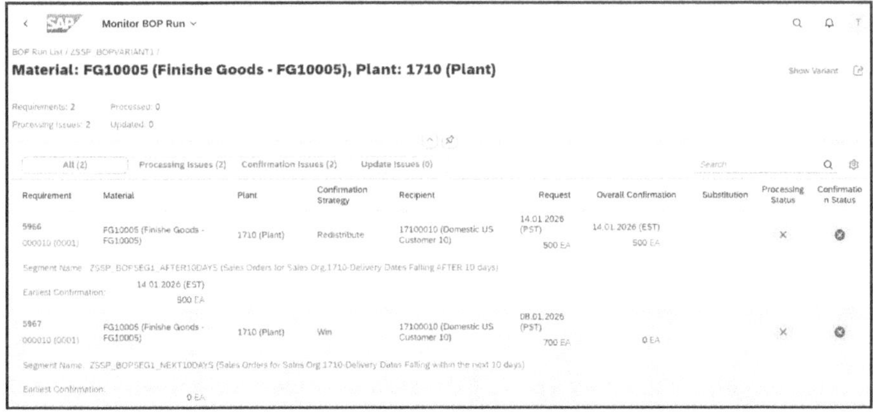

Processing Status "X" – Alert (WIN or GAIN Strategy)

Processing Status "X" indicates an alert raised during the BOP run, typically caused by the WIN or GAIN strategy. This occurs when the strategy cannot be executed as expected (for example, when full confirmation is not possible).

Processing Status **"X"** helps identify strategy-related exceptions and provides flexibility to resolve them either **automatically** (via fallback variants) or **manually**.

How to Fix This Exception-

There are three possible resolution options:

1. Manually run another BOP variant
    - Create and execute a different BOP variant, such as one using the Improve strategy, to resolve the issue.
2. Use a Fallback Variant

- Configure a fallback variant in the main BOP variant.
- When the exception is raised, the fallback variant (for example, with the Improve strategy) is executed automatically.

3. Manual Correction
   - Resolve the issue manually by using Release for Delivery to reallocate the quantity.

Minimal BOP Run Using Only One Strategy (Redistribute)

You *can* create a minimal Backorder Processing (BOP) run in SAP S/4HANA aATP by using only one confirmation strategy, such as Redistribute.

Why Redistribute Works for a Minimal Setup - The Redistribute strategy allows the system to freely adjust confirmations for all selected requirements — improving, keeping equal, or worsening them based on overall availability. Because Redistribute does not rely on prioritization segments (like Win, Gain, or Lose), it can function alone, making it an ideal choice for a simple, low-complexity BOP activation.

When you execute a Backorder Processing (BOP) run in S/4HANA that includes **only the *REDISTRIBUTE* strategy**, the system performs a very narrow and controlled type of reallocation. This setup is often used during testing because it enables you to validate basic BOP behavior quickly **without configuring the full WIN/GAIN/LOSE priority structure**.

- The system collects all relevant open requirements in the segment
- It recalculates availability for each item
- It redistributes available stock across the selected requirements
- There are **no prioritization rules**; all items follow the same logic
- A single strategy is easier to configure and test

BOP Process Flow

SAP Backorder Processing (BOP) handles unconfirmed sales order items through scheduled runs that re-evaluate availability using segments and strategies.

1. Sales Order Creation- During sales order creation or save, advanced ATP (aATP) and Product Availability Check (PAC) execute immediately,

proposing dates for unconfirmed items or leaving none, resulting in "Not yet processed completely" status.

2. BOP Run Execution- The scheduled background job, such as /SAPATP/RERUN_AVBL_CHK, selects backorders based on BOP Segments (criteria like order type, customer priority, plant) and applies Sorting/Sequence for prioritization (e.g., by priority class or revenue).

3. Confirmation Strategies- BOP applies strategies including Win (retain high-priority confirmations), Gain (pull from lower priority), Redistribute (rebalance across orders), Fill (partial confirmation), and Lose (deconfirm low-priority items), re-running ATP and update confirmations across orders.

- You define BOP segments and variants that select which orders are in scope and which confirmation strategy applies, then schedule BOP runs in the background.
- After a run, confirmations on impacted orders are updated automatically, and business users can review the changes in the monitoring apps to see how stock has been redistributed.

Key Setup for Mass Confirmation

- Define BOP segments/variants in Fiori app "Manage Backorder Processing" (sorting defines sequence).
- Schedule periodic BOP runs (daily/hourly) to process accumulated backorders based on current stock situation.

This ensures controlled, priority-based mass ATP rather than automatic per-order processing.
BOP periodically re-evaluates all relevant open requirements against the current supply situation and can change confirmations (quantities and dates) for existing orders.
It works with confirmation strategies such as Win, Gain, Redistribute, Fill, and Lose, which let you prioritize certain customers, order types, or channels and pull stock away from low-priority orders toward high-priority ones.

Sequence / Execution Logic in BOP Run

- During a BOP run, requirements are grouped (via “segments”) and assigned a confirmation strategy.

- The run execution follows the priority order: first Lose (i.e., release those confirmations), then Win, Gain, Redistribute, Fill, and Improve (if used) per the defined priorities.
- Freed quantities from "Lose" (and possibly from other strategies) become available and then reallocated according to higher-priority strategies (Win, Gain, etc.) to maximize fulfilment for more important orders.
- The result is a re-confirmation (or adjustment) of order confirmations (dates/quantities) based on latest supply, priorities and business rules.

aATP and BOP Strategies Integration

aATP performs initial availability checks during sales order creation using standard rules, while BOP strategies activate later in scheduled runs to re-prioritize and adjust confirmations across backorders.

Initial aATP Check (No Strategies)- When creating a sales order in SAP S/4HANA with aATP, the system performs an immediate availability check using standard rules (checking group, scope of check, product/location substitutions, supply creation) and confirms quantities/dates based on current stock and supply - no BOP strategies apply at this stage.

BOP Strategies Execution (Activation and Processing)- Unconfirmed or partially confirmed items become backorders but remain untouched until a scheduled BOP run executes program –
/SAPATP/RERUN_AVBL_CHK
The BOP variant then dynamically classifies these backorders into segments using predefined criteria (order type, customer priority, plant, etc.) and applies strategies like WIN, GAIN, or REDISTRIBUTE in sequence to re-run ATP and adjust confirmations across orders.

BOP re-runs aATP logic per strategy, overriding initial confirmations based on business prioritization.

Complete Workflow

1. Order Entry: aATP confirms what's available now (first-come-first-served basis).

2. BOP Run: Selects backorders → matches to segments → sorts by priority → executes strategies to prioritize/reallocate (e.g., high-priority WIN keeps/steals supply from LOSE).

3. Result: Confirmations update automatically; no manual strategy assignment during order creation—BOP overrides initial confirmations based on business rules.

17 Release for Delivery (RFD)

What is “Release for Delivery” (RefDy) in SAP S/4HANA aATP
Release for Delivery is an advanced ATP functionality within *Advanced Available-to-Promise (aATP)* in SAP S/4HANA:

- It’s an interactive process (often done via an SAP Fiori app) where supply or material availability confirmed during the ATP check can be manually reviewed and adjusted *before* those sales order confirmations are released to delivery creation (e.g., picking/packing) in SD.
- The app helps supply planners/fulfillment specialists:
  - visualize limited availability situations,
  - consider the financial impact of fulfilling or not fulfilling certain orders,
  - redistribute confirmed quantities across orders,
  - protect high-priority orders,
  - *then finally release order items for delivery processing.*
- This happens *after* backorder processing (BOP) or standard ATP checks, providing manual override and prioritization capabilities just before delivery creation.

Key Points-

In standard S/4HANA aATP, the Release for Delivery step (RefDy) does not automatically create deliveries; it only prepares and prioritizes order items for delivery creation. The Fiori “Release for Delivery” app can optionally trigger delivery creation from within the same UI, but this is an interactive convenience, not an automatic behavior of RefDy itself.

- In standard aATP, “Release for Delivery” (RefDy) itself does not create deliveries as part of the ATP/BOP logic. It prepares and prioritizes order items (reallocation, who should get what, when), and then you or a follow-up function trigger delivery creation.
- In newer S/4HANA releases, the Fiori “Release for Delivery” app offers a button/option to create deliveries directly from the app. That is a UI convenience on top of the RefDy proposal, not an automatic background step of RefDy as an ATP method.

- RefDy does not automatically create deliveries; it only prepares (primes) sales orders for the next step in SD, where deliveries are created in a separate process.

How "Release for Delivery" compares to Classic ATP/"ATP v2"
First, let's clarify the classic ATP context:

- Classic ATP (ECC / SAP ERP / ATP v2 style)
  - Checks inventory and planned receipts during sales order creation and *automatically confirms* schedule lines if stock is available.
  - Confirmation happens *in the order document*, with simple backorder queue logic and first-come/first-served matching of supply to demand.
  - There is limited manual intervention at the confirmation stage (aside from Backorder Processing at night or via SD transaction) and *no refined UI or pre-delivery release step* for supply prioritization.
  - You *can* split deliveries based on availability, but there's no built-in interactive step to reassign availability before delivery.

This older logic is sometimes referenced as ATP "V2" (in SAP cloud or simplified ATP modes) - though SAP doesn't officially call it that in the core product; it essentially means *basic availability check + simple confirmation logic*.

Advanced ATP (aATP) Release for Delivery-

Feature	Classic ATP (ATP / ATP v2)	Advanced ATP (aATP) w/ RefDy
Availability check	Yes, basic stock & receipts	Yes, real-time, multi-constraint
Backorder processing	Limited	Enhanced BOP with prioritization
Manual intervention before delivery	Not interactive	Yes — via Fiori *Release for Delivery*
Prioritization of orders	Mostly system/automated	Manual reassign + business impact view
Alternative optimization	Minimal	Supports allocation and real-time decisions
UI	SAP GUI screens	Modern Fiori apps with simulations

Outcome	Sales order confirmation only	Confirmation + pre-delivery release decisions

What's new with RefDy vs classic ATP:

1. Planners/specialists *decide which confirmed quantities are released to logistics* based on priority/impact before delivery creation.
2. Manual redistribution of stock across orders in shortage situations.
3. Financial impact can be estimated if some requirements *cannot be met.*
4. Fiori UX for better visualization and decision support.

In contrast, classic ATP just confirms what's available at sales order entry and then delivery creation proceeds normally — with no interactive prioritization step.

Why this matters-

1. Better control in shortage - You can flex supply to the orders that matter most.
2. Prioritize high-value customers/orders before delivery.
3. Reduce costly delivery cancellations or overrides later in the process.

Classic ATP lacks this refined prioritization and timing control.

Configuration/Setup Steps-

Configuration/Setup steps for enabling and using the SAP Fiori Release for Delivery (RefDy) app in SAP S/4HANA aATP. This covers what you must do from a configuration and authorization perspective, so the app works correctly in your system.

1. Activate Advanced ATP in Backend

Before anything else, make sure aATP features are active in your S/4HANA system (usually done during aATP setup):

- Ensure your ATP check (PAC), Backorder Processing (BOP), and related services are configured.
- In some systems, activate relevant features via SFW5 (Switch Framework) - e.g., S4H_AATP and, if needed, SAP APO compatibility blocks.

Note: This step is standard for enabling ATP functionality generally - not specific only to RefDy.

2. Configure Order Fulfillment Responsibilities (Required)

The Release for Delivery app depends on Order Fulfillment Responsibilities to determine which materials/orders a user can see and act on.

Fiori App-

Configure Order Fulfillment Responsibilities

Fiori App: *Configure* ***Order Fulfillment Responsibilities*** (App ID F2246)

Steps (functional configuration):

1. Open the *Configure Order Fulfillment Responsibilities* app.
2. Click Create.
3. Enter Selection Criteria — typically based on Product, Plant, Customer, Sales Org, etc.
4. Assign the responsible User ID who will run Release for Delivery.
5. Save/Publish the configuration.

This ensures users only see applicable materials/orders in the Release for Delivery app.

Use Create –

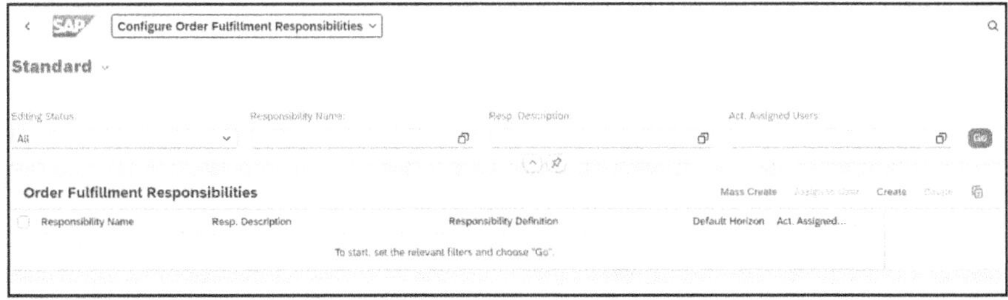

For testing, I have created a Responsibility named 'SSP' with selection conditions: Plant and Product Group. In Release for Delivery, the Sales Document is selected, and the user is assigned accordingly.

General Information –

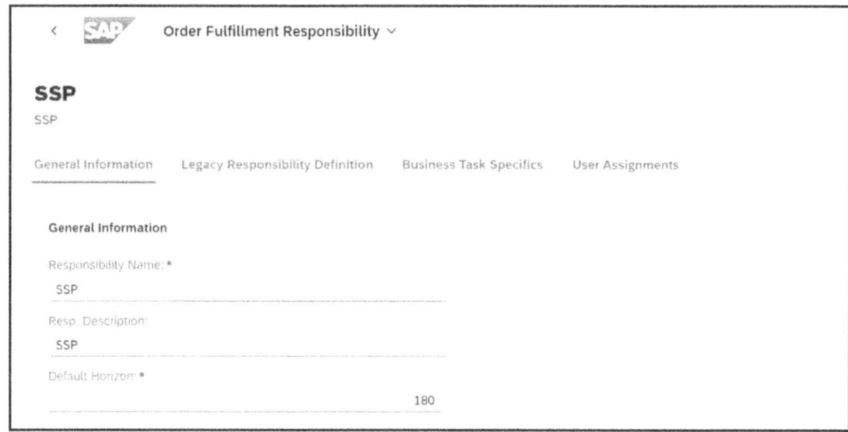

Legacy Responsibility Definition-

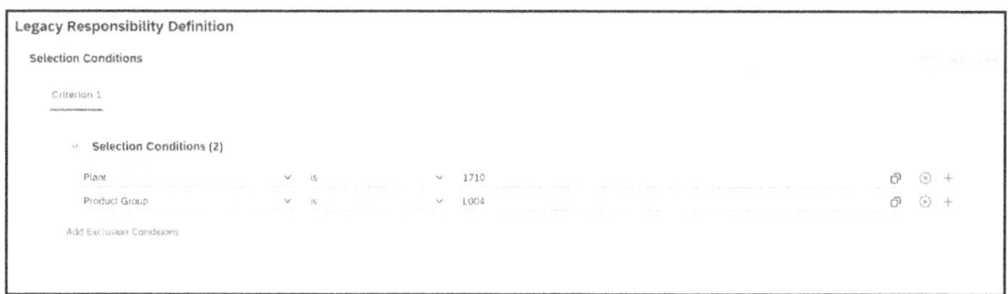

Business Task Specifics & User Assignments-

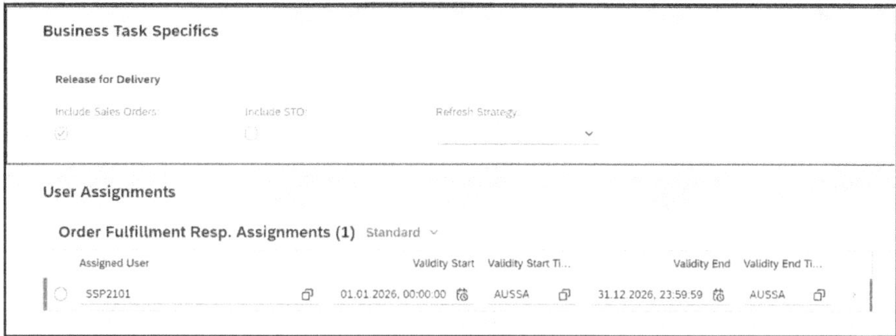

3. Using the Fiori App – Release for Delivery

Once the above configuration is complete:

1. Open Release for Delivery (App ID F1786).
2. Select criteria (e.g., Product/Plant/Date) to display prepared confirmations.
3. Review availability and redistribute confirmations if needed.
4. Mark items and Release for Delivery (status change).
5. From the app, you may optionally trigger delivery creation depending on implementation and authorizations — but the core function is release, not delivery creation.

Fiori App-

This will allow you to make allocation/release decisions.

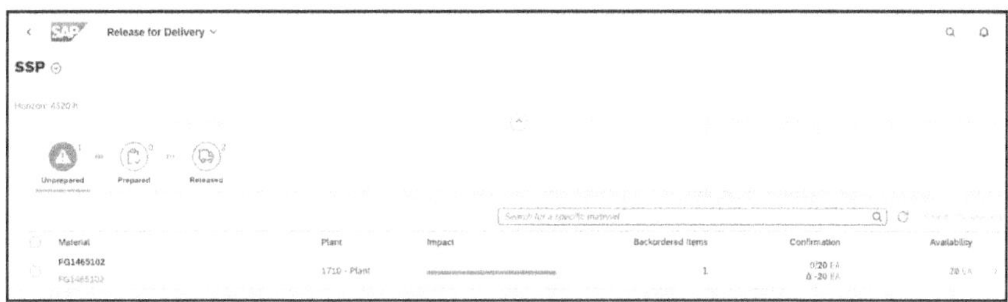

When you select a schedule line, the app often shows a bar chart or visual representation of quantities. Typical indicators include:

1. Confirmed - Quantity that has been ATP confirmed and is eligible for release.
2. Available - Stock that is physically available in the plant but not yet confirmed to this order.
3. Over confirmed - Quantity that was confirmed beyond available stock (e.g., due to safety stock or backorder logic).
4. Unconfirmable - Quantity that cannot be confirmed due to insufficient stock, allocation to higher-priority orders, or other constraints.

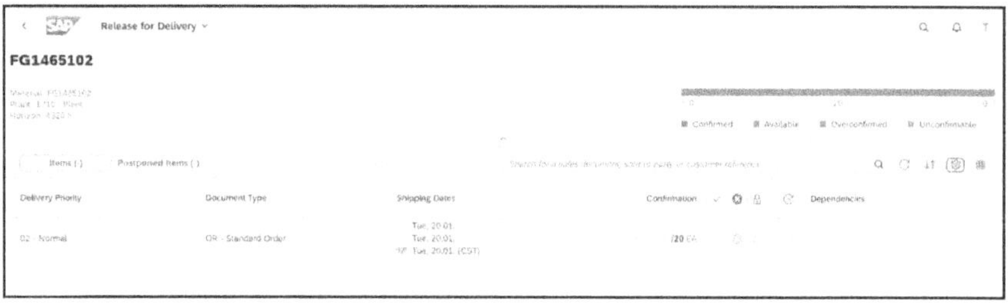

Release for Delivery App – Confirm / Unconfirm / Protect Icons

Icon	Action	Effect in RefDy
Confirm ☑	Release schedule line	Line is ready for delivery; included in allocation
Unconfirm ✗	Cancel release	Line is removed from release; stock can be reallocated
Protect	Lock line	Line's allocation cannot be changed in redistribution or BOP

Example – Release for Delivery (Reallocating Confirmed Quantity)

Business Scenario-

A high-priority customer order must be fulfilled urgently, but no ATP quantity is available because stock has already been confirmed for another sales order.

In case of an urgent requirement, the "Release for Delivery" app can be used to reallocate confirmed quantities from an existing sales order to a backordered sales order.

- Material ABC1001 has ATP quantity = 50 units.
- You create the 1st sales order (100001) with 50 units → the order is confirmed.
- You then create a 2nd sales order (100002) with 20 units → this unconfirmed order goes to backorder because no ATP quantity is available.

Later, there is an urgent requirement to fulfill the 2nd sales order (100002). Using the "Release for Delivery" app, you can:

- Reassign or reallocate quantity (for example, 20 units) from the 1st sales order (100001).
- Confirm those 20 units for the 2nd (urgent) sales order (100002).

As a result:

- The 2nd sales order (100002) is confirmed for 20 units
- The 1st sales order (100001) will now have its confirmed quantity reduced accordingly (for example, from 50 → 30 units)

Result After Release for Delivery-

Sales Order	Ordered Qty	Confirmed Qty
100001	50 units	30 units
100002	20 units	20 units

18 Important Notes

In this section, I have provided additional key information related to aATP, highlighting the most important points and notes.

1. Available to Promise Overview

Basic ATP calculation is primarily based on the three key elements – stock, receipts (supply), and requirements (demand). In addition to these core elements, the system uses data from the various master and organization objects – such as material master, customer master, plant, shipping points, and transportation settings – to control the scope, rules, and timing of the ATP check, enabling the system to determine accurate availability quantities and confirmation dates. The quantity calculation itself always comes from stock, receipts, and requirements.

Important Note – the system does not calculate availability directly from master data, but it uses master and organizational data to control how the ATP calculation is performed.

As the ATP calculation depends on key elements, any missing or incorrect data in these elements can lead to inaccurate ATP dates.

In ATP, quantities come from stock, receipts, and requirements – but the Material Master defines the rules that control how ATP interprets and confirm those quantities.

Common ATP Issues Caused by Material Master: -

1. ATP not triggered -> Wrong Availability Check Group
2. Wrong Confirmation Date -> Incorrect Replenishment Lead Time (RLT)/Lead Times
3. ATP ignores receipts -> Wrong Procurement Type

2. Warehouse Capacity

In aATP (Advanced Available-to-Promise), warehouse capacity refers to the system's ability to check whether the warehouse has enough resources to handle and process the product on time not just stock, but also the capability to pick, pack, and ship the order.

Warehouse capacity typically includes:

1	*Picking Capacity*	Do we have enough available picking resources (people, equipment, time slots) to pick 1,000 pieces?
2	*Packing Capacity*	Is there enough packing capacity to process this order on the planned date?

3	*Staging / Loading Capacity*	Can the warehouse stage and load the materials onto a truck for shipment on time?
4	*Working Time / Operational Capacity*	Does the warehouse have operational hours to process the order for that date?

Why Warehouse Capacity Matters in SAP S/4HANA aATP-
Even when material stock is available, the warehouse must have sufficient physical processing capacity (picking, packing, dock doors, labor hours) to fulfill orders on time.

Warehouse Capacity Example Summary

Scenario:	5,000 units of material stock available, but warehouse capacity limited to 1,000 units/week. Customer orders 1,500 units → aATP rejects due to warehouse overload (dock/labor constraint). Line item gets partial ATP confirmation (1,000/1,500), not complete rejection. Warehouse capacity creates realistic delivery proposals, not order blocks.
Key Point:	Material availability alone isn't enough - aATP checks physical processing capacity to ensure realistic delivery promises.
Business Value:	Prevents over-promising on warehouse handling capacity, avoiding delivery delays and service level failures.

Bottom Line: Warehouse capacity ensures aATP doesn't just check "Do we have the product?" but also "Can the warehouse ship it on that date?" - This prevents the classic disconnect between material availability and physical fulfillment capacity.
Simple Definition - Warehouse capacity in AATP means the warehouse's ability to pick, pack, and ship products on time, based on available operational resources.

Key Benefits

- Prevents over-promising warehouse handling (not just material)
- Protects service levels for priority customers within capacity limits
- Realistic dates considering full supply chain constraints

3. aATP Document Types-
In **SAP S/4HANA Advanced Available-to-Promise (aATP)**, several key document types are involved in the order fulfillment process where AATP functionality plays a critical role:

These document types form the core around which **aATP checks and confirmations** revolve, enabling complex, **real-time, and customer-prioritized order promising** within SAP S/4HANA's supply chain processes.

#	Document Types	Details
1	*Sales Orders (Standard Sales Order Document Types)*	Primary documents where AATP performs availability checks at the **schedule line level** to confirm delivery dates and quantities based on material availability, supply plans, and allocations.
2	*Stock Transport Orders (STO)*	Used to move stock between plants or storage locations. AATP ensures the availability and timely delivery of stock transfers.
3	*Production Orders / Production Planning Documents*	AATP checks materials required for production orders to confirm the availability of **components and finished goods** to promise.
4	*Purchase Orders (POs)*	Availability of purchased materials is considered in AATP calculations to propose delivery commitments based on expected receipts.
5	*Backorder Processing (BOP) Segments / Documents*	AATP uses special backorder categories to manage and reprioritize open orders according to business strategies such as **WIN, GAIN, REDISTRIBUTE, FILL, and LOSE**.
6	*Product Allocation (PAL) Documents*	Document allocation rules and quotas that control how limited stock is assigned to **customer groups or sales channels** under AATP.
7	*Alternative-Based Confirmation (ABC) Documents*	Manage **alternative sourcing or product substitution** scenarios in the confirmation process.

4. How aATP works in Sales Order processing-

In SAP S/4HANA, AATP (Advanced Available-to-Promise) works as an intelligent order fulfillment engine during sales order processing by providing realistic confirmations for requested quantities and dates based on comprehensive checks and advanced logic.

aATP (Advanced Available-to-Promise) determines **how much** of a material can be delivered and **when**, based on real-time stock and supply chain conditions.

In Sales Order processing, AATP operates mainly at the **Schedule Line level** and drives the confirmed delivery dates.

Order Creation & Trigger:
When a Sales Order is created with key details such as Material, Quantity, Plant, and a Requested Delivery Date, the system automatically triggers the AATP (Advanced Available-to-Promise) check.
AATP runs at the schedule line level and begins evaluating real-time availability and supply conditions for ATP-relevant materials.
During the aATP ATP Check in SAP S/4HANA, the system follows a multi-layered sequence that progressively validates availability through basic checks, business rules, capacity constraints, and scheduling before proposing final confirmed dates.

Important Note: Every order interaction (create/change) runs the complete configured sequence from PAC through BPS, ensuring confirmations always reflect current reality across all constraints. Only configured advanced functions execute; unconfigured steps (like PAL with no objects) are skipped silently.

During VA01/VA02, users see immediate ATP results (PAC+PAL+etc.). BOP runs separately to optimize the total order book via background jobs. BOP does NOT run during every order interaction (create/change).

Correct Sequence During Order Create/Change (VA01/VA02)
****Interactive ATP Check (PAC → BPS):****

- PAC (Product Availability Check)
- PAL (Product Allocation)
- Supply Protection
- Capacity Allocation
- ABC (Alternatives)
- BPS (Business Process Scheduling)

Important Note-

** ***BOP** – Excluded from interactive checks***

aATP is triggered during sales order creation when:

a. The material is AATP-relevant - In the material master (MRP3 view): Availability Check field is assigned to an AATP-relevant check. (i.e., the material uses AATP instead of Classic ATP). The Availability Check field in the material master (MRP3 view → Availability Check) determines which ATP engine is used:

- Classic ATP
- Advanced ATP (AATP)

Availability Check Value - 02 (AATP) and SR (Standard ATP). When a material is assigned to 02, the system does NOT use Classic ATP & it uses the aATP framework.
If a material is not assigned to an aATP check, aATP will NOT trigger.
You can have both ATP and AATP in the same S/4 HANA system. BUT:

- One material can use only one type of ATP
- One Sales Order item can run only one ATP type
- A material assigned to 02 (AATP) cannot run Classic ATP

Material	Availability Check Value	ATP Type Used
Material A	02	aATP
Material B	SR	Classic ATP

Same system → different materials → different ATP engines.

The Availability Check field in the material master and its corresponding configuration are critical for activating and controlling the aATP check during sales order creation in SAP S/4 HANA.

Importance of the Availability Check Field for aATP

- Material Master Setting:
  The "Availability Check" field (also called Checking Group) in the MRP 3 view or Sales: General/Plant view of the material master specifies the checking group that governs the availability logic used for that material. This field determines whether and how availability checks are performed for sales orders and other transaction types.

- Activation of AATP Check:
  For aATP to be triggered during sales order processing, the Availability Check field must be set to a checking group configured with Advanced ATP enabled. This means in SAP Customizing, the checking group assigned to the material must have AATP activated (e.g., with a value "Active" in the Advanced ATP column).

b. A plant is determined - aATP requires a **plant** to determine stock and supply.
When the plant is entered or determined automatically (via master data or configuration), AATP becomes active.
aATP requires a plant to execute stock and supply determination, activating only after plant determination completes.
Plant = aATP prerequisite. No plant = no stock check = no confirmations. Once plant determined (auto/manual), full aATP sequence executes immediately.

Plant Determination → aATP Activation Sequence

Plant determination is triggered in the **Sales Order** in the following sequence:

1	*Customer-Material Info Record (CMIR):*	Preferred plant is selected.
2	*Customer Master:*	Default delivering plant is used if CMIR is not available.
3	*Material Master:*	First storage location plant is considered.
4	*Manual Entry:*	Users can override the system-determined plant.

If a plant is found: aATP is activated and PAC is executed (Stock + Receipts − Requirements). **If no plant is found:** aATP check is not performed.
If the **aATP check is not performed** (because no plant is determined), here's what typically happens in SAP Sales Order processing:

1	*No Availability Check:*	The system cannot verify whether the requested quantity can be delivered on the requested date.
2	*No PAC Calculation:*	Since PAC (Product Availability Check) relies on Stock + Receipts − Requirements, it won't run.
3	*System Cannot Commit Quantities:*	The sales order line may remain unconfirmed, or confirmation depends on default rules (like backorder processing).
4	*Potential Manual Intervention:*	A planner or user may need to manually assign a plant or adjust delivery dates to process the order.
5	*Impact on IBP / Planning:*	Downstream planning (e.g., supply-demand matching, allocation, and delivery scheduling) cannot accurately account for this order until a plant is assigned and aATP is activated.

In short: No aATP check → no automated confirmation → risk of unconfirmed or delayed order fulfillment.

c. The schedule line category allows ATP checking

aATP-Relevant Schedule Line Category Is Determined

During sales order creation, the system determines a schedule line category that is set to perform availability check.

If the schedule line category is configured with:

- Availability Check = "Yes" then aATP is triggered.

d. Quantity/date is entered

This starts with the **aATP availability check** at the **schedule line level**.

aATP kicks in when and Any of these actions trigger the availability check.

- The material is entered
- The quantity is entered or changed
- The requested delivery date has been entered
- The line item is saved

5. System Runs Availability Check (aATP Logic)

aATP evaluates real-time supply elements, such as:

Stock Available	• On-hand stock • Safety stock • Blocked or reserved quantities
Future Supply	• Production orders • Purchase orders • Stock transfer orders (STO) • Planned orders
Constraints	• Warehouse capacity • Production capacity • Transportation lead times
Business Rules	• Allocations (PAL) • Supply Protection • Alternative-Based Confirmation (ABC) • Backorder Processing (BOP) logic

aATP combines all this to determine the **earliest possible delivery date**.

Practical Example-

Order 1000 units requested 01-Feb-2026:

PAC sees:

Stock: 400 units

+ PO inbound: 300 units

+ Prod Order: 500 units

- Safety stock: 100 units

- Existing SO: 200 units

= 900 units available → Passed to PAL

PAL checks: Customer group limit 1200 → 900 (900 confirmed Month Feb 2026)

During ATP (Available-to-Promise) check in SAP S/4HANA (including aATP), the stock check depends on the scope of check defined in the checking rule. Essentially, SAP evaluates different types of stock and receipts to see if the requested quantity can be delivered. Here's a detailed breakdown:

i. Types of Stock Checked During ATP

Stock Type	Description	Example / Use in ATP
Unrestricted-Use Stock	Stock available for any customer order.	Usually, the first stock considered for confirmation.
Quality Inspection Stock	Stock under inspection, not yet released for sales.	Only considered if configuration allows.
Blocked Stock	Stock reserved for special purposes or blocked.	Usually **not considered** for ATP.
Stock in Transit	Stock moving between plants or storage locations.	It can be included if you want to promise incoming stock.
Restricted-Use Stock	Stock reserved for a specific purpose (e.g., customer consignment).	Can be considered if allowed in the checking rule.

ii. Stock vs. Receipts

During ATP, SAP doesn't just check **existing stock**—it also considers **incoming stock** or **receipts**, depending on the scope:

- **Purchase Orders (POs)** – Incoming deliveries from suppliers.
- **Production Orders (PP)** – Planned or ongoing manufacturing orders.

- **Planned Independent Requirements (PIRs)** – Demand-based planned production.
- **Stock Transfers** – Material moving between plants or storage locations.

iii. Scope of Check & Stock

- The **scope of check** determines **which stock types are included**.
- Example:
    - Scope = "Plant stock + Purchase Orders + Production Orders" → ATP considers current stock plus receipts from POs and production.
    - Scope = "Plant stock only" → ATP only considers unrestricted-use stock at the plant.

iv. Key Points

- **Unrestricted-use stock is always checked first.**
- **Other stock types (inspection, restricted, in-transit)** are optional depending on configuration.
- **Planned receipts** extend the confirmed delivery date if stock is insufficient.
- **Product allocations and substitutions** can further affect confirmation.

6. How to manage this scenario in SAP S/4 HANA aATP

Orders created in VA01 are not confirmed immediately. BOP runs aATP checks starting with PAC, ensuring accurate order confirmation.

- Sales orders (VA01) are not automatically confirmed upon creation.
- Backorder Processing (BOP) triggers the aATP checks for these orders.
- BOP execution sequence:
    1. Basic Availability Check (PAC)
    2. Followed by subsequent aATP checks based on system logic
- This ensures accurate order confirmation, considering stock, receipts, and requirements.

This scenario is standard SAP aATP design. VA01 creates unconfirmed orders (no ATP confirmation), then BOP batch run applies full aATP sequence with priority strategies.

BOP Run Sequence (Complete aATP)

BOP Processing Steps:

1. Select orders (via BOP Variant filters)
2. PAC (Product Availability Check)
3. PAL (Product Allocation Check)
4. Supply Protection
5. Capacity Allocation
6. ABC (Alternative-Based Confirmation)
7. Apply BOP Strategies (Win/Gain/Fill/Lose)
8. BPS (Business Process Scheduling)
9. Update confirmed quantities/dates

Practical Example

Day 1 – VA01 (No ATP):	SO#1001 (Priority Customer) -> 0 confirmed SO#1002 (Normal) -> 0 confirmed Stock: 800 units available
Day 2 – BOP Run (Variant: High Priority First):	BOP Segment 1 (Win Strategy): SO#1001 → 800 confirmed BOP Segment 2 (Lose Strategy): SO#1002 → 0 confirmed Result: Only priority orders confirmed by business rules • Business priority rules applied consistently • Full aATP sequence (PAC→BPS) during BOP

7. Sources of Stock for Order Confirmation

Stock for confirmation may come from Purchase Order Goods Receipts (PO GR) and Customer Returns.

- GR from Purchase Orders represents the primary and planned future supply.
- Customer Returns provide an unexpected stock boost.
- Both sources dynamically increase confirmable quantities during ATP checks.

8. Planned Delivery Time vs. RLT (Replenishment Lead Time) with (Procurement Type – E, F and X)

- Definitions-

Term	What it is	Where maintained	Used for	Covers
Planned Delivery Time (PDT)	Time (in days) the vendor needs to deliver material after a PO is sent	Material Master → MRP 2 view	External procurement (buy from vendors)	PO processing, vendor manufacturing/handling, transportation to plant
Replenishment Lead Time (RLT)	Total time to replenish material, from demand to availability	Material Master → MRP 3 view	MRP planning, especially for complex procurement	PDT, goods receipt processing, in-house production time (if applicable)

- Procurement Type Impact-

Procurement Type	PDT Usage	RLT Usage	MRP Behavior
E – In-house Production	Not relevant	Used (calculated from in-house production time)	MRP uses production lead times (routing, work centers)
F – External Procurement	Used	Optional; can override PDT	MRP normally uses PDT; RLT, if maintained, can override detailed lead-time calculation
X – Both (In-house + External)	Used for external option	Used	RLT takes priority if maintained; otherwise, system calculates RLT from PDT (external) + production time (internal)

- Key Rules of Thumb-
  - External procurement (F): PDT is critical.
  - In-house production (E): RLT / production times are critical.
  - Mixed procurement (X): RLT dominates if maintained; otherwise, MRP calculates based on a combination of PDT and production times.

- Example Comparison Table-

Procurement Type	Planned Delivery Time	Replenishment Lead Time	MRP Result / Behavior
E – In-House	10 days	12 days	PDT ignored. RLT (production time + GR processing) used by MRP.
F – External	10 days	12 days	PDT is used. RLT = PDT + GR processing. MRP uses RLT for planning.
X – Both	10 days	12 days	RLT depends on procurement method. PDT used only if MRP selects external; otherwise RLT based on in-house production time.

9. How AATP related MD04 and CO09 at the Sales Orders

In SAP, AATP (Advanced Available-to-Promise) uses the same basic data foundation as classic ATP (what you see in MD04/CO09), but it interprets and prioritizes it differently for sales orders.

Utilize the *Monitor Product Availability / Availability Overview (CO09)* app rather than relying solely on the *Monitor Stock Requirements List (MD04)*. In the Availability Overview app, the availability review is based on the defined scope of check and displays only the stock, demand, and supply elements that are relevant for ATP. The header of the app shows the applicable availability check and checking rule. You can also select the *Scope of Check* button to view the detailed scope-of-check settings.

MD04 shows what is planned and CO09 shows what can be promised; The differences are explained below.

1. Monitor Stock Requirements List MD04 is a planning and analysis tool, and it shows the time-phased supply and demand situation for a material. MD04 is for "How does supply and demand look over time?".

2. Utilize Monitor Product Availability CO09 is an ATP simulation tool, and it shows what quantity can be confirmed and on what date. CO09 is for "What can i promise to the customer, and when?"

- **Roles of MD04 and CO09-**

MD04 (Stock/Requirements List)	• Shows all supply and demand elements (stock, planned orders, purchase orders, sales orders, deliveries, etc.) by *requested* dates and quantities. • It is an MRP/production view: planners see *what is requested when*, not what is actually confirmed by ATP. • MD04 "Available Qty" is not calculated using ATP scope-of-check logic; it is a simple MRP-based netting of requirements and receipts.
CO09 (ATP Details / Availability)	• Shows the ATP picture by *confirmed* quantities and dates, using the defined ATP checking rule and scope-of-check. • It is a sales/ATP view: SD uses this logic when confirming schedule lines in the sales order. • CO09 is where you see how much is still available-to-promise after existing confirmations.

- **How this ties into aATP and sales orders**
  - When a sales order is saved with aATP active:
    - aATP reads stock and receipts (and possibly planned future receipts, safety stock, etc.) similarly to classic ATP, but then applies additional logic such as alternative confirmations, product substitutions, or advanced rules (e.g., allocation, rules-based ATP).
    - The final *confirmed* quantities and dates that aATP decides are stored in the sales order schedule lines (VBEP), just like classic ATP.
  - These stored confirmations then:
    - Reduce the ATP in CO09 for future checks (CO09 will show less remaining ATP once an order is confirmed).
    - Appear as requirements in MD04 with their requested/requirement dates (and sometimes also with their confirmed dates, depending on how you view the elements), driving MRP.

- **Conceptual relationship at sales-order level**
  - MD04 is MRP-focused:
    - Shows the sales order requirement as "demand" for production/procurement, usually by requested date.
    - aATP does not "live" in MD04, but the result of aATP (the confirmed schedule lines) becomes one of the MRP elements visible there.
  - CO09 is ATP-focused:
    - Shows the *ATP calculation* that aATP (or classic ATP) relies on to confirm the sales order.
    - The sales order confirmations created by aATP immediately impact the ATP quantities in CO09.
  - aATP itself is an enhanced ATP engine:
    - Uses the same underlying requirements/receipts that feed MD04 and CO09.
    - Writes its results back to the sales order; those results are what you later recognize in CO09 as consumed ATP and in MD04 as requirements.

10. Core date relationships in ATP - Backward & Forward Scheduling

In sales orders, backward and forward scheduling are used to determine delivery and goods issue dates based on availability and lead times.

1. **Backward scheduling** -> Can we meet the customer's requested date?
2. **Forward scheduling** -> If not, when is the earliest we *can* deliver?

- **Backward Scheduling-**

When a requested delivery date (RDD) is entered in the sales order, the system normally does backward scheduling:

Step#1 – Goods issue date (GI date) = Requested delivery date−Transportation time

The system subtracts the transit/transportation time from the RDD to get the date when the goods must leave the plant.

Step#2 – Loading date = Goods issue date −Loading time
From the GI date, the system subtracts loading time (per shipping point/route) to get the loading date.

Step#3 – Material availability date (MAD / product availability date)
Material availability date=Loading Date−Picking & packing time
From the loading date, the system subtracts picking/packing time to get the date on which stock must be available for picking. This is the key date for the ATP quantity check.

These formulas are adjusted against factory and route calendars (working days, goods issue days), so dates may shift if a calculated date falls on a non-working day.

Note: Unless a customer calendar is in place, a customer request date can be on any calendar day and is not influenced by our calendar.

- **Forward Scheduling (when requested delivery rate cannot be met)-**

If ATP finds that stock is not available on the MAD from backward scheduling, the system switches to forward scheduling:
It finds the earliest date with sufficient ATP quantity, sets this as a new material availability date, then Adds picking/packing, loading, and transportation times to propose a new confirmed goods issue date and confirmed delivery date.

In SAP sales order schedule lines (accessed via VA01/VA02/VA03 under the Schedule Lines tab by clicking the shipping button), the dates represent key milestones in the backward scheduling process, calculated from the customer's requested delivery date using the route schedule and transit times from the shipping point/plant to the customer.

- **Date Significance-**

These dates ensure logistical feasibility and ATP confirmation:

Delivery Date:	Customer receives goods; primary customer-facing date and main driver for backward scheduling.
Goods Issue Date:	Stock leaves the plant; triggers inventory reduction, accounting postings, and billing relevance.
Loading Date:	Goods are physically loaded onto transport vehicles; precedes goods issue.

Transportation Planning Date:	Time to plan/notify carriers or arrange freight; allows lead time for booking.
Material Availability Date:	Latest date material must be ready in unrestricted stock for picking/packing (determined by ATP check).

- **Calculation Sequence-**

SAP uses strict backward logic starting from the Requested Delivery Date (customer input):

1. Delivery Date (confirmed via ATP) – Starting point.
2. Goods Issue Date = Delivery Date minus Transit Time (from route).
3. Loading Date = Goods Issue Date minus Loading Time (plant-specific).
4. Transportation Planning Date = Loading Date minus Transp. Planning Time (shipping config).
5. Material Availability Date = Loading Date minus Pick/Pack Time; ATP confirms qty here first.

This sequence drives delivery creation (VL01N picks the earliest feasible schedule line) and integrates with MD04/CO09 for availability.
These dates flow into the outbound delivery, where they control the timing of picking, loading, and posting goods issue. The **Goods Issue (GI) Date** represents the planned date on which goods are physically issued from the warehouse or plant. It is a key logistical date that drives picking, packing, and shipping activities and is calculated based on logistics lead times.

11. How the material determination is different than ABC Substitution

Material Determination vs. ABC Substitution (S/4HANA aATP 2023)

- Material Determination is an SD function that replaces a material based on predefined business rules, before ATP, and without checking availability. It is best suited for planned substitutions like phase-in/phase-out or customer-specific replacements.

- ABC Substitution (AATP) is an availability-driven function that runs during the ATP check or BOP, proposes alternative materials/plants only when supply is insufficient, and supports 1:n substitution, percentage splits, and date shifts.

Material Determination decides what material is sold; ABC Substitution decides how demand is fulfilled when availability is constrained.

12. aATP Settings

Here's a clear, practical breakdown of aATP settings showing which require Transport Requests and which do not. This is *very* useful for projects, and troubleshooting.

aATP uses a hybrid model: core logic is transported via IMG, while operational behavior is controlled through key settings maintained directly in each system.

Simple Rule to Remember-

- IMG = Transport required
- Fiori / App / Business control = No transport

i. aATP Settings WITHOUT Transport Request

(Key Settings / Business Settings)

These are maintained **directly in each system** (DEV, QA, PRD):

- Product Allocation (PAL) activation
- Backorder Processing (BOP) variants
- aATP check control at product level
- Supply assignment priorities
- Confirmation strategy controls
- Alternative-based confirmation (ABC) settings
- ATP categories activation
- Key figures used in aATP checks

Key Points-

- *Treated as key settings*
- *No TR required*
- *Must be manually aligned across systems*

ii. aATP Settings WITH Transport Request

(IMG / Configuration)

These are maintained via **SPRO / IMG** and **require transport**:

- Checking rules
- Availability check control
- ATP scope of check
- Requirement class & requirement type
- Strategy groups

- Business process integration settings
- Integration with SD / PP / MM
- General aATP framework configuration

Key Points-

- *Transported via TR*
- *System-consistent*
- *Typical customizing objects*

13. aATP Integration with other modules

Advanced Available-to-Promise (aATP) is tightly integrated with multiple SAP modules to enable accurate, priority-based order fulfillment.

aATP acts as the real-time bridge between sales demand and supply execution by integrating SD, MM, PP/PP-DS, EWM, TM, and IBP in S/4HANA. aATP pulls real-time data from multiple modules for accurate confirmations:

aATP Integration with Other Modules (Table)

SAP Module	How It Integrates with aATP	Role in aATP
SD (Sales & Distribution)	Triggers aATP during sales order creation/change and delivery processing	Provides demand and customer requirements
MM (Materials Management)	Supplies on-hand stock, purchase orders, STOs, subcontracting stock	Provides procurement and inventory supply
PP (Production Planning)	Provides planned orders and production orders	Enables manufacturing-based availability
PP-DS	Supplies capacity-checked production dates	Enables realistic, finite ATP confirmations
EWM (Extended Warehouse Management)	Considers warehouse stock and allocation constraints	Ensures warehouse-executable confirmations
TM (Transportation Management)	Provides transportation lead times and routing	Supports realistic delivery promise dates
IBP (Integrated Business Planning)	Feeds demand forecasts and allocation inputs	Aligns planning with execution
FI / CO	Impacted indirectly through order fulfilment and revenue timing	**Financial impact**, no direct ATP logic

aATP Integration with SAP TM for BPS (S/4HANA 2023 FPS02+)

- BPS ↔ TM: Transportation scheduling delegation

Key Enhancement: BPS can delegate transportation scheduling to SAP TM master data, eliminating duplicate route/transit time maintenance.
TM provides: Route duration + Carrier calendar + Means of Transport
→ BPS returns precise delivery date to aATP confirmation

Data Flow Example:

Sales Order (SD) → PAC → No stock
↓
SBC → PP/DS → Creates planned order
↓
BPS → TM → Adds transport duration
↓
Confirmation → Customer sees realistic date

14. aATP at Delivery Level in SAP S/4 HANA

In S/4HANA, aATP does not stop at the sales order level. It is also executed at the delivery level using a separate checking rule, ensuring realistic and executable confirmations.
aATP at delivery level (Checking Rule "B") revalidates stock and allocations during delivery creation to ensure only physically and logically available quantities are delivered.

i. Delivery-Level aATP Execution

- aATP works at delivery level, not only at sales order item level.
- Delivery-level ATP uses Checking Rule "B".
- This allows SAP to:
    - Revalidate availability at the time of delivery creation
    - Consider latest stock, allocations, and priorities

ii. Delivery Creation & PGI Flow

Delivery creation is performed using:

- VL01N / VL02N
- Followed by Post Goods Issue (PGI)

During delivery creation, aATP rechecks:

- Available stock
- Product allocations (PAL)

- Supply protection constraints

This prevents over-commitment between order confirmation and physical delivery.

iii. Triggers for Delivery-Level aATP

aATP with checking rule "B" is triggered during:

- Delivery Due List (VL10*)
- Delivery Creation (VL01N)
- Delivery block removal or reprocessing

These triggers ensure that only currently available stock is used for delivery creation.

iv. Checking Rule Difference (Very Important)

Document Level	Checking Rule	Purpose
Sales Order	A	Sales-order-level ATP check
Delivery	B	Delivery-level ATP recheck

Key point:

- Checking Rule "B" has separate product allocation (PAL) consumption
- This avoids double or incorrect allocation usage between SO and Delivery

v. Configuration

- Define scope of check with Delivery Checking Rules – B. This controls:
  - Stock types considered
  - Allocations
  - Receipt elements
  - Delivery-relevant availability logic

vi. 2023 Enhancements (Fiori)

Delivery Creation Monitor (Fiori App)

New capabilities include:

- Monitor delivery-relevant sales orders
- Adjust ATP confirmations
- Create deliveries directly from the app
- Real-time aATP recalculation during delivery grouping

This ensures delivery quantities are always aligned with current availability.

vii. Business Scenarios-

Scenario 1: Stock Reduced After Sales Order Confirmation	• SO confirmed earlier using rule A • Stock consumed by higher-priority orders • During delivery creation, rule B rechecks availability • Result: Delivery quantity reduced or split

Scenario 2: Product Allocation Control at Delivery	• PAL consumed at SO level
	• Separate PAL consumption at delivery ensures:
	  o No over-delivery
	  o Fair allocation enforcement

Scenario 3: Delivery Block Removal	• Order was delivery-blocked • Block removed later • aATP rechecks availability before delivery creation

Key Points-

1. Delivery-level aATP is standard in SAP S/4HANA and uses Checking Rule "B".
It cannot be switched off via a single on/off setting; however, its behaviour can be controlled through configuration, specifically via the scope-of-check for Checking Rule B.

2. In SAP S/4HANA, aATP is executed by default during delivery creation, and this is standard system behaviour. There is no single switch called *"Delivery-level aATP"*. Instead, delivery-level aATP is controlled through the checking rule and scope-of-check configuration.

3. Delivery documents automatically use Checking Rule B, which is standard SAP behaviour. You cannot change the checking rule at the delivery item level, but you can control what Checking Rule B evaluates, such as:

- Whether stock is checked
- Whether product allocation is checked
- Whether receipts are checked
- Whether reservations are considered

If stock and allocation checks are removed, delivery-level aATP is effectively neutralized.
Indirect deactivation (not recommended)

4. You can make delivery-level aATP behave like a "no check" scenario by:
    - Defining a custom scope of check for Checking Rule B
    - Excluding all relevant supply elements

This approach is generally not recommended.

5. What you CANNOT do -
    - Completely switch off delivery-level ATP using a single flag
    - Bypass Checking Rule B without modification
    - Use Checking Rule A at the delivery level

6. Why SAP recommends keeping delivery-level aATP active
    - Stock situations can change between sales order confirmation and delivery creation
    - Prevents over-delivery
    - Protects product allocation and supply protection logic

15. Project Essentials – Migration from Classic ATP to Advanced ATP (aATP) in SAP S/4HANA 2023

A migration from classic ATP to advanced ATP (aATP) in S/4HANA 2023 is a functional transformation project, not just a technical switch. It needs structured preparation, design, and phased enablement.

Vision and Assessment	• Clarify why you move to aATP: need for PAL, BOP, ABC, Supply Protection, TM/EWM/PP-DS integration, or better Fiori analytics. • Assess current use of classic ATP: checking groups/rules, scope of check, rescheduling, substitutions, user exits, CIF/APO dependencies. • Decide target scope for phase 1: PAC only, or PAC + PAL + BOP, etc., to avoid a "big bang" of all aATP features.
Landscape and Pre-requisites	• Ensure S/4HANA 2023 system is in place with required FPS/SPS for the aATP features you plan to use. • Confirm simplification items: classic APO/gATP processes planned for decommissioning; understand where aATP replaces APO scenarios. • Align with SD, TM, EWM, PP-DS teams for cross-module dependencies (transportation scheduling in BPS, supply creation, warehouse constraints).

Conceptual Design: Classic ATP → aATP	Map classic ATP behavior to aATP capabilities: • Classic ATP check → aATP Product Availability Check (PAC) with checking group/rule and scope of check in Fiori. • Simple allocations (manual rules, Z-tables) → Product Allocation (PAL) objects and sequences. • Manual redispatch/rescheduling → Backorder Processing (BOP) with segmentation (Win/Gain/Redistribute/Fill/Lose) and Fiori control. • Manual plant/material substitution → Alternative-Based Confirmation (ABC) with substitution strategies and control profiles. • "Always reserve stock for VIPs/Channels" → Supply Protection where needed.
Configuration Workstream	Key activities in S/4HANA 2023: • Define/check aATP-relevant checking groups in material master and checking rules for sales, deliveries, STOs. • Maintain scope of check in aATP apps (supplies/demands, stock types, safety stock handling, special stocks). • Set up PAL (objects, sequences, allocation plans) for constrained products/markets; load historical or planned allocations. • Configure BOP variants (segments, filter/sort, fair-share rules) and scheduling (batch vs online). • Configure ABC (substitution strategies, characteristic combinations, plant/material hierarchies). • Activate and tune explanation/analytics apps (ATP explanation, BOP result monitor, PAL situation, etc.).
Data and Master-Data Preparation	• Clean and harmonize material masters (MRP views, ATP check group, strategy groups, plants, storage locations). • Classify products/customers/channels for PAL, BOP segmentation, and Supply Protection (e.g., key accounts, regions). • Prepare substitution master data: cross-plant availability, alternative materials, and transportation/production constraints. • If moving from APO/gATP, decide which rules/allocations will be re-implemented versus dropped or simplified.

Integration Design	• SD: sales order and delivery check strategies, scheduling, partial confirmation rules, delivery creation integration with aATP. • TM: integration of transportation scheduling with BPS/aATP (handover of transit times, calendars, lanes). • EWM: stock visibility at warehouse/storage type level for PAC; impact of waves/tasks on perceived availability. • PP-DS / PP: use of supply creation (SBC), planned orders, production capacity in confirmation logic. • IBP or other planning: how demand/supply plans influence allocations and priorities.

Transition Strategy (Coexistence and Cutover)	• Decide coexistence model: run classic ATP and aATP in parallel for different materials or plants, then phase out classic ATP; S/4 allows selective activation via check groups.
	• Define cutover steps:
	  ◦ Freeze APO/gATP interfaces (if any).
	  ◦ Switch customizing for selected plants/materials to aATP.
	  ◦ Perform initial PAL uploads and validate allocations.
	  ◦ Schedule first BOP runs and validates results against legacy logic.
	• Plan rollback and feature toggles (e.g., revert aATP for specific materials if issues appear).

Testing and Validation	• Unit tests per component: PAC, PAL, BOP, ABC, Supply Protection, explanation tools. • End-to-end scenarios: quote-to-cash with shortages, priority customers, substitutions, and cross-plant fulfilment. • Volume/performance tests for high-order environments using mass BOP and parallelized PAC. • Regression tests to ensure no negative impact on non-aATP processes (billing, finance integration, logistics execution).
Change Management and Training	• Train SD planners, CSR teams, and supply chain planners on new Fiori apps (e.g., BOP cockpit, PAL maintenance, ATP explanations).

	• Update business rules: who defines priorities, how often BOP runs, when PAL is adjusted, escalation paths for shortages. • Provide "before vs after" examples to explain difference between classic ATP dates and new aATP confirmations.
Typical Project Deliverables	• Migration strategy and scope document (classic ATP vs aATP mapping). • End-to-end aATP solution design (PAC, PAL, BOP, ABC, integrations). • Configuration and development specs (incl. exits/BAdIs replacements, if any). • Test scripts, cutover plan, and support playbook for hyper care.

Avoiding Data Mapping Errors in Classic ATP to aATP Migration
Approximately **60% of ATP migration issues stem from incorrect data mapping**, including mis-mapped allocation objects, inconsistent checking groups, and missing substitution rules. Precise alignment of Classic ATP data structures with aATP objects is critical to ensure accurate availability checks and reliable order confirmations after go-live.

16. Common Pitfalls in Classic ATP to aATP Migration (SAP S/4HANA 2023)

Migration projects frequently fail due to functional gaps, incomplete testing, and underestimated business impacts when transitioning from Classic ATP to Advanced ATP (aATP). Understanding and addressing these pitfalls early is critical to a successful go-live.

1. Incomplete Feature Mapping

Pitfall:	Assuming Classic ATP rescheduling jobs are equivalent to aATP Backorder Processing (BOP), while overlooking segmentation logic differences (Win/Gain/Lose).
Impact:	Orders confirmed correctly in Classic ATP are not reprioritized as expected during BOP runs.
Fix:	Document all Classic ATP user exits and rescheduling variants, then map them explicitly to aATP Product Availability Check (PAC) sequences, BOP filters, and fair-share rules.

2. Product Allocation (PAL) Data Overload at Cutover

Pitfall:	Loading full-year allocation quantities without period-based sequences; consumption sequences do not reflect historical demand patterns.
Impact:	Immediate over- or under-consumption of allocations post–go-live.
Fix:	Roll out PAL in phases (Phase 1: constrained SKUs only). Use the *Initial Fill* Fiori app and historical sales data to seed realistic allocation consumption.

3. Scope-of-Check Misalignment

Classic ATP	aATP (PAC)	Common Gap
V_V2 (Sales)	Stock categories	Safety stock exclusion
OVZ9 static configuration	Fiori-based dynamic checks	In-transit stock timing
No EWM visibility	Bin-level stock	Warehouse task delays

Fix: Run the *ATP Impact Analysis* Fiori app prior to migration to validate scope-of-check equivalence between Classic ATP and aATP.

4. BOP Performance Surprises

Pitfall:	Executing BOP online (instead of batch) for volumes exceeding 50,000 orders, with no parallel processing configured.
Impact:	Sales order creation freezes exceeding 30 seconds.
Fix:	Schedule /SAPAPO/BOP_EXEC as a background job and limit segmentation to a maximum of 3–5 segments.

5. Substitution Strategy Gaps

Pitfall:	Migrating Classic plant substitution logic directly into Advanced Business Configuration (ABC) without control profiles or ranking logic.
Impact:	Customers receive incorrect alternative products or plants, with no fallback strategy.
Fix:	Define a clear ABC substitution chain, for example: Plant 1 → Plant 2 → Material 2, with min/max quantities and priority defined at each step.

6. Integration Cutover Sequence Errors

❌ **Incorrect:** Switch SD checking groups → Activate TM/BOP

☑ **Correct:**

1. Activate PAC
2. Activate PAL
3. Enable TM delegation
4. Activate BOP and ABC

7. User Training Deficiencies

Pitfall:	SD users continue using VA05 rescheduling instead of the BOP Cockpit Fiori apps.
Impact:	Manual workarounds bypass aATP logic, undermining the new design.
Fix:	Mandatory training on key apps, including *ATP Explanation* (to trace PAC decisions) and *Reschedule Orders* (BOP Cockpit).

8. Rollback Planning Underestimation

Pitfall:	No documented procedure to revert to Classic ATP in case of critical issues.
Impact:	Extended business disruption if aATP must be temporarily disabled.
Fix:	Document the checking group switch-back process and retain Classic ATP customizing for at least three months post–go-live.

Pro Tip

Create an **aATP Readiness Scorecard** before migration. Recommended minimum thresholds:

- ≥ 80% Product Allocation (PAL) coverage
- < 5 seconds PAC response time
- 100% BOP scenario test coverage

Meeting these criteria significantly increases the likelihood of a stable and predictable aATP go-live.

19 Introduction to SAP IBP (Integrated Business Planning)

Here, I want to give an overview of IBP that will help you understand its core concepts and how it is used as a planning tool.

SAP IBP (Integrated Business Planning) is a cloud-based planning solution that helps companies align demand, supply, inventory, and financial plans in real time.
SAP IBP enables better planning, faster decision-making, and improved supply chain coordination using advanced analytics and simulations.
SAP IBP is a supply chain **planning** solution that plans across the supply chain network and SAP IBP supports **end-to-end supply chain planning**.
SAP IBP integrates with SAP S/4HANA to enable supply chain planning.

Supply Chain Network – Overview
The supply chain network is a strategic, data-driven system connecting customers, retailers, distribution centers (DCs), manufacturing units, and suppliers, coordinating the flow of goods from raw materials to finished products delivered to customers based on their demand. The supply chain process typically begins with customer demand, progressing through retail stocking checks, DC inventory verification, manufacturing production based on raw materials availability, supplier sourcing, and back through distribution until the customer receives the product.

SAP Advanced Available-to-Promise (aATP) fits within this supply chain network and process primarily at the order fulfillment and inventory allocation stages. aATP includes capabilities such as product allocation, supply protection, alternative-based confirmation, and supply creation, enhancing real-time inventory visibility, prioritizing high-value customer orders, and responding dynamically to supply constraints across the entire supply chain network. It ensures that inventory is used optimally, key customers' needs are protected even during shortages, and alternatives for sourcing or delivery are offered to reduce delays. Thus, aATP sits at the crucial intersection where customer demand meets supply chain execution, enabling businesses to confirm orders accurately and reliably within the SAP ecosystem, particularly in solutions like SAP S/4HANA and SAP IBP.

In summary, the supply chain process involves steps from demand planning to procurement, manufacturing, and distribution. aATP enhances the process by providing intelligent, rule-based order promising and allocation capabilities that optimize supply fulfillment in the network, making it a vital part of modern supply chain operations.

This explanation aligns with your example of customer order flow from customer demand to retail, DC, manufacturing units, and suppliers and positions aATP as the system that manages the availability and allocation of products to meet that demand efficiently.

When we talk about a Supply Chain Network, it always starts with customer demand and follows the supply chain elements based on the business-to-business setup.
A Supply Chain Network represents the interconnected flow of demand, supply, materials, and information across multiple entities involved in delivering a product to the end customer. It typically starts with customer demand and spans across retailers, distribution centers (DCs), manufacturing plants, and suppliers.

Demand flows from downstream to upstream - (Customer → Retailer → DC → Plant → Supplier)
Supply flows from upstream to downstream - (Supplier → Plant → DC → Retailer → Customer)

Demand flow from the downstream to upstream – starting from the **customer to the retailer**. If the retailer does not have sufficient stock to fulfill the demand, the requirement moves to the **Distribution Center (DC)**. If the DC also lacks stock, it passes the requirement to the **manufacturing plant**, where the product is produced.

To produce finished goods, the plant procures **raw materials from suppliers/vendors**,
These entities—**Suppliers, Plants, DCs, Retailers, and Customers** - are called **nodes** in the supply chain network. These nodes work together to ensure customer demand is fulfilled efficiently while balancing inventory, capacity, and cost.

Flow of Demand and Supply-

- **Demand / Requirements** always flow from **downstream to upstream**
- **Supply** flows from **upstream to downstream**

Demand/Requirements Flow:
Suppliers ← Plant ← DC ← Retailer ← Customer

Supply Flow:
Suppliers → Plant → DC → Retailer → Customer

Organizations serve multiple customers and operate with multiple plants and DCs, managing the supply chain manually is not practical. This complexity requires integrated supply chain planning tools to forecast demand, plan supply, optimize production, and manage transportation effectively. In short, a well-designed supply chain network ensures:

- Timely fulfillment of customer demand
- Optimal use of production and logistics capacity
- Cost-efficient and reliable end-to-end operations

Example:

If a customer demands 100 units, the requirement flows upstream to trigger production. Once produced, the supply flows downstream to fulfil customer demand.

To fulfill customer demand effectively, the organization must ensure that the **entire supply chain cycle from downstream to upstream and back is properly set up**.

This is not limited to a single customer. Organizations typically deal with:

- Multiple customers with varying demand
- Multiple DCs and plants
- Complex, multi-node supply chain structures

Supply Chain Planning Tools-

Managing demand and supply manually is not feasible due to uncertainties such as:

- Demand volume
- Supply availability
- Capacity constraints

Therefore, organizations rely on supply chain planning tools. Multiple tools are available in the market, including the following:

1	**Anaplan**
2	**JDA**
3	**Kinaxis**
4	**SAP APO**
5	**SAP IBP**

In the ERP market, approximately 70% is dominated by SAP products. Hence, the focus here is on SAP-based planning solutions.
SAP provides two major planning tools:

SAP APO	An on-premises solution available for the last two decades
SAP IBP	A cloud-based planning solution

When SAP products are used end to end, system connectivity is seamless. However, when non-SAP tools are used, custom integrations are required.

SAP APO Modules (End-to-End Planning)
SAP APO (Advanced Planning and Optimization) supports end-to-end supply chain planning through the following five core major modules:

1	*Demand Planning (DP)*
2	*Supply Network Planning (SNP)*
3	*PP/DS (Production Planning & Detailed Scheduling)*
4	*TP/VS (Transportation Planning / Vehicle Scheduling)*
5	*gATP (Global Available-to-Promise)*

1. Demand Planning (DP)
Demand Planning is a forecast-based planning process that uses historical sales data to estimate future customer demand. It is typically performed for long-term horizons, such as 6 months to 1–2 years, to predict expected sales volumes.
The primary objective of demand planning is to predict future demand based on past sales performance. It helps organizations understand how much demand they are likely to receive over an upcoming period, enabling better production, inventory, and supply planning decisions.

Example Forecast:
If demand planning is done for one year, the forecasted demand may look like this:

- January: 100 units
- February: 150 units
- March: 200 units
- April: 250 units
- … and so on for all 12 months

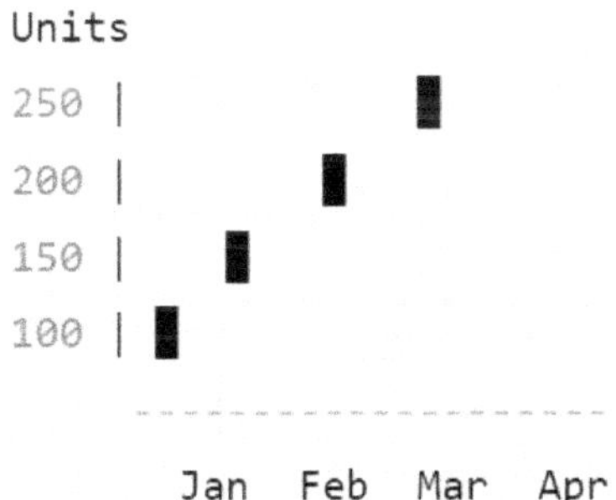

This forecast represents the expected customer demand for each month over the planning horizon.

2. Supply Network Planning (SNP)
Supply Network Planning (SNP) is a mid-term supply planning process that focuses on how to fulfill forecasted demand across a supply chain network. Unlike demand planning, which predicts how much customers will buy, SNP determines how the supply chain can meet that demand efficiently.
SNP helps answer critical questions such as:

- Where to produce the product – which manufacturing plant should produce it.
- How much to produce at each location – allocating production quantities to plants.
- Which distribution center (DC) should supply which customer – optimizing logistics and delivery.
- How products should flow through the network – ensuring materials and products moves efficiently.

The goal is to balance demand, supply, and network constraints (e.g., capacity, transportation limits, inventory levels) to create a feasible mid-term supply plan.

Example Scenario

Forecasted demand:

- January: 100 units
- February: 150 units
- March: 200 units
- April: 250 units

SNP planning might include:

- Assigning production quantities to specific plants based on their capacity.
- Determining sourcing relationships between plants and distribution centers.
- Assigning customer demand to appropriate DCs.
- Planning material flows to minimize costs and delays.

The output is a rough-cut supply plan, providing a high-level view of:

- Production requirements
- Distribution assignments
- Transportation needs

This mid-term plan serves as an input for detailed planning, such as production schedules or daily operational plans. The output of SNP is a rough-cut supply plan, providing a high-level view of production, distribution, and transportation requirements. SNP bridges the gap between long-term demand planning and short-term execution planning, guiding the supply chain to meet forecasted demand efficiently.

3. PP/DS (Production Planning & Detailed Scheduling)

Once a rough-cut supply plan is created through SNP, it is converted into an actual production plan using PP/DS, which handles detailed scheduling. PP/DS is a short-term planning process that transforms rough-cut plans into specific production schedules.

Example:

- Suppose 10 orders need to be produced tomorrow for 3 products (P1, P2, P3).
- PP/DS determines the exact production sequence – for example, produce P1 first, then P2, and finally P3.

PP/DS handles:

- Order sequencing – deciding the order in which products are manufactured.
- Machine capacity – ensuring production fits within available resources.
- Production priorities – managing which orders are more urgent or critical.

This process ensures that the supply plan from SNP is practically executable on the shop floor.

4. TP/VS (Transportation Planning / Vehicle Scheduling)

TPVS is used to manage transportation efficiently by optimizing delivery schedules. For example, suppose a customer requires 300 units:

- Today, you produce 200 units.
- Tomorrow, you will produce 100 units.

Instead of sending two separate deliveries, TPVS can help decide whether to:

- Ship based on production (200 units today, 100 units tomorrow), or
- Wait and send all 300 units together instead of making two trips

By consolidating shipments, TPVS reduces transportation costs while meeting customer demand.

Key Benefits:

- Optimizes transportation and delivery planning
- Reduce costs by consolidating shipments

Note-

All four planning steps—Demand Planning, SNP, PP/DS, and TPVS - are internal processes within the organization. They focus on managing production, supply, and transportation efficiently and do not involve direct interaction with customers.

These steps ensure that the supply chain operates smoothly, from forecasting demand to scheduling production and optimizing deliveries, before the products reach the customer.

5. gATP (Global Available-to-Promise)

gATP is a customer-facing tool that interacts directly with customer requirements. It:

1. Takes customer orders or requests.

2. Check the supply chain to see if the requested quantity is available.
3. Verifies available stock and production capacity.
4. Propose a delivery date based on availability and supply constraints.

Unlike SNP, PP/DS, or TPVS, which are internal planning processes, gATP directly connects the organization to customer demand, ensuring reliable order fulfillment.

Transition from SAP APO to SAP IBP-

On top of these planning modules, SAP introduced tools to enable a more systematic and collaborative planning process, known as S&OP (Sales & Operations Planning).

In S&OP, key stakeholders from different functions—such as sales, supply chain, operations, logistics, and finance—come together to:

- Discuss demand and supply scenarios
- Review capacity constraints
- Consider logistics capabilities
- Aligning with financial targets

In short, S&OP is a demand and supply review process where capacity is balanced with logistics and financial considerations to arrive at a finalized, consensus-based demand plan.

Later, SAP renamed and expanded S&OP into SAP IBP (Integrated Business Planning). As part of this transformation:

- SAP began moving planning functionalities from SAP APO
- Some modules were integrated into SAP S/4HANA
- Other advanced planning capabilities were migrated and enhanced within SAP IBP

In short - SAP started moving planning functionalities from SAP APO. Certain modules were integrated directly into SAP S/4HANA, while other advanced planning capabilities were migrated and enhanced within SAP IBP.

SAP IBP now serves as a strategic and tactical planning platform, supporting end-to-end planning across demand, supply, inventory, and S&OP processes.

SAP introduced S&OP (Sales & Operations Planning) as a collaborative planning process, which later evolved into SAP IBP (Integrated Business

Planning). As part of SAP's strategic shift, planning capabilities from SAP APO were gradually redistributed:

- Some functionalities were integrated into SAP S/4HANA
- Advanced planning and S&OP capabilities were moved to SAP IBP (Cloud)

Module Mapping-

From SAP APO:

- DP, SNP, and S&OP functionalities were moved to SAP IBP.
- PPDS, TP/VS, and gATP functionalities were moved to SAP S/4HANA.

S/4HANA	SAP APO (On-Prem)	SAP IBP (Cloud)
	DP (Long Term Planning)	**IBP Demand**
	SNP (Mid Term Planning)	**IBP Supply**
ePPDS	PPDS (Short Term Planning)	—
TM	TP/VS	—
aATP	gATP	—
—	S&OP	**IBP S&OP**
—	—	IBP Inventory
—	—	IBP SCCT

- *TP/VS = Transportation Planning / Vehicle Scheduling)*
- *SCCT = Supply Chain Control Tower*
- *S&OP = Sales & Operations Planning*
- *gATP = Global Available to Promise*

Next Steps – SAP IBP

With the foundation in place, the next focus should be on understanding SAP IBP architecture and its capabilities.

SAP Integrated Business Planning (IBP) provides cloud-based supply chain planning tightly integrated with SAP S/4HANA execution systems. This "at a glance" covers IBP's architecture, core capabilities, and operational planning cycle.

- **SAP IBP Architecture**
- **SAP IBP Capabilities**
- **SAP IBP Operational Planning Process**

1. SAP IBP Architecture

SAP Integrated Business Planning (IBP) is a cloud-based, HANA-powered platform for end-to-end supply chain planning that combines demand, supply, inventory, S&OP, and response planning in one integrated architecture.

SAP IBP runs as Software-as-a-Service (SaaS) on the SAP HANA in-memory platform, delivering real-time planning across global supply networks. Its modular design includes:

Planning Areas:	Central data models that combine time-series data (forecasts, inventory levels) and order-based data (specific customer orders, planned receipts).
Modules:	Demand Planning, Response & Supply Planning, Inventory Optimization, Sales & Operations Planning (S&OP), Demand-Driven Replenishment (DDMRP), and Supply Chain Control Tower.
Integration Layer:	Standard connectors to S/4HANA (via OData/ODP), ECC, non-SAP ERPs, and external data sources. Data flows include master data (products, customers, plants), transactions (sales orders, inventory), and planning results back to execution.

The architecture supports multi-tenant cloud deployment with embedded analytics, AI/ML forecasting engines, and what-if scenario simulation capabilities.

Supply Chain Control Tower

Exceptional Handling and Business Network Collaboration

Sales and Operations Planning

Strategic and Tactical Decision Processes

Demand	Inventory	Demand-Driven Replenishment	Response & Supply
Statistical Forecasting, Consensus Planning, Demand Sensing	Multi-Stage Inventory Optimization	Demand-Driven Material-Requirements Planning (DDMRP)	Unconstrained & Constrained Supply Planning, Allocations & Deployment Planning, Order Rescheduling

IBP Platform

Analytics and Web-based Planning UI, Microsoft Excel Planning Frontend, Job Scheduling, Data Integration, Version Planning and Simulation, User Management and Authorizations, Data Realignment, ...

SAP HANA

SAP IBP – Architecture Diagram (Standard)

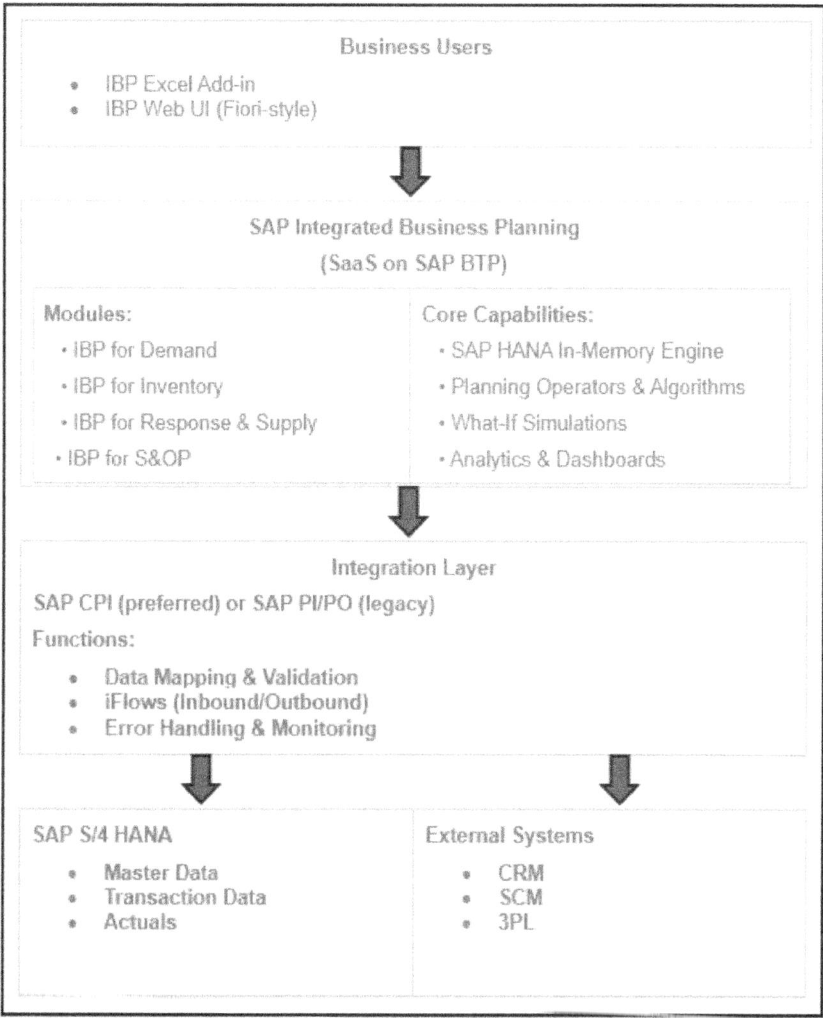

Key Architecture Notes-

- IBP is 100% cloud (SaaS) running on SAP HANA
- SAP CPI is the recommended middleware
- Data integration is mainly batch-based
- Excel Add-in is the primary planning interface
- S/4HANA remains the system of record

Typical Data Flow

- Master & transactional data → S/4 → CPI → IBP
- Planning results (forecast, supply plan) → IBP → CPI → S/4
- External signals can feed demand planning

Functional components-

Commonly described IBP components (all share the same technical platform and data model):

IBP for Demand	Statistical forecasting, demand sensing, promotion uplift, and consensus demand planning.
IBP for Response & Supply	Constrained supply planning, prioritization, and optimization across network (plants, DCs, customers).
IBP for Inventory	Multi-echelon inventory optimization and safety-stock setting.
IBP for S&OP	Scenario planning, financial alignment, and executive reviews.
Demand-Driven Replenishment (DDMRP)	Decoupling points, buffers, and demand-driven planning.
Supply Chain Control Tower	End-to-end visibility, alerts, and KPIs over the same data model.

Data model and planning areas-

Planning area & time-series model	• Central configuration object holding key figures (e.g., forecast, supply, inventory), attributes (product, location, customer), and planning levels. • Time-series and order-based planning supported; same HANA model drives Excel and web UI.
Master & transactional data	• Product, location, customer, resources, BOMs, transport lanes replicated (often from S/4HANA). • Historical sales orders, stock, and orders used for forecasting and planning runs.

Integration with S/4HANA and other systems-

Standard integration content	Preconfigured scenarios for S/4HANA using SAP Cloud Integration (CI-DS/CI-IT): master data, time-series key figures, and planning results (e.g., planned orders) exchanged regularly.
APIs & non-SAP	OData/REST and flat-file options for other ERPs, external data providers, or data lakes.

Deployment characteristics

- Cloud only (no "on-prem" IBP):
    - Single IBP tenants can support multiple ERPs.
    - Quarterly innovations delivered by SAP; customers control when to switch configuration, not the codebase.

2. SAP IBP Capabilities

SAP IBP offers specialized modules for end-to-end planning. Key ones include sales and operations planning (S&OP) for aligning financial and operational goals, forecasting and demand management with AI-driven predictions, and inventory planning using multi-echelon optimization to balance stock levels.

Key Features-

- Real-time scenario simulation and what-if analysis to test demand or supply changes.
- Advanced demand sensing and statistical models for accurate short- and long-term forecasts.
- Response and supply planning with constrained optimization and order prioritization.
- Supply chain visibility through control towers, alerts, and network collaboration.

Core capabilities at a glance

IBP addresses end-to-end supply chain planning challenges through specialized modules:

Module	Primary Function	Key Features
Demand Planning	Statistical forecasting and sensing	AI/ML algorithms, lifecycle management, promotion effects, hierarchical forecasting
Response & Supply	Constrained network planning	Multi-stage supply planning, capacity constraints, heuristics/optimizer
Inventory Optimization	Multi-echelon stock optimization	Safety stock calculation, service level balancing, push/pull strategies
S&OP	Integrated business planning	Volume/value planning, scenario comparison, financial integration
Control Tower	Visibility & exception management	Order/shipment tracking, risk alerts, collaboration workspaces
DDMRR	Demand-driven replenishment	Buffer profiles, dynamic reorder points, visual signals

3. SAP IBP Operational Planning Process

IBP follows a rolling monthly planning cycle aligned with S&OP cadence:

1. Data Integration (Weekly/Daily): Master data and transactions load into planning areas via scheduled jobs. Key inputs include sales history, current inventory, open orders, and planned receipts.
2. Demand Planning Cycle:
    - Statistical forecasts generated automatically using historical patterns.
    - Planners apply overrides for known events (promotions, one-offs).
    - Demand sensing adjusts short-term forecasts using recent actuals.
    - Consensus demand plan approved and released to supply planning.
3. Supply & Inventory Planning:
    - Constrained supply planning across the network proposes feasible production/purchase plans.
    - Inventory optimization sets target stock levels per location/product.
    - Results published back to S/4HANA as planned orders or supply proposals.
4. S&OP Review Process:
    - Cross-functional teams review base/upside/downside scenarios.
    - Demand-supply gaps identified and resolved through trade-offs.
    - Executive S&OP meeting approves final plan with financial targets.
5. Execution & Monitoring:
    - Plans released to S/4HANA for order promising (aATP integration).
    - Control Tower monitors deviations with automated alerts.
    - Response planning handles disruptions through "what-if" simulations.

Key Benefit: IBP creates a single source of truth that bridges strategic planning (S&OP), tactical planning (supply/demand), and operational execution (S/4HANA aATP), reducing manual reconciliation and improving supply chain responsiveness.

- **IBP (Integrated Business Planning)** is the modern SAP planning solution and is gradually replacing **SAP APO**.
- **SAP APO** is expected to **sunset around 2027**.
- **MRP (Material Requirements Planning)** is a planning tool that generates **receipts against demand**.
- Before running MRP, **PIRs (Planned Independent Requirements)** are generated by **Demand Planning**.

MRP- SAP's Standard Execution Tool

MRP is an execution tool within S/4HANA ERP that generates procurement proposals. It is not a planning solution and not a standalone product.

MRP (Material Requirements Planning) is SAP's standard execution planning tool that generates procurement proposals to cover material shortages. In ATP/aATP context, MRP creates the future supply elements (planned orders/PRs) that aATP consumes during real-time checks.

MRP Definition & Core Function

MRP compares demand vs supply and creates receipts:
Demand (SO + PIR + Reservations)
- Existing Supply (Stock + PO + Prod Orders)
= Net Requirements → **Procurement Proposals**

MRP Role in ATP/aATP Ecosystem

- MRP -> Creates “Planned Orders/PRs”
- aATP -> “Consumes” those planned orders as future supply during PAC

Flow:
1. Sales Order Created → Demand Recorded (MD04)
2. MRP Live run (MD01N) → Generates Planned Orders
3. aATP PAC sees: Stock + “MRP Planned Orders” -> Confirms Customer

Standard MRP Process (4 Steps)
1. Netting: ** Stock + Incoming - Requirements = Shortage
2. Lot Sizing: ** Apply lot size rules (EX, FX, weekly)
3. Scheduling: ** Backward/Forward from requirement date
4. Procurement: ** Planned Order → PR/PO/Prod Order

MRP Types in S/4HANA

MRP Types → Planning Strategy:
PD-Standard MRP (deterministic), VB-Lot-size planning, M0-Manual reorder point, M1-Deterministic/Period (PIR coverage). **Live MRP (MD01N):** Real-time, single-material execution

Practical ATP Integration Example

Day1: Sales Order 1000 units created -> No stock -> **Not confirmed**

Day2: MRP Run -> Creates planned order 1000 units

Day3: Same Sales Order -> aATP PAC sees **Planned Order** -> **1000 confirmed**

MRP Execution Triggers

Transactions:

- MD01N: MRP Live (single plant/multi-material), - MD02: Classic MRP (total planning)
- Fiori: "MRP Cockpit" apps

Scheduling: Nightly batch + real-time on-demand

Key Point:

MRP feeds aATP. Without MRP generating planned orders/PRs, aATP would only see existing stock + firm supply -> Limited confirmations. MRP ensures future supply visibility in real-time ATP checks.

For more details about MM/MRP, please refer to my book: *"Easy SAP S/4HANA MM".*

Search on Amazon.com

All ▾ easy sap mm

End Note for an SAP aATP Book

As we conclude this book, remember that mastering SAP begins with a solid understanding of its core modules and processes. aATP settings and advanced scheduling form the backbone of efficient availability checks and order promising. By carefully configuring these fields and understanding their impact, you can ensure smoother operations and better decision-making in real-world business scenarios.

I encourage you to explore SAP hands-on, experiment with different settings, and continuously learn through practical application. SAP is a tool, but its true power lies in enabling businesses to operate smarter, faster, and more effectively. Let this knowledge guide your journey toward becoming a confident SAP professional.

SAP aATP is a critical component of the SAP S/4HANA system, helping organizations streamline availability-to-promise processes, supply chain visibility, and order fulfilment. As this book has highlighted, SAP aATP integrates deeply with other key modules to provide a comprehensive solution that drives operational efficiency and business agility. Mastering SAP aATP requires an understanding of its core processes from Product Availability Checks (PAC) and Basic Supply Control (BSC) to Backorder Processing (BOP), scheduling agreements, and advanced integration scenarios along with key configuration settings such as checking rules, Product Allocations (PAL), scheduling schemas, and alternative-based confirmations. The modern SAP S/4HANA platform further enhances aATP's capabilities, offering flexible workflows, real-time data processing, and advanced integration options with TM, PP/DS, and EWM.

As businesses face increasingly complex supply chains and customer expectations, SAP aATP stands out as a vital tool for ensuring accurate material availability, realistic delivery dates, priority fulfilment, and seamless collaboration across order-to-cash processes.

With the knowledge and practical insights shared in this book, readers are well-equipped to leverage SAP aATP effectively to meet organizational goals, optimize order fulfilment operations, and support informed decision-making. Continued learning and hands-on experience will be key to fully unlocking the potential of SAP aATP, making it indispensable to any enterprise's digital transformation journey.

This closes the book with a reflective yet forward-looking message, emphasizing the importance and value of SAP aATP.

About the Author: Sunil Patil

SUNIL PATIL is a senior SAP Sales and Distribution (SD), Material Management (MM) and aATP consultant with extensive experience in numerous implementation projects. He has years of hands-on experience across a wide range of industries, including manufacturing, electronics and high-tech, food, medical, healthcare, and small to midsize enterprises. Sunil is an SAP Certified Consultant with a strong background in end-to-end SAP implementations, integration, and process optimization. His deep understanding of business processes and technical configurations has helped organizations streamline operations and achieve measurable efficiency improvements.

SAP Certifications:

- SAP Certified Application Associate – SAP S/4HANA Sales
- Cloud-Sales Implementation
- Utilities with SAP ERP 6.0
- Logistics Execution and Warehouse Management with ERP 6.0 EHP5
- Solution Consultant SCM – Order Fulfilment with mySAP ERP

SUNIL PATIL is a seasoned SAP professional and certified consultant specializing in Sales & Distribution and aATP. With decades of experience across a variety of industries, including Fortune 500 enterprise implementations, upgrades, roll-outs, and ongoing support, he brings a comprehensive and practical perspective. Sunil has been involved from project kick-off through to go-live, handling requirement gathering, fit/gap analysis, and blueprinting. Beyond his core modules, his expertise spans SD, MM, FSCM, RAR, and FICO, enabling him to convert business requirements into optimized SAP solutions that drive efficiency and value.

www.ingramcontent.com/pod-product-compliance
Lightning Source LLC
Chambersburg PA
CBHW071528030726
47598CB00001B/44

* 9 7 8 1 9 7 2 0 2 0 0 0 5 *